THE ELEMENTS OF RELATIVITY

The Elements of Relativity

David M. Wittman

University of California, Davis

OXFORD
UNIVERSITY PRESS

Great Clarendon Street, Oxford, OX2 6DP,
United Kingdom

Oxford University Press is a department of the University of Oxford.
It furthers the University's objective of excellence in research, scholarship,
and education by publishing worldwide. Oxford is a registered trade mark of
Oxford University Press in the UK and in certain other countries

First Edition published in 2018
Impression: 1

Published in the United States of America by Oxford University Press
198 Madison Avenue, New York, NY 10016, United States of America

British Library Cataloguing in Publication Data
Data available

Library of Congress Control Number: 2017962192

ISBN 978–0–19–965863–3 (Hbk.)
ISBN 978–0–19–965864–0 (Pbk.)

DOI 10.1093/oso/9780199658633.001.0001

Printed and bound by
CPI Group (UK) Ltd, Croydon, CR0 4YY

To Vera, whose love and support has helped me grow so much.

Preface

Relativity is one of the triumphs of twentieth-century physics, but physics curricula tend to reserve it for advanced students. The message to students is; if you survive everything else, you may move on to relativity. This is a missed opportunity to engage a much larger audience of nonspecialists. Given that many general education students are genuinely curious about relativity, and that special relativity can be understood with remarkably little math, it is a shame that we do not teach it more widely. Furthermore, in relativity a rich set of interesting consequences can be deduced from just a few axioms and a lot of disciplined thinking. This makes relativity an excellent topic for a college course for nonspecialists! I urge my fellow faculty to offer this type of course more often— it is rewarding for students and faculty alike.

My take on such a course is that while it should explicitly *not* require any previous physics knowledge, it *should* offer training in disciplined thinking. (This could potentially make it a first course for physics majors as well.) Instructors should provide thinking tools that are accessible to beginners, while making clear that students must be willing to put in the hard work to practice those thinking tools; this is a college physics course, not a broad survey. While making the subject approachable and avoiding unnecessary complications, we should pursue a thorough understanding, avoid shrinking from difficult concepts, and require students to apply what they learn to new situations. Such a course should follow Einstein's exhortation to "Make things as simple as possible, but not simpler."

I wrote this textbook to make a course like that possible. A key question in designing that course is: should it be limited to special relativity (where a complete understanding is possible) or should it try to address general relativity? Knowing that students are intensely curious about black holes and the Big Bang, I have done my best to help them develop the relevant thinking tools there too, while relaxing the goal of complete understanding. Instructors can therefore use this book in various ways. If your highest priority is systematic understanding, you may wish to proceed sequentially and slowly, leaving the last few chapters for interested students to read on their own. If black holes are a must-do, consider skipping some of the details of earlier chapters. (I have marked some sections with an asterisk to indicate candidates for skipping; I also skip some mathematical details in sections that cannot be skipped conceptually.) Similarly, while my approach is to require a minimum of math in homework problems, there are many opportunities to use more math if that suits your audience. If you would rather emphasize scientific literacy and media consumption skills, there are links in those areas as well.

Having taught the course several times, I gradually learned how to present the ideas to students effectively:

- Spacetime diagrams are introduced very early, in the context of Galilean relativity. Students avoid cognitive overload by becoming familiar with worldlines, events, and so on *before* they wrestle with any of the ideas of special relativity.

- The emphasis more generally is on graphical understanding. Although equations are necessary for rigor, beginners do better when they can *see* the essentials of the situation rather than try to extract them from an equation.

- Accelerated frames are also introduced prior to special relativity, so students are not blindsided by the twin paradox.

- General relativistic thinking tools are presented as a natural evolution of special relativistic thinking tools, so general relativity seems less like a separate and forbidding domain of knowledge.

- Thinking tools are presented quite explicitly as tools. This makes relativity more accessible, but equally importantly it cultivates metacognitive skills. Students who may think "I'm not good at this" are explicitly given the tools to practice and *become* good at it.

- Thinking tools that go beyond relativity, such as symmetry, are also emphasized. My goal is to make these thinking tools so familiar that students may begin to apply them outside the context of the course.

- Research shows that students benefit from revisiting topics and making connections between different topics. I therefore allow understanding to unfold in layers rather than attempt to force complete understanding of a topic on a single chapter. The Einstein velocity addition law is a case in point: Chapter 5 merely provides a mental picture to make such a law seem intuitively possible, Chapter 6 first shows how it works graphically, Chapter 8 provides more graphical velocity addition practice, and Chapter 9 explains the law mathematically.

In summary, I believe this book enables students with no physics background to *understand* relativity rather than just read a description of it, and enables more faculty to offer general education courses on relativity. I hope you will find this book stimulating and rewarding.

David Wittman
Davis, California
January 20, 2018

Acknowledgments

This book grew from a course I taught, so I start by thanking people related to that course. I never would have tried teaching such a course without Will Dawson as a teaching assistant; I knew I could trust him to do a great job developing discussion activities, guiding students, challenging my facile explanations, and helping me shape the course. Jeff Hutchinson performed just as ably in later iterations of the course; he also corrected mistakes and suggested clarifications in an initial draft of this book. I also thank the students who suffered through initial drafts of this book. The honors students who took the course in fall 2013 deserve special mention for constructive exchanges that stimulated me to rewrite the draft almost from scratch. Four students from that course—Shuhao David Ke, Olga Ivanova, Dean Watson, and Adam Zufall—read a completely new draft in fall 2014 and provided useful feedback. Adam Zufall in particular provided specific and insightful feedback and debated some changes with me.

Outside the context of the course, Chuck Watson deserves special mention for reading every single line of two separate drafts and providing extremely thorough and thoughtful feedback. Chuck provided a greatly needed wake-up call that many sections of an earlier draft were inadequate for beginners, and if the book now makes sense to them they should thank Chuck for that. Chuck also caught numerous typos and awkward or ambiguous passages, often suggesting better phrasing than I came up with myself. My wife, Vera Margoniner, also read some versions of each chapter and help me clarify the presentation of many points. I also thank Steve Carlip for providing expert advice on a variety of points in the later chapters. It is a cliché to write that any remaining errors are my own, but now I know that the cliché is absolutely true. Each of these people kept me on track at some point, but I may have veered off track through numerous revisions since then.

I thank Adam Taylor for the handful of stylish drawings you see in this book; the less stylish ones are my own. I also thank Vivian Ellinger for making it possible to work with Adam.

I owe the biggest thanks to my family—Becca, Linus, and most of all Vera—for supporting me through such a large time commitment. My parents, Linus and Bonnie Wittman, not only raised me well but also gave freely of their time and energy in the past few years so I would have more time to focus on the book. Thanks, geysers!

Contents

Guide to the Reader

Teachers rarely give their students explicit instruction in how to read. The assumption must be that they learned how to read as children, so we have nothing to add at this point. I believe this consigns many students to ineffective study habits; many have never thought explicitly about reading strategies. The fact is that you should not read a book of ideas straight through like a novel. You should be engaging in a conversation with the book, identifying the key points and arguing back until you come to terms with them.

True learning does not happen quickly and easily, so budget plenty of time for each chapter and perhaps skim its sections first to help you budget wisely. Then read one section at a time and give yourself time to really think about the concepts. Consider taking a break between sections—reading too much in one sitting will reduce your comprehension of the later parts.

When you finish reading a section, thoroughly consider the *Check Your Understanding* question before moving on. Use your performance on that question to rate your level of understanding, and keep track of which sections you will need to reread. When rereading a section, focus on the paragraphs and figures that seem most important or most relevant to your difficulties rather than rereading uniformly from start to finish. If a point is still unclear after rereading, make a note to discuss the question with another student or with the instructor to clarify the concept. Then, make sure to reread the relevant point again after that discussion to check your new understanding. This is crucial because listening to a clear explanation does not necessarily make it stick in your mind, even if you feel strongly at the time that it will (a phenomenon psychologists call the fluency illusion).

At the end of a chapter, check the list of key concepts in the chapter summary and ask yourself if you understand them completely. The chapter summary will come in handy when reviewing or rereading, but do not fool yourself into thinking that reading—or even memorizing—the summary alone is a substitute for deeply engaged reading. A great way to process the ideas is to close the book and attempt to write down the major points yourself. Writing boosts your learning by engaging a different set of brain circuits.

Chapter end matter includes both exercises and problems, and the distinction is extremely important. Exercises are straightforward procedures that help you rehearse concepts and skills. Problems do not come with a well-defined procedure; you really have to *think* about the solution. Take rock climbing as an analogy. If you want to be a good rock climber, you need strong arms so you repeatedly do chinups, which are straightforward and trivial to describe, but still take practice.

Confusion alert

These are posted to sharpen the distinction between two similar concepts or words, or between the physics and everyday meanings of a word, or in other situations where miscommunications are common. These alerts inoculate you against the most common sources of miscommunication in discussing relativity so keep them in mind not only as you read, but also as you discuss relativity with others.

Think about it

If you find the main text on a page relatively clear and easy to digest, you *should* be simultaneously engaged in relating it to earlier points and to personal experience; these notes help prompt this engagement, and help answer questions such as "But how is that consistent with...?" In your first time through any given section, you may have cognitive overload just processing the main text. In that case, it may be better not to dwell on a *Think About It* note, but make sure to reread the section later when you are able to process it on this deeper level.

But you *also* need to practice climbing on real obstacles; this is where the *skill* of rock climbing is. Exercises are necessary, but problems develop higher-level skills. Problems force you to apply your thinking tools to new and unfamiliar situations. This can be the most difficult part of learning—but also the most rewarding, because it builds true understanding.

★*Optional sections.* One of the beauties of science is that everything is connected, but this is also one of the difficulties of teaching science: where to stop? Sections marked with asterisks (and boxes, which are smaller) are not absolutely necessary to understanding the main thread of the course. Students who feel comfortable with the main thread will benefit from making these additional connections, but students who need to focus fully on the main thread may skip these boxes at first. Of course, instructors will vary in their opinion of what is optional, depending on the length of the term and the level of student preparation; students are advised to rely on their instructor for detailed guidance. Readers who are teaching themselves should pay close attention to the asterisks; if an optional topic seems more confusing than enriching, refocus on the main thread and return to the optional topic later as desired.

A First Look at Relativity

Relativity is a set of remarkable insights into the way space and time work. The basic notion of relativity, first articulated by Galileo Galilei (1564–1642), explains why we do not feel the Earth moving as it orbits the Sun and was successful for hundreds of years. By the turn of the twentieth century, however, it became apparent that Galilean relativity did not provide a complete description of nature, particularly at high speeds such as the speed of light. In 1905, Albert Einstein (1879–1955) discovered unexpected relationships between space and time that allow relativity to work even at high speeds; this is now called special relativity. Soon after, Hermann Minkowski (1864–1909) found a way to express these relationships in terms of the geometry of a single unified entity called spacetime. Einstein initially resisted this point of view but eventually adopted it and pushed it much further in his 1915 general theory of relativity, which explained gravity itself in these geometric terms. The insights of general relativity are abstract and help us understand extreme phenomena such as black holes, but they also have everyday consequences: general relativity is used by smartphones everywhere to locate themselves in Earth's gravitational field with the help of the Global Positioning System.

We begin with Galilean relativity.

1.1 Coordinates and displacement

Understanding relativity requires, first of all, clear language for *describing* motion. Imagine an unmoving camera capturing a series of images of a bicyclist going by from west to east:

Assuming we care only about the overall position of the bike at any given time we can boil this information down to a **motion diagram**, which represents the position of the bike at any given time as a dot:

The Elements of Relativity. David M. Wittman, Oxford University Press (2018).
© David M. Wittman 2018. DOI 10.1093/oso/9780199658633.001.0001

To help you recognize the order in which the positions were recorded, the older dots are progressively more faded. This motion diagram eliminates details such as how the pedals were turned or when the rider drank from the water bottle, but it captures the essence of the motion. By simplifying the bike down to a featureless dot we have adopted a **particle model** of the bike. The word *particle* will appear often in this book, indicating that we do not care about the details of the object performing a particular motion. For variety, we may also refer to named objects or characters in motion, but the particle model is still implicit unless otherwise noted.

By stripping out other details, the motion diagram helps us focus on a particle's position. The *change* in a particle's position from one time to another is called its **displacement** during that time. Displacement is distinct from position; if an object never moves during an experiment, it has no displacement, but it definitely has a position! Although they are distinct concepts, we measure displacement and position in the same units, the most common being meters (abbreviated to m) and kilometers (km). We will also occasionally use feet and miles for variety.

To quantify position and displacement, imagine a tape measure anchored at the west edge of the scene and stretched to the east:

The bicycle's displacement between two snapshots is then the tape measure reading in the second snapshot minus the tape measure reading in the first snapshot. To avoid cumbersome phrases such as "the tape measure reading in the second snapshot" we give the tape measure reading a shorter name; physicists like to use x. By itself, x will refer to a tape measure reading at any time. Subscripts will refer to specific tape measure readings; for example, x_1 is the tape measure reading in the first snapshot and x_2 is the tape measure reading in the second snapshot. Furthermore, the symbol Δ (the upper-case Greek letter "delta") will indicate a *change* in any quantity; for example, Δx is the change in x. So the displacement between the first two snapshots is $\Delta x = x_2 - x_1 = 3.5 - 0.5 = 3$.

By the simple act of placing the tape measure, we have defined a **coordinate system**. A coordinate system consists of an **origin** (the start of the tape measure), a direction (numbers increase to the east, for example), and a scale (meters, feet, inches, or whatever is most convenient). Note that our choice of origin does *not* affect the displacement we compute; had we anchored the tape measure 10 m more to the west, then *each* of the two readings that determine the displacement would be 10 m larger, and their *difference* would not be affected. So, in physics problems we are free to set the origin where it is most convenient. We may encounter problems where a proper choice of origin makes the answer easier to calculate, but the physical result cannot *depend* on the choice of origin. For example, we can choose to measure the height of a tennis ball in terms of height above ground or height above the net. This choice affects the numerical value of the height of the ball and of the height needed to clear the net, but does *not* change the answer to the question, "Did the ball clear the net?" This is just one example of **coordinate independence**, a key idea that will appear in additional forms throughout the book.

Check your understanding. In some countries the first floor of a building is understood to be the floor you walk in on, while in others it is understood to be the floor immediately above that one. For each of the following statements, assess whether it is a position or a displacement, and whether it is coordinate-dependent or -independent: *(a)* Alice's office is on the fifth floor; *(b)* Bob's office is on the third floor; *(c)* Alice's and Bob's office are two floors apart.

1.2 Velocity

If displacement tells us how far the bicycle moved, velocity tells us how quickly it executed this motion. To compute this, we need to introduce an additional coordinate, **time**, which is measured by clocks and denoted by t. The *difference* in time between measurements of the bike's position is denoted Δt, and **velocity** is defined as $v \equiv \frac{\Delta x}{\Delta t}$. The $\equiv$ symbol (read "is defined as") is used here to reinforce the notion that this is a *definition* rather than a conclusion. A definition is a relationship stronger than mere equality; for example, $v = 2$ m/s may be true in some particular situation, but we would never write $v \equiv 2$ m/s. The definition $v \equiv \frac{\Delta x}{\Delta t}$ is useful because it provides a recipe for quantifying the *rate of change* of the position x. The displacement Δx alone cannot distinguish, for example, between the motion of a snail and a sprinter in a 100 m race. The distinction lies in the sprinter completing the displacement in a small Δt (thus yielding a large $\frac{\Delta x}{\Delta t}$) while the snail requires a large Δt (thus yielding a small $\frac{\Delta x}{\Delta t}$).

The direction of motion is inherent in the idea of velocity. If the coordinate system for the 100 m dash is a tape measure stretched from start to finish, someone who runs in the wrong direction has a *negative* velocity because x_2, the runner's position at time t_2, is *less* than x_1, the runner's position at time t_1. This makes $\Delta x = x_2 - x_1$ negative, which in turn makes $v = \frac{\Delta x}{\Delta t}$ negative. In this one-dimensional coordinate system velocity is a single number, with the direction of motion encoded by the presence or absence of a minus sign in front of the number. With coordinate systems that describe two or more dimensions (e.g., a map that extends north-south as well east-west), the full specification of velocity requires a bit more care, and we defer that to Section 1.4.

Velocity appears on a motion diagram as follows. Each dot on a motion diagram indicates an **event**, which is defined by its time as well as its position. In principle, we can label each dot in a motion diagram with the time it was recorded, but it is more convenient to simply record snapshots at regular time intervals so that Δt is the same between any two successive snapshots. Then the motion diagram is

> **Confusion alert**
>
> *Velocity* is one of several words—including *acceleration, energy,* and *momentum*—that have specific meanings in physics but are used loosely in everyday speech, so take care to understand each physics definition as it arises.

a visual representation of velocity as well as position and displacement; with Δt constant, any variations in Δx *must* be due to variations in v and vice versa. In this particular motion diagram,

we see that Δx is the same (+3 m) between any two successive snapshots, so the velocity here is constant. In fact, we will study constant velocity for the remainder of this chapter because there is much to say even in this simple case.

Because $v \equiv \frac{\Delta x}{\Delta t}$, velocity can have units of meters per second (m/s), kilometers per hour (kph), or miles per hour (mph) for everyday things such as cars, or kilometers per second (km/s) for extremely fast things such as spaceships. In the motion diagram we have studied, if the units of distance are meters and the camera takes a snapshot once each second ($\Delta t = 1$), the velocity of the bike is +3 m/s.

I list these units to help you relate velocity to everyday experience, but physicists find it helpful to focus less on the specific units and more on what they *mean*. We will often do abstract things like compare the velocity of some object to the velocity of light, to see if they are of comparable size. But if they are of comparable size in one system of units, then they are of comparable size in *any* system of units. So, in a very important sense, units will not matter in much of this book; what matters is that velocity is a displacement divided by a time. That said, sometimes attaching specific units to an abstract idea does help you understand the idea. Feel free to take any abstract statement or idea in this book and take it for a test drive in the units of your choice.

Check your understanding. What is the velocity of a rocket that moves 10 km eastward in 0.5 s? What is the velocity of a car that moves the same 10 km in 10 minutes? Compare the two velocities in the same units.

1.3 Galilean velocity addition law

Let us call our bicyclist Alice. In Section 1.2 we used a motion diagram to determine Alice's velocity as +3 m/s (or 3 m/s to the east). This really means 3 m/s to the east *through the coordinate system we defined*. You probably assumed the coordinate system and attached camera were fixed to the road, thus interpreting Alice's velocity as 3 m/s to the east *relative to the road*. But not all coordinate systems are fixed to the road. Imagine that the motion diagram data were actually recorded by a camera in the helmet of a second cyclist, Bob, who is himself moving relative to the road as shown in Figure 1.1. If this is the source of the motion diagram data we saw in Section 1.2, we must specify that Alice moves at +3 m/s *relative to Bob*.

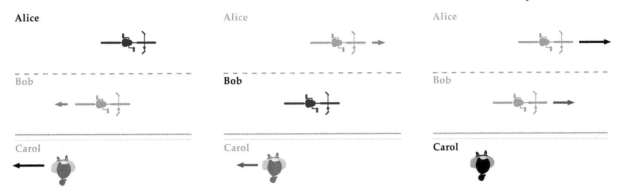

Figure 1.1 *Alice, Bob, and Carol move at different velocities. Arrows represent the velocities of each character as measured in coordinate systems attached to Alice* (left panel), *to Bob* (middle), *and to Carol* (right). *Any two characters always measure each other as moving at the same speed but in opposite directions; arrows have been shaded to help you match equal and opposite velocities. The challenge in this section is to see how relative velocities of two characters, say Alice and Carol, could be deduced from measurements in a third frame such as Bob's.*

This example demonstrates that the meaning of a velocity depends on the coordinate system—also known as the **frame of reference**, or simply **frame**—in which it is measured. When you read the phrase "Bob's frame" picture an entire coordinate system attached to Bob; in this frame Bob's velocity is always zero.

Practice thinking in different frames of reference for a moment. Figure 1.1 adds a third character, Carol, who is fixed to the road. The middle panel shows that in Bob's frame Alice is moving to the east but Carol is moving to the *west*. To see why, imagine the view from a camera attached to Bob: Alice is passing him even as he passes Carol. For more practice, imagine the view from Alice's frame as in the right panel of Figure 1.1. In this frame, Bob has a westward velocity, and Carol has an even larger westward velocity. We will use subscripts to track the frame in which a velocity is measured: v_{AB} is Alice's velocity as measured in Bob's frame (also stated as "relative to Bob"), v_{BC} is Bob's velocity relative to Carol, and so on. Note that any two frames always have equal and opposite velocities relative to each other: if Carol measures Bob moving east at 5 m/s relative to her ($v_{BC} = 5$ m/s), then Bob must measure Carol as moving *west* at 5 m/s relative to him ($v_{CB} = -5$ m/s).

Let us return to Bob's frame to ask a fundamental question. If we know what Bob measures for Alice's velocity *and* what he measures for Carol's velocity, can we deduce what Carol and Alice measure for their velocity relative to each other? Most people have a strong intuition on this question, based on everyday experience. To use a money analogy, if Alice has $3 more than Bob and Bob has $5 more than Carol, then Alice clearly has $8 more than Carol. Why would we not do the same with velocities? If we know that $v_{AB} = +3$ m/s and $v_{BC} = +5$ m/s, how can it not be the case that $v_{AC} = +8$ m/s? Abstracting away from specific numbers, this intuition suggests that velocities add according to $v_{AC} = v_{AB} + v_{BC}$.

Think about it

Each coordinate system includes *time* as a coordinate; without this coordinate we could describe locations but not motion. Time is measured by clocks rather than rulers, so thinking of time as a coordinate may take some effort initially—but this effort will pay off in the long run.

Confusion alert

Avoid using terms such as *left* and *right* when describing directions of motion. These terms cause confusion because they depend on the direction a person is (or imagines) facing, whereas all participants agree on the meaning of terms such as *east* and *west*.

Think about it

The type of addition used in the Galilean law is referred to as **linear**. To illustrate that other kinds of addition are possible, consider a stack of pillows: because the lower pillows compress, the height of the stack is less than the sum of the heights of the pillows separately. The addition of pillow heights is sub-linear.

This is called the **Galilean velocity addition law**. You can make sense of the subscripts in this equation by thinking of Bob as a middleman: the left side of the equation cuts out the middleman and predicts the result of a direct velocity measurement between the other two parties.

The Galilean law makes intuitive sense, but intuition is often flawed—velocity measurements are based on rulers and clocks, and do not necessarily behave like money. Science demands a two-pronged strategy here: identify the assumptions behind our intuition so we can present a clearly defined *model* of nature, and perform experiments to determine whether nature actually follows this model. Experiments do show that the Galilean law works very well for everyday velocities—but not for very large velocities. This section unravels the model behind *why* it works for everyday velocities, so we can better understand (in later chapters) why it does *not* work in all situations.

Ready to unravel the assumptions? We are asked to predict v_{AC} (Alice's velocity through Carol's coordinate system) given a measurement of Alice in *another* coordinate system. If we assume, as did Galileo, that clock velocities do not affect their time measurements, then the time Δt between any two events is the same regardless of the coordinate system, and we can write $v_{AC} = \frac{\Delta x_{AC}}{\Delta t}$ without putting any subscript on the Δt to specify the frame of the clocks involved. This seems reasonable, but keep in mind that this is an *assumption* about the behavior of clocks, to be revisited in later chapters.

Next, we predict Δx_{AC} (Alice's displacement through Carol's frame) knowing only Δx_{AB} (Alice's displacement through Bob's frame) and Δx_{BC} (Bob's displacement through Carol's frame). If we assume (as did Galileo) that ruler velocities do not affect their distance measurements, then meters of displacement measured in Bob's frame are completely interchangeable with those measured in Carol's frame. Again, this is an *assumption*, not a conclusion, and evidence will forced us to revise this assumption in later chapters. But for now, this assumption allows us to add displacements as if they had been measured by the same ruler: $\Delta x_{AC} = \Delta x_{AB} + \Delta x_{BC}$. Under these assumptions, then, $v_{AC} = \frac{\Delta x_{AC}}{\Delta t} = \frac{\Delta x_{AB} + \Delta x_{BC}}{\Delta t}$. We can rewrite this last quotient as $\frac{\Delta x_{AB}}{\Delta t} + \frac{\Delta x_{BC}}{\Delta t}$, which, under the interchangeable-time assumption. is the same as $v_{AB} + v_{BC}$. This completes the proof that $v_{AC} = v_{AB} + v_{BC}$, provided that our assumptions about rulers and clocks are correct.

You are probably not surprised that velocities add this way—anyone who has walked on a moving sidewalk or train has experienced it. Yet, if all velocities add this way there will be profound implications:

Nature should have no speed limit. In principle, there is no upper limit to the speed we can achieve by concatenating an arbitrarily large number of velocity additions, such as firing a bullet from a missile launched from a moving train and so on. A more practical way to achieve such high speeds would be in space, where there is no air resistance, using an engine to provide a long series of small boosts rather than a few dramatic boosts. But the practical details are less important than the logical conclusion that the Galilean model must allow arbitrarily high speeds.

Today it is common knowledge that nature *does* have a speed limit—the speed of light—so you already know that one or more of Galileo's assumptions must be wrong. In later chapters we will discover how and why they are wrong.

The laws of motion are the same in any constant-velocity frame. If all frames have equally valid distance and time measurements then there is nothing special about a frame fixed to your portion of the surface of the Earth. Imagine yourself inside a smoothly moving train or airplane. If you drop an object, it does not fly backward as it would if it were stuck to the frame of the Earth; it simply falls straight down relative to the moving vehicle. Galileo was the first to notice this: inside any laboratory (which is just a more concrete word for frame of reference) moving at constant velocity, the laws of motion are the same as on the "stationary" ground. He argued that if the laboratory's motion has no effect on experiments inside, there is no reason to declare one laboratory "stationary" and the other "moving"—we can only say that they are moving relative to each other. We are tempted to reserve the words "stationary" or "at rest" for labs fixed to the surface of the Earth, but the insight here is that *even Earth need not be stationary*—we would not feel or measure anything different on a stationary Earth versus an Earth moving at constant velocity. Galileo's insight into relativity helped overcome a persistent objection to the idea that Earth orbits the Sun: that if Earth moved, we would feel it.

Today it is easy to view those who argued "if Earth were moving people would feel it" as ignoramuses, but their experience was rife with situations in which motion *is* felt. Consider running or horseback riding: you feel the wind in your face and *dropped objects do fly backward*. We now attribute this to air resistance because we can contrast the feeling of riding in a car with the windows open (or in a convertible with the top down) versus with the windows closed (or the top up). Seventeenth-century citizens never saw air resistance turned off, nor could they easily imagine the emptiness of space that allows Earth to move forever without resistance. A second reason behind the widespread "if you are moving you feel it" belief is that everyday life is full of *variations* in velocity, which *can* be felt; the laws of motion are *not* the same in frames that change their velocities. We defer more discussion of this important point to Chapters 2 and 4.

You might think that glancing out the window of a smoothly moving laboratory is enough to tell you whether it is moving, but in fact this only tells you whether it is moving *relative to the Earth*. This is the origin of the word *relativity*: we can determine the relative velocities of laboratories, but there is no such thing as an absolute velocity.

If the laws of motion are the same in any constant-velocity frame, then perhaps *all* the laws of physics are the same in any constant-velocity frame. This in fact seems to be the case, as no counterexample has ever been found. Experiment alone can never *prove* this conjecture because we can never do all possible experiments. But because no exceptions have been found, we take this conjecture as a working hypothesis and deduce further consequences that are then tested by new experiments. This process of *hypothetico-deductive reasoning* is the backbone

Think about it

The velocity of Earth in fact varies over time, but these variations are too small to notice in everyday life.

of science. Relativity is a wonderful arena for hypothetico-deductive reasoning because many fascinating and testable consequences can be deduced from a few basic principles.

Check your understanding. Aboard a train moving eastward at 90 kph, a bicyclist rides toward the rear of the train. The speedometer on the bike reads 20 kph. *(a)* What is the velocity of the bike relative to the ground? *(b)* The bicyclist sees an ant on the bike, crawling at 0.1 kph toward the rear of the bike. What is the velocity of the ant relative to the ground?

1.4 Velocity is an arrow

Imagine that, in previous sections, Alice was riding her bicycle away from home. She now returns home:

From one snapshot to the next, displacement is now negative; for example, $x_2 - x_1 = -3$. The velocity between those snapshots is also negative because its numerator (the displacement $x_2 - x_1$) is negative. The sign of the displacement or velocity tells us which direction the bicycle moved. Displacement and velocity are called **vector** quantities because they describe a direction as well as a size—think of them as arrows rather than numbers. A quantity that is simply a size with no associated direction, such as two cups of flour, is called a **scalar** quantity.

In many situations, the size—also called the **magnitude**—of a vector is more important than the direction. For example, a 100 kph wind is dangerous regardless of its direction. Physicists therefore have a special word just to describe the magnitude of the velocity vector: **speed**. In the one-dimensional motion diagrams in this chapter, a speed of 3 m/s can correspond to one of only two velocities: 3 m/s to the east (+3 m/s) or to the west (−3 m/s). In this special case you can infer the direction from the sign on the number, but this will not be possible for motions in two or three dimensions.

Vectors are often described with a magnitude and a direction: for example, 100 m to the southwest or 50 kph to the north-northeast. But we will more often describe a vector by breaking it down into **components** that align with the coordinate system. For example, a 13 m displacement to the north-northeast may break down into 5 m to the east and 12 m to the north (Figure 1.2). Keep in mind that either method is simply a way to describe an arrow. You should always think of a vector, such as displacement or velocity, as an *arrow* rather than a number or list of numbers.

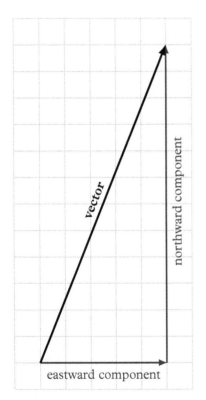

Figure 1.2 *A vector broken down into perpendicular components.*

The velocity addition law $v_{AC} = v_{AB} + v_{BC}$ (and all other equations involving vectors) work with either type of vector description. For those who prefer literally drawing arrows, there are many internet resources demonstrating graphical vector addition and subtraction. Those who prefer components should simply apply the equation separately to each component. Practicing vector laws with a single component therefore saves time and is usually enough for a solid understanding. Most examples in this book deal with motion in one dimension, which is depicted as toward the right ("east") or left ("west") of the page. When the direction of motion is perfectly clear from the context, a simple number such as $v = 10$ m/s may be an acceptable way to specify velocity. "Velocity is an arrow" then helps us see that $v = -20$ m/s would be an arrow twice as long *in the opposite direction*.

Let us practice the velocity addition law with velocities of different signs in one dimension. Draw this out for yourself: a car moves east with a velocity relative to the road of $v_{CR} = 100$ kph and in the frame of the car, a fly flies from front to back at 10 kph. Because the fly is moving to the *west* in the frame of the car and we have decided that eastbound velocities are positive, the velocity of the fly as observed by the car must be written as $v_{FC} = -10$ kph. The velocity of the fly relative to the road therefore must be $v_{FR} = v_{FC} + v_{CR} = -10 + 100 = 90$ kph. A variation on this example: a child in the back seat of this car throws a toy forward at 5 kph (in the frame of the car). The velocity of the toy relative to the road is then $v_{TR} = v_{TC} + v_{CR} = 5 + 100 = 105$ kph.

Check your understanding. Consider a one-dimensional motion. *(a)* Can an object with negative velocity have a positive position $(x > 0)$? *(b)* Can an object with positive velocity have a negative position $(x < 0)$?

1.5 Symmetry and the principle of relativity

The experimentally determined laws of physics do not seem to depend on where the experiment is done; one part of the universe is like any other as far as the laws of physics are concerned. When parts of a whole are similar to each other we say that the whole has **symmetry**. For example, the appearance of a wheel is unchanged by turning it (rotational symmetry), and human beings have a particular kind of symmetry in which the left half is similar to the right half (bilateral symmetry). A useful way to think about symmetry is that it allows something to remain unchanged when you perform an action on it, such as rotation (for a wheel) or reflection in a mirror (for a human being). This helps us use symmetry as a thinking tool in more abstract situations—for example, we can ask if the laws of motion remain unchanged when we change the velocity of the laboratory.

Symmetry is a big idea in physics for both practical and profound reasons. Practically, symmetry provides a valuable tool for solving problems, as we shall see throughout the book. More profoundly, the symmetries we observe in the laws of physics are clues to the nature of the universe we live in. The vast amount

Think about it

Real human beings are not exactly symmetric; for example, the heart is on the left side. In this case physicists say the symmetry is "broken"—it is not perfect but it still provides a useful approximate description.

of experimental evidence supporting the **principle of relativity**—the idea that the laws of physics are the same in any constant-velocity frame—implies a new, nonobvious "velocity symmetry" of the universe. This principle by itself is too broad to allow us to deduce any details such as how to add velocities. If we assume that Galileo's velocity addition law is correct, we can fill in many more details; the resulting set of conclusions is called **Galilean relativity** (Chapters 3 and 4). Nature, however, seems to add velocities in a way that approximates the Galilean law at low speeds while prohibiting speeds above 299,792.458 m/s. Starting in Chapter 5 we will see how this observation, combined with the principle of relativity, allows us to deduce a more nuanced set of conclusions called **special relativity**. We will then see how understanding special relativity forces us to think about gravity in a new way; the resulting model of gravity is called **general relativity**.

Check your understanding. In what ways are each of these objects at least approximately symmetric? *(a)* a pinwheel; *(b)* a ball; *(c)* a sea star.

CHAPTER SUMMARY

- Coordinate independence: because the physical result in any situation cannot depend on your choice of coordinate system, you are free to define whatever coordinate system you find convenient.

- Galilean velocity addition law: if Alice moves through Bob's coordinate system with velocity v_{AB} and Bob moves through Carol's coordinate system with velocity v_{BC} then Alice moves through Carol's coordinate system with velocity $v_{AC} = v_{AB} + v_{BC}$.

- Principle of relativity: the laws of physics are the same in any constant-velocity frame of reference.

→ STUDY ADVICE

Research has shown that the best way to study is to practice *retrieval* at spaced intervals. For each of the concepts in the chapter summary, write down now what you remember without looking back at the text; the effort of trying to remember may be difficult but this effort is good for your learning. (Of course, you are encouraged to look back at the text to check and refine your list *after* writing down everything you remember.) Tomorrow, practice retrieving these concepts from memory without looking at your list and repeat again within a few more days. It may feel awkward to put things in your own words but research has also shown that *generating* your own statements is key to learning. This system is much more effective than other forms of studying, such as highlighting and rereading, and will help you *target* your rereading to where it is most needed.

Chapter-end exercises and problems will be crucial for strengthening your understanding, but retrieval practice is important for getting the big picture and seeing the connections between different concepts.

The study advice here is based on *Make It Stick: The Science of Successful Learning* by Peter C. Brown, Henry L. Roediger III, and Mark A. McDaniel. This book helped me as an instructor understand the poor results some students were getting despite studying hard: some study habits are not only ineffective but also mislead the student into thinking he or she has mastered the material. All students could benefit from consulting this book.

◤ CHECK YOUR UNDERSTANDING: EXPLANATIONS

The end of each chapter contains model responses to the *Check Your Understanding* questions at the end of each section in that chapter. Many questions do not have a single correct answer; the model response to such questions should be considered just one example of a range of potentially correct responses.

1.1 (a) position, coordinate-dependent; (b) position, coordinate-dependent; (c) displacement, coordinate-independent. Note that positions are by their nature coordinate-dependent. (If you responded "coordinate-dependent" to (c) because you were thinking of coordinate systems marked off in various units such as floors, meters, or feet: this is understandable, but the main point of this exercise was simply to understand that the displacement in floors does not depend on how we label the floors.)

1.2 20 km/s to the east and 1/60 or 0.0167 km/s to the east. (I provide a decimal answer here in case you used a calculator, but I recommend keeping fractions as fractions; this more clearly exposes certain relationships and avoids the issue of rounding.)

1.3 (a) 70 kph to the east; (b) 70.1 kph to the east.

1.4 (a) Yes. (b) Yes. More generally, the position of a particle at a given time tells us nothing about the *rate of change* of its position, and vice versa.

1.5 (a) rotation, but only by a specific amount about a specific axis, leaves the pinwheel unchanged; (b) rotation by any amount about any axis; (c) the sea star has the symmetry described in part (a) plus mirror symmetry—which the pinwheel does *not* have.

? EXERCISES

1.1 State in your own words the distinction between position and displacement.

1.2 State in your own words the distinction between velocity and speed.

1.3 What are the benefits of modeling an object as a featureless particle? What is lost in this model?

1.4 Explain how a motion diagram is made.

1.5 In what sense does it matter where you put the origin of a coordinate system? In what sense does it not matter?

1.6 Identify each of the following as a vector or a scalar: *(a)* 100 kg; *(b)* 2 blocks north; *(c)* a 50 kph wind from the west; *(d)* 1 cup of rice; *(e)* a top speed of 200 mph.

1.7 Train A moves at 200 kph to the east, while on a parallel track train B moves 200 kph to the west (these velocities are relative to the ground). *(a)* What is the velocity of train B relative to train A? *(b)* What is the velocity of train A relative to train B? *(c)* Does the relative velocity change after they pass each other?

1.8 What, in your own words, is the principle of relativity?

1.9   A piece of cargo falling off a truck on a highway is extremely dangerous because it moves at high speed relative to the vehicles on the highway. How is this consistent with the principle of relativity, which would seem to predict that loose cargo will not fly backward from the truck?

➕ PROBLEMS

Subsequent chapter endings will list problems as well as exercises. The difference is that an exercise follows an established procedure (sometimes as simple as summarizing an established point in your own words) while a problem deepens your understanding by forcing you to navigate new and more open-ended situations. There is value in each: you need to practice the basics before you can use them in new ways and new situations. Because problems often go deeper than a single chapter, there are none here— but be aware that rehearsing established procedures (i.e., completing exercises) is only half the battle at most. The greatest challenge for physics students is applying a concept to new situations, and for a good reason—this is the true test of understanding.

You can increase your chance of success in navigating new situations by making sure you practice each new tool or skill as it arises, before you encounter the new situation. Navigating a new situation *while* attempting to catch up on mastering tools and skills is a recipe for cognitive overload. This is why I have separated exercises and problems.

Acceleration and Force

2

So far, we have considered coordinate systems moving at constant velocity relative to each other. Now imagine Alice driving a car at constant velocity relative to the road and Bob driving in the next lane, *not* at constant velocity relative to the road. Bob and Alice must measure nonconstant velocities relative to each other, but can Bob argue that he is the one traveling at constant velocity and Alice is the one whose velocity is changing? Your intuition may say that the road frame is a good arbiter of this dispute, but why? In this chapter, we will discover a profound difference between constant-velocity frames and other frames by thinking about these questions.

2.1 Acceleration

An object not traveling at constant velocity is said to be **accelerating**. If Bob is accelerating relative to the road, his motion diagram as recorded by a camera fixed to the road may look something like this:

Between snapshots, Bob moved 2 m, then 3 m, and finally 4 m. If the snapshots were taken once per second, then his velocity v was 2 m/s between the first two snapshots, then 3 m/s, then 4 m/s. His velocity increased by $\Delta v = 1$ m/s each second. The change in velocity per unit time is also known as the **acceleration**. Mathematically, acceleration is defined as $a \equiv \frac{\Delta v}{\Delta t}$, so it has units of m/s per second or m/s^2. Although this is pronounced "meters per second squared" it is best understood as "meters per second (change in velocity) each second."

Imagine that in the road frame Alice's velocity is a constant 2 m/s to the east. In the first second of this story she is therefore moving at the same velocity as Bob (i.e., she measures him as moving at zero velocity relative to her), in the next second she measures Bob as moving 1 m/s to the east relative to her, and in the final second of the story she measures him as moving 2 m/s to the east relative to her. So she agrees with the road frame that Bob's acceleration is 1 m/s^2. This is a hint that there is a universal standard of acceleration. Try the argument with hypothetical observers traveling at *any other* constant velocity relative to the road: they will

> **Think about it**
>
> Practice applying the logic in the opening paragraph: why must Alice and Bob measure nonconstant velocities relative to each other?

> **Confusion alert**
>
> For reasons that will become clear shortly, *acceleration* in physics also includes cases where the velocity *decreases* with time—what we would call *deceleration* in everyday life.

always measure Bob's acceleration to be 1 m/s². This is remarkable, because these hypothetical observers do not agree on Bob's displacement or velocity at any time. Yet somehow all these disparate coordinate systems manage to agree on his *rate of change of velocity*. Perhaps Bob *really is* accelerating at 1 m/s² in some absolute sense, not merely relative to each of these frames.

Before stating this conclusion with any confidence, we consider in Section 2.2 a closely related question: what does Bob have to *do* to accelerate?

Check your understanding. The text states that if Bob accelerates at 1 m/s² as measured in Alice's constant-velocity frame, then *any* constant-velocity observer (regardless of their velocity) must also observe him to accelerate at 1 m/s². *(a)* Explain to a hypothetical fellow student why this must be true. *(b)* Build on this to explain why different constant-velocity observers agree that all their relative velocities are indeed constant.

2.2 Acceleration, force, and mass

Everyone knows that to get a stationary object moving (i.e., accelerate it so that its velocity differs from zero) you have to give it a push. Everyone also "knows" from everyday experience that if you stop pushing, the object will slow down and stop. Aristotle (384 BCE–322 BCE) generalized this by teaching that the natural state of motion of any object is to be at rest, and our everyday experience is so salient that this went unquestioned for millenia. But this generalization is mistaken, because everyday experience is limited. If you take special care to reduce any rubbing (also known as *friction*) of the moving object against anything else—think of an air hockey puck—you find that an object in motion has a much greater tendency to *stay* in motion. The more friction you eliminate, the less slowing you observe— and in outer space, things really can continue forever because friction is not an issue. Velocity changes only through interactions with other objects, which we call *forces*.

Even without air hockey tables, seventeenth-century thinkers such as Galileo and Descartes began to grapple with these ideas. Isaac Newton (1643–1727) was the first to articulate clearly that an object maintains constant velocity unless it is acted upon by a force. This is **Newton's first law of motion**. Technically, a constant-velocity trajectory implies no *net* force rather than no forces at all. Consider an evenly matched tug-of-war: each team exerts a large force on the rope but these forces—being of equal size and opposite direction—cancel each other out and leave no net force, so the rope does not accelerate. From our observation that the rope is not accelerating, we can conclude only that there is not *net* force on the rope; we would be quite mistaken if we concluded that neither team is exerting any force. Thus, Newton's first law is best stated: *an object maintains constant velocity unless acted upon by a net force.* We may drop the word "net" when only one force is present, but the "net" concept is always part of the law.

Confusion alert

If an object bounces off a wall without slowing, this still qualifies as a change of velocity because the direction of motion changed.

Confusion alert

Objects with "constant velocity" include those at rest, which have constant zero velocity.

Let's get more specific about the velocity change caused by any given net force. We can investigate this empirically: apply forces of different sizes and directions to an air hockey puck or a ball. You will see that the resulting acceleration is in the same direction as the applied force, and the amount of acceleration is directly proportional to the magnitude of the force. We can write this relationship as $a \propto F$, where the $\propto$ symbol is pronounced "is proportional to." Now if you compare the effect of a given force on balls of different masses, you will find that the acceleration is also *inversely* proportional to the mass; we write this as $a \propto \frac{1}{m}$. If you experiment further you will also find that no other variable affects the acceleration. Therefore, the relationship between acceleration, force and mass can be summarized in one simple equation: $a = \frac{F}{m}$. This is **Newton's second law of motion**. It is more commonly written $F = ma$, but the $a = \frac{F}{m}$ form helps you think of acceleration as the *result* of applying a force, the effect of which is diluted by the mass—think of this as the inertia—of the object.

This law is astoundingly useful. When we observe something accelerate—that is, move in any way other than a straight line at constant speed—we can *infer* that a force was applied even if we do not see directly who or what applied the force. For example, the Moon's motion around Earth is not a straight line so it must be experiencing a force; we will see in Chapter 16 how Newton realized that this force must be the same force that pulls apples toward the ground. Furthermore, if we measure the amount of acceleration—easily done by recording positions and times—we can infer the net force if we know the mass of the object, or the mass of the object if we know the net force. We will see later how a chain of such reasoning allows us to infer the mass of just about anything in the universe.

But what is **mass**? Everyone has a general sense that the mass of an object corresponds to the "amount of stuff" in it: a full bucket of water is undoubtedly more massive than a partially filled bucket. Mass is *not* just size, because we all agree that a small lead ball has more mass than a much larger balloon. We can sense that the lead ball has more "stuff" packed into a smaller volume, but how can we quantify the amount of "stuff"? The answer is to stop thinking of Newton's second law as an empirical pattern based on some intuitive notion of mass, and turn it into a *definition* of mass: apply a force to your test particle, measure the resulting acceleration, and take the ratio, $m = \frac{F}{a}$. This makes mass synonymous with resistance to acceleration, or inertia. The kilogram (abbreviated to kg) is our unit of mass based on the International Prototype Kilogram, a specially made piece of metal stored in a climate-controlled vault in France (proposals for a more stable and reproducible definition of the kilogram are being considered). When we determine the mass of an object we are comparing—through a chain of intermediary comparison masses—our object's inertia to that of the International Prototype Kilogram.

Mass *is* inertia.

Check your understanding. Explain why an empty train can accelerate more quickly than it can when fully loaded.

Think about it

Newton's second law implies the first: $F = ma$ implies that $a = 0$ if and only if $F = 0$.

Think about it

In Chapter 12 we will discover that by this definition the mass of an object depends not just on the masses of its parts but on how those parts are arranged, thus making "amount of stuff" an even less useful way of thinking about mass.

Confusion alert

Mass is not the same thing as weight; we will address this in Chapter 15.

2.3 Accelerating frames and fictitious forces

We saw in Section 2.1 that *all* constant-velocity frames agree on Bob's acceleration (1 m/s^2 to the east), and measure each other as moving at constant velocity. But what prevents Bob from claiming that *he* is the one moving at constant velocity and the other observers are accelerating to the west at 1 m/s^2 relative to him? Certainly, that is what he measures given his coordinate system, so why—other than majority rule—should his measurement be considered less valid than those of the other frames?

It turns out that accelerating frames are objectively different: they violate Newton's first law. Objects with no net force on them appear to accelerate, and objects *with* net force on them may not accelerate. To see this, picture a coordinate system spanning the interior of an accelerating car. In this frame, passengers maintain constant positions (they do not accelerate), but they *do* feel net forces, such as the force of the seat on your back as the car accelerates forward. This force objectively exists, as we can see by putting a force gauge—a spring—between your back and the seat back. When the car accelerates, it is clear to *all* observers that the spring compresses; that is, there is a force at work. If the car returns to constant velocity, all observers agree that the spring returns to normal. The acceleration of the frame itself causes objectively measurable forces on objects fixed to that frame, so Newton's first law cannot work in accelerating frames.

Conversely, in accelerating frames free objects *do* accelerate. In Figure 2.1, the shaded area represents the path of a car turning left. As measured in the constant-velocity frame of the page, a free object such as a phone on the left side of the dashboard continues straight ahead along the dashed line—a textbook example of Newton's first law. But as measured in the accelerating frame, the phone slid to the right, and this violates Newton's first law because force gauges on the phone measure nothing during this slide. A "force gauge" can be almost anything; for example, the screen of your phone flexes and may even crack when a force is applied to it. Picture this in the scenario of Figure 2.1—the phone is completely safe during its slide across the dashboard, and force is applied only when it comes to "rest" against the far side of the car. The accelerating frame thus turns Newton's first law completely backward. In the constant-velocity frame of the page, Newton's first law works as well as ever: the phone experiences force only when the far side of the car finally forces it off the constant-velocity path.

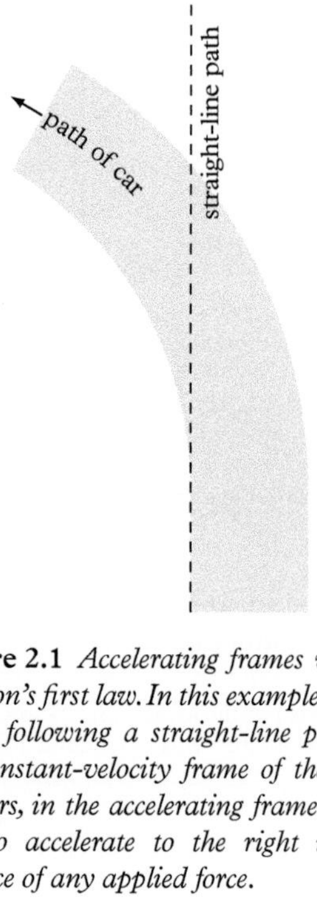

Thus, by testing Newton's first law all observers can agree on which frames are accelerated and which are constant-velocity. Subway trains make great laboratories for practicing the distinction because they start and stop often and accelerate fairly quickly. While the train is accelerating, passengers who stand must brace themselves against falling backward, they must be very careful when they walk, and if they left a phone on the floor it would quickly scurry all the way to the back of the car. All these effects vanish when the train reaches constant velocity—regardless of what velocity that is.

Figure 2.1 *Accelerating frames violate Newton's first law. In this example, a free object following a straight-line path in the constant-velocity frame of the page appears, in the accelerating frame of the car, to accelerate to the right in the absence of any applied force.*

Imagine instead a train that accelerates without end. On this train passengers never experience constant velocity, so they find it difficult to appreciate Newton's first law. For them, the tendency for objects to slide toward the back of the train is simply a permanent feature of life on the train. They may suggest that the back of the train simply exerts an attractive force, like gravity, on everything in the train. Physicists call this a **fictitious force**: a force invented by accelerated observers to make Newton's first law appear to work in their frame. "Centrifugal force" is a familiar example. Passengers in the car depicted in Figure 2.1 would surely explain the slide of the phone as caused by "centrifugal force" but really there is no such force, just the tendency of free objects to continue on constant-velocity paths.

A good mental picture for an acceleration detector is a cup of coffee: the coffee sloshes whenever the lab changes velocity. We will use this coffee-sloshing test as an icon for objective tests of acceleration that do not require position and time measurements.

Check your understanding. Alice and Bob are on a spinning merry-go-round when Alice rolls a ball toward Bob. What kind of path does the ball follow in the merry-go-round frame? In the ground frame?

> **Think about it**
>
> The coffee cup illustrates why the thought experiments of Chapter 1 specified "smoothly" moving trains and ships. Real conveyances are subject to brief accelerations, also known as bumps.

2.4 Inertial frames

Constant-velocity frames are called **inertial frames** to emphasize that Newton's first law is valid in these frames. Experimenters in a laboratory can test whether, for example, a rolling ball maintains constant velocity; if not, the laboratory does not respect Newton's first law and thereby does not constitute an inertial frame. Thus, laboratories can easily determine whether they are accelerating *without* having to measure their motion relative to any other frame. In contrast, the *only* way a laboratory can determine its velocity is by measuring a velocity relative to something else, because there is no absolute test of velocity. Any two inertial frames always measure constant velocity relative to each other. If two frames do not measure constant relative velocity, at least one must be noninertial.

In laboratories on Earth, Newton's first law is obeyed to a very good approximation in the north-south and east-west directions (think of an air hockey table) but in the vertical direction freely moving objects always accelerate downward at 9.8 m/s^2. We call this *gravity*, but it bears some remarkable parallels to simply living in an accelerating frame. We will revisit and extend this idea in Chapter 13.

Check your understanding. A dog named Jack has a ball in his mouth and runs toward his stationary human companion. Jack stops suddenly when he reaches his human but first drops the ball while still at full speed, so the ball rolls far beyond. Specify which of the following frames are inertial: a frame attached to Jack; a frame attached to the ball; a frame attached to the human.

Box 2.1 Newton's third law

Newton's third law of motion is one of the most widely misunderstood concepts in physics because the usual wording, "for every action there is an equal and opposite reaction," is misleading in multiple ways. Newton's third law is not a leading character in this book though, so I will confine my rant to this box.

The law really says that a force is always an interaction between *two* objects; you will never find an object pushing itself in isolation. For example, a car "pushes itself forward" by pushing *back against the Earth*. If this seems abstract, imagine the car on a road made of loose logs; the car is pushed forward only as the logs are pushed backward. If the logs are instead fixed to the Earth they are still pushed backward along with the Earth, but their acceleration $a = \frac{F}{m}$ is immeasurably small because m is now the enormous mass of the Earth.

In the interaction between two objects (call them A and B) the two forces, that of A on B and that of B on A, must be in opposite directions and of the same size. (Otherwise, we could violate Newton's first and second laws by forming an object consisting of A plus B that accelerates *without* an external net force applied.) These two forces are called "action" and "reaction" in the usual wording, but this is misleading because the latter force was not *caused* by the first; instead they are two sides of the same coin. Furthermore, the usual wording seems to imply that nothing can ever happen because every action cancels itself out. But the true meaning is only that an interaction between A and B cannot result in an acceleration of "A plus B."

You can verify this law empirically by pushing against a partner or object with a bathroom scale to measure the force you are applying. Now insert a second bathroom scale facing the other way to measure the force with which the partner or object pushes back. Try it not only when you are stationary but also when you slide in response to the force (try wearing socks without shoes). If the scales are accurate the two readings are always equal.

CHAPTER SUMMARY

- If an object is accelerating, observers in *all* constant-velocity frames agree on its acceleration. In physics, *acceleration* includes all forms of changing velocity, whether speeding up, slowing down, or changing direction.

- Constant-velocity (also known as inertial) frames respect Newton's first law: objects accelerate only when acted upon by a net force. In these frames, the mere observation that an object has accelerated is enough to infer the existence of a net force.

- Accelerating frames do *not* respect Newton's first law. In these frames objects follow accelerating paths in the absence of forces, or show no acceleration despite the clear presence of forces.

- Therefore, we have a foolproof way of distinguishing accelerating frames from inertial frames. Unlike velocity, the acceleration of a frame is measurable without reference to other frames. Simply put, acceleration is not relative.

- To find the mass of an object, exert a net force on it, measure the acceleration, and take the ratio $m = \frac{F}{a}$. This relationship is also written as $F = ma$ and is called Newton's second law.

CHECK YOUR UNDERSTANDING: EXPLANATIONS

2.1   (a) An acceleration of 1 m/s^2 means that Bob's velocity *changes* by 1 m/s each second; it has no bearing on his *initial* velocity. Different constant-velocity observers will measure different initial velocities for Bob, but when that velocity changes each of those observers registers a change compared to their baseline measurement. (b) Instead of Bob consider a hypothetical Carol whose acceleration is 0 m/s^2. According to part (a) all constant-velocity observers will measure her acceleration to be zero; that is, they will agree that she is moving at constant velocity. By extension, no constant-velocity observer can find any other constant-velocity observer to be accelerating.

2.2   When fully loaded the train has more mass, but its engine is the same so it can apply no more force than before. The ratio $a = \frac{F}{m}$ therefore decreases when fully loaded.

2.3   In the ground frame the ball is free of forces once it leaves Alice's hand, so it follows a straight path. In the merry-go-round frame this must be a curved path, because the frame attached to the merry-go-round spins relative to the ground.

2.4   The frame attached to Jack is the only *non* inertial frame, because Jack accelerates. (Reminder: a motion you may consider as "decelerating" counts as accelerating in the physics sense of changing velocity.) If you considered that air resistance eventually slows and stops the ball then a frame attached to the ball is also non inertial.

? EXERCISES

2.1   A donut is attached to a string and swung in a circle. After some time, the string cuts through the soft donut and the donut is no longer pulled by the string. What kind of path will the donut promptly begin to follow?

2.2   *Mythbusters* investigated whether a bullet could be made to follow a curved path by swinging the gun as it is fired. What do you think they found, and why?

2.3   A hula hoop is cut at one point and the ends are separated slightly, making a nearly-circular "blow-gun." A marble is inserted and shot out the blow-gun. What path does the marble follow after leaving the gun?

2.4   Alice measures Bob's velocity relative to her and finds that it is not constant. *(a)* What can you conclude about Bob's measurement of Alice's velocity relative to him? *(b)* Can we conclude that Bob does not define an inertial frame? Explain your reasoning.

2.5   A common type of accelerated frame is a rotating frame. Consider observers aboard a merry-go-round moving at constant speed. Is this an accelerated frame? Explain why or why not.

2.6   A rocket appears to accelerate itself without pushing against any other object (not even against the air, because it works in empty space as well). How does it do this?

＋ PROBLEMS

2.1 Even with the most powerful engine, a locomotive can accelerate at most about 4 m/s^2 (if it tried to accelerate more than that, its wheels would slip on the track). What is the maximum acceleration of a fully loaded train (including the locomotive) with a mass 100 times that of the locomotive? Explain why locomotives are built to be very massive.

2.2 Consider a railroad locomotive applying a constant force and accelerating all its attached cars. At some point cars start falling successively off the back, steadily decreasing the total mass of the train. (a) What happens to the acceleration of the connected part of the train over time? Explain why, in terms of Newton's laws of motion. Neglect friction and air resistance. (b) Graph the acceleration and velocity as a function of time. Make sure the plots are vertically aligned with each other.

2.3 A boomerang follows a curving path after it is thrown. This must indicate a force pushing on the boomerang, but an object cannot push on itself. How do you resolve this apparent contradiction?

2.4 A bathroom scale works by reading the compression of a spring due to your body. (a) Explain why this is not measuring your mass, even if it is measuring something closely related to your mass. (b) If you really had to measure mass, how would you do it? *Hint:* how do scientists measure masses of atoms and molecules?

2.5 You work in a library and need to move two frictionless carts full of books. Cart 2 is twice as heavy as cart 1. You push cart 2 with the same constant force that you push cart 1. You cannot run in a library, so whenever you reach a velocity of 2 m/s you stay at that velocity. (a) Draw a plot of velocity versus time for cart 1, and then on the same plot add another line for cart 2. Be sure to label each line with its cart number! (b) Under the velocity plot, make an acceleration plot with the time axis lined up with the velocity plot. Again, draw and label one line for cart 1 and another line for cart 2. (c) How must you adjust your pushing over time to keep the carts from going faster than 2 m/s? (d) Now let us admit that there is friction; how must you adjust your pushing over time to keep a cart moving at 2 m/s?

2.6 A rope slides off a table as shown in Figure 2.2. Note that only the weight of the part of the rope dangling off the edge of the table provides a force to move the rope, but that the mass of the *entire* rope must move if any part moves. Neglect friction and air resistance. (a) Using Newton's laws of motion, explain what happens to the acceleration of the vertical part of the rope over time. (b) Assuming that the initial velocity is zero, graph the qualitative behavior of acceleration and velocity as a function of time. Make sure the plots are vertically aligned and consistent with each other.

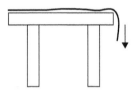

Figure 2.2 *A rope slides off a table.*

2.7 You are assigned to mark the outside of a train so that when the train accelerates constantly the marks pass a ground-based observer at equal time intervals. Describe how you must arrange the *distance* intervals between marks.

2.8 The surface of the Earth is a rotating frame, so particles moving from one point to another along the surface should not follow a straight line. (a) How is this evident in wind patterns in your hemisphere? (b) Describe and explain how low-pressure weather systems behave differently in the hemisphere opposite to yours. (c) Explain why your answer to part (b) does *not* apply to the direction of water swirling down a drain. *Hint:* sketch Earth as if you are looking down on a merry-go-round, and sketch the winds and your sink to scale.

2.9 Use Newton's third law to explain why astronauts on spacewalks need tethers.

2.10 Explain how a centrifuge works using the concepts in this chapter.

2.11 The *Mythbusters*, in the episode *Unarmed and Unharmed*, need to simulate the force of a bullet *hitting* a gun. Watch the episode and explain in detail how Jamie uses Newton's third law. Furthermore, explain why Jamie's solution does not provide an exact replica of the desired force.

3

Galilean Relativity

With the concepts of acceleration and force established in Chapter 2, we are now ready to dig into relativity. Because we are still assuming the Galilean velocity addition law is fully correct, the body of reasoning and conclusions presented in this chapter is called **Galilean relativity**.

3.1 Motion in two (or more) dimensions

Galileo studied projectiles and found that motion in the up-down direction is completely independent of motion in the forward direction. This independence is true for any two (or more) perpendicular directions you care to examine. Consider an eastward-moving marble on a table, and give it a push or tap in the direction of north so it begins moving northeast. Its velocity has changed both direction and size, so the relationship between the old and new velocities may be difficult to discern at first. But the relationship is easier to see if we think of the velocity as consisting of two independent components: an eastward component that is unaffected by north-south taps, and a northward component that was intially zero but became nonzero as a result of the northward tap. To prove this, give the ball an equal-size southward tap after some time, and the original velocity is restored:

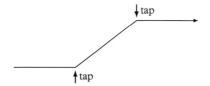

This may seem unremarkable at first, but bears further scrutiny. A motion diagram recording only the east-west position as a function of time will be *unaffected* by the north-south taps. If the motion diagram is unaffected, then *no* analysis based on that motion diagram—such as the marble's east-west velocity and acceleration at any point in time—can be affected. Any laws of physics we derive by studying the east-west motion diagram will be the same as if no action took place in the north-south direction.

Likewise, the north-south component of motion is unaffected by the east-west motion component. A description of the north-south component alone would be "an initially stationary marble was set in motion northward by the first tap

The Elements of Relativity. David M. Wittman, Oxford University Press (2018).
© David M. Wittman 2018. DOI 10.1093/oso/9780199658633.001.0001

and then stopped dead in its tracks by the second tap." This makes perfect sense in terms of Newton's laws of motion, with no need to reference the east-west component at all. The laws of motion are true independently in perpendicular directions.

Section 1.4 briefly introduced the notion of describing a vector (any "arrow" quantity such as displacement, velocity, or acceleration) in terms of components aligned with the coordinate system. The marble-tapping exercise illustrates why this is so useful: the behavior of each component is independent of the others.

If you have worked with vectors before, you also know that the component description of a vector simplifies many calculations. For example, adding a three-dimensional Δv to a three-dimensional v is as simple as adding the components separately. Still, the magnitude-and-direction description of a vector remains powerful conceptually because a vector is most easily pictured as an arrow. In this book, we will marry the two descriptions as follows. To keep things as simple as possible, we will generally add vectors only along a single direction (e.g., the velocity of a passenger walking along a moving train). This enables us to think in terms of arrows while calculating with a single component. The independence of components then implies that the understanding we gain by thinking about one component can be generalized, if and when we are ready to tackle problems that require multiple components.

Check your understanding. A cart carries a spring-loaded mechanism that shoots a marble straight up. If the cart is stationary, it is clear that the ball will fall back down into the cart. What happens if the cart is moving at constant velocity when the marble is released: does the ball fall back down behind the cart, into the cart, or ahead of the cart? Search for "ballistics cart" videos on the internet to confirm your prediction.

3.2 Projectile motion

Now consider a bomb released from an airplane that moves horizontally at constant velocity, and assume no air resistance. Recall the experimental fact that objects released near the surface of the Earth always accelerate downward at 9.8 m s^{-2} (Section 2.4). The independence of motion in perpendicular directions means that, after leaving the airplane, the bomb continues with the same *horizontal* velocity but in the vertical direction accelerates downward due to gravity. Figure 3.1 shows a motion diagram of the story, with the bomb released on the third dot to clearly establish the pre- and post-release motions. The horizontal distance is the same between any two successive bomb dots because the horizontal velocity is constant, but the vertical distance increases rapidly with time just as with any freely falling object.

Meanwhile, the airplane continues at the same horizontal velocity and zero vertical velocity. The horizontal positions of the airplane and the bomb are

> **Think about it**
>
> Velocity is a vector *because* displacement is a vector: when we divide displacement by Δt to obtain velocity, direction is preserved because the time interval has no direction. Acceleration is a vector for similar reasons. In fact, all vectors are based in some way on the displacement vector, and thereby have similar mathematical properties.

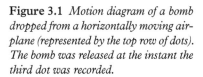

Figure 3.1 *Motion diagram of a bomb dropped from a horizontally moving airplane (represented by the top row of dots). The bomb was released at the instant the third dot was recorded.*

therefore identical; only the vertical positions differ. So, *relative to the airplane, the bomb falls straight down with the usual acceleration due to gravity.* Generalizing from the airplane, we see that in any constant-velocity laboratory or reference frame near the surface of the Earth, dropped objects accelerate straight downward at 9.8 m/s^2 just as they do in a laboratory fixed to the Earth. We need not have an actual moving laboratory to make use of this thinking tool. Galileo first developed this tool when studying the trajectories of cannonballs fired at some angle between horizontal and vertical. He vastly simplified the calculation by realizing that in a hypothetical laboratory moving horizontally with the cannonball's constant horizontal velocity, the cannonball must go straight up and down. This implies two separate but simple calculations: a horizontal calculation based on constant speed, and a vertical calculation based on the simple well-known case of an object launched straight up. Together these components completely specify the trajectory, and do so far more elegantly than a laborious calculation based on angles.

Check your understanding. A hunter sits in a tree waiting for a good shot. He sees a monkey eating a banana in another tree at exactly his height. He aims the gun perfectly horizontally and fires. At the instant the bullet leaves the gun, the monkey drops the banana. What does the bullet hit: the monkey, the banana, or something else? Why?

3.3 Principle of relativity

The principle of relativity (Section 1.5) is a conjecture that the laws of physics are the same in any inertial frame. The previous two sections certainly support this notion. The independence of vector components implies that any law of physics deduced from studying one component will remain true regardless of the state of motion expressed by the other components. But the principle of relativity goes beyond this to conjecture that frame-independence pertains to *all* the laws of physics, not just those of motion. We cannot *prove* this experimentally because we cannot test all laws in all frames at all times. But so far no real evidence has ever contradicted the principle of relativity, so it remains a solid basis for deducing additional nonobvious consequences, which can then be further tested. Throughout this book we shall see abundant evidence that these nonobvious consequences really happen. This amply supports the idea that the principle of relativity is a fundamental part of how nature really works.

The principle of relativity implies that there are no preferred inertial frames: any inertial frame is as good as any other. In real life, however, it certainly *seems* as if there are preferred frames. Objects rolling or sliding "freely" across flat ground, for example, slow down and come to rest in the frame of the ground and the air, so that frame seems special. This is compatible with the principle of relativity if we recognize that such objects are not really free, but are acted upon by forces of friction and air resistance. Are these forces real, or are they fictitious

---

Think about it

Mythbusters demonstrated the independence of vertical and horizontal motions for the very high horizontal speed of 500 m/s achieved by a bullet. They fired one bullet horizontally and simultaneously dropped another from the same height. Both hit the floor at the same time.

forces invented by physicists merely to explain violations of relativity? Real forces always leave evidence *in addition to* the observed changes in velocity. In the case of friction, the sliding object and the ground heat up. With air resistance—or more generally, forces caused by motion through a medium—we may observe the wake left in the medium by the object, or perhaps the deformation of the object into a teardrop shape. This proves that such forces are real.

Practice the habit of thinking of *all* motion as relative motion; the redundant adjective *relative* will be dropped from the nouns *motion*, *velocity*, and *speed* throughout most of this book. In everyday life we often describe motion without specifying what the motion is relative to; the context usually implies a particular reference frame such as the Earth. Thought experiments in this book often follow this convention, precisely so you can practice the necessary skill of framing a story. If a rocket is described as moving at 1 million kph to the east, practice thinking instead *there is some frame in which the rocket moves at 1 million kph to the east.*

Check your understanding. In the television series *Futurama*, a robot named Bender falls asleep in a spaceship's torpedo tube and accidentally gets fired in the forward direction. When the crew attempts to speed up to rescue him, they are unsuccessful; the captain says "It's no use—we were going full speed when we fired him, so he's going even faster than that." Does this make sense, given that the ship has engines and plenty of fuel? Explain your reasoning, and remember that there is no friction in space.

CHAPTER SUMMARY

- It is useful to think of motion in terms of perpendicular components. These components are completely independent: Newton's laws of motion apply separately to each component, and what happens to any one component has no effect on the other components.

- We gain insight into trajectories by thinking in a frame that moves along with one component; for example, a cannonball appears to go straight up and down (exactly as if it were vertically launched) in a frame moving at the appropriate constant horizontal velocity.

- The principle of relativity states that Newton's laws of motion (and all other laws of physics) are equally valid in any inertial frame.

- If the principle of relativity seems not to apply in some situation, look for the reason that some frame *appears* to be preferred. Motion through a medium such as air, for example, causes resistive forces that are not easily seen and therefore make the air appear to constitute a preferred frame. But such forces always leave frame-independent evidence, however subtle, that they are at work.

◢ CHECK YOUR UNDERSTANDING: EXPLANATIONS

3.1 Even while in the air the marble maintains the same horizontal velocity as the cart, so the marble is *always* in the horizontal position required to re-enter the cart.

3.2 The bullet will hit the banana because both fall with the same acceleration. Search for "Monkey and a Gun" video demonstrations to see this play out in real time. Real bullets are so fast that they have little time to fall a noticeable distance before hitting the target—but they do fall.

3.3 This does not make sense if the ship has engines. At the instant of launch we can consider the ship at rest regardless of its speed relative to, say, the nearest planet. Now if Bender's launch gave him velocity v relative to the ship, the ship can simply use its engines to accelerate until it reaches the velocity required to overtake him. In the absence of friction, additional pushes forward *always* increase the forward velocity.

? EXERCISES

3.1 To practice thinking about relatives and absolutes, *(a)* identify a real-life situation unconnected to physics where only relative amounts matter, and *(b)* identify a real-life situation unconnected to physics where only absolute amounts matter. Example: (a) a cookie recipe tastes good if the relative amounts of flour, sugar, and so on are correct, regardless of the size of the batch; (b) eating ten cookies in a sitting is probably unhealthy regardless of how many cookies others around you are eating.

3.2 This chapter asks you to practice mentally inserting the adjective *relative* whenever you see the nouns *motion, velocity,* or *speed*. Why not for *acceleration*?

3.3 Watch the video http://www.youtube.com/watch?v= yPHoUbCNPX8 and explain it in terms of Galilean relativity and its velocity addition law.

3.4 A gun 1 m above the ground fires a bullet horizontally. Simultaneously, a bullet 1 m above the ground is dropped. *(a)* Do they hit the ground at the same time? Consider the situation both with and without air resistance. *(b)* Find a video on the web in which this experiment was actually done. Provide the link and describe the results. Was air resistance a factor?

3.5 A classic physics demonstration called the *drop and shoot* releases one ball straight down, while another is simultaneously shot out the side at the same height. Which, if any, of the balls hits the ground first? Explain your reasoning.

3.6 Where does a bomber pilot release a bomb, directly over the target or well before that point? Assume a "dumb" bomb that simply falls without any course corrections.

✚ PROBLEMS

3.1 California Jones is crossing Death Canyon on a zipline, travelling horizontally at constant velocity and carrying a valuable crystal skull. He approaches a rival who is standing on the zipline, eager to grab the crystal skull. Jones decides to throw the skull up over the bad guy and catch it as it comes down on the other side. *(a)* At what angle (as perceived by Jones) should Jones throw the skull to make sure he catches it, and why? Describe the skull's motion from his point of view. Neglect any effect of air resistance. *(b)* Does your answer change if Jones and the skull are equally slowed by air resistance? Explain your reasoning.

3.2 The two rockets deep in space (Figure 3.2) are engaged in battle while moving at the same (very rapid) constant velocity toward the top of the page. They are armed with torpedos that can be aimed in any direction. You are the captain of the ship on the left, where the torpedo officer aims the torpedo directly to the right. The first officer says that the torpedo will miss, because by the time it gets to where the enemy rocket is now, the enemy rocket will have moved ahead. Instead, the first officer says, the torpedo should be aimed ahead and to the right. How should you aim the torpedo? Why? Keep in mind there is no friction in space.

Figure 3.2 *Two rockets moving at the same very large constant velocity toward the top of the page.*

3.3 You are in a spaceship far from any planet, traveling in a straight line at constant velocity. Your ship has no speedometer like a car does, but you do have a radar speed gun that you can point at planets or spaceships when you see them. *(a)* How can you know at any point in the future if you are still traveling at constant velocity, even if your speed gun breaks? *(b)*

Another spaceship flies by; your speed gun says that ship is moving at a constant 20,000 mph. By this standard (i.e., as read by your radar gun), what is your velocity? *(c)* How would the person in the other spaceship describe your motion? *(d)* At some point the Intergalactic Highway Patrol stops you and shows their speed gun reading of your ship: 30,000 mph, far above the speed limit. How can you argue your way out of a speeding ticket?

3.4 An American football player throws the ball to another player very far away. You are sitting in the audience. As you see it, when the ball leaves his hand, the ball is moving mostly to the right but also upward. *(a)* draw the ball's *horizontal* velocity and acceleration *vs* time. Make sure your three plots are vertically aligned so that it is easy to see the relationships between the three quantities at any given time. *(b)* Do the same for the ball's *vertical* velocity and acceleration *vs* time. *(c)* Now imagine there is a camera on a track alongside the field that follows the ball so that the ball and the camera always share the same horizontal position. Draw the ball's vertical velocity and acceleration *vs* time as seen by this camera (again, align the plots vertically). *(d)* Describe the path of the ball as seen by someone watching on TV, as broadcast through the camera described in part (c).

3.5 Why do airplanes take off and land against the wind?

3.6 (For those who are comfortable with vectors.) *(a)* Explain how boats can sail upwind. (*Hint:* it is important to understand how the boat responds to pushes in various directions, before you start thinking about the wind.) *(b)* Now use your understanding of relative motion to explain why sailing is actually faster upwind than downwind.

Time is what keeps everything from happening at once.
—Ray Cummings, *The Girl in the Golden Atom*

[For Tralfamadorians] all moments, past, present and future, always have existed, always will exist. The Tralfamadorians can look at all the different moments just that way we can look at a stretch of the Rocky Mountains, for instance. They can see how permanent all the moments are, and they can look at any moment that interests them. It is just an illusion we have here on Earth that one moment follows another one, like beads on a string, and that once a moment is gone it is gone forever.
—Kurt Vonnegut, *Slaughterhouse Five*

Rose Tyler (stunned): I can see everything, all that is, all that was, all that ever could be. The Doctor: That's what I see, all the time. And doesn't it drive you mad?
—*Doctor Who*

Reasoning with Frames and Spacetime Diagrams

4

Soon we will enter a high-speed world where Galilean relativity breaks down. To prepare for that, we now practice the skill of thinking in different frames. Practicing this in our familiar low-speed world will help us avoid cognitive overload when we enter the more counterintuitive high-speed world. In this chapter we will examine two problems that illustrate the process of thinking in different frames.

4.1 The river and the hat

Imagine you are rowing a boat on a river. Your constant rowing speed is 3 kph relative to the water, which yields 4 kph relative to the land when rowing downstream and only 2 kph relative to the land when rowing upstream. While rowing upstream you lose your hat, which moves downstream at the same speed as the water, but you do not notice until 1 hour later. At that point, you turn around and row downstream to retrieve your hat. How long does it take to reach your hat?

One way to solve this problem is to think about positions in the land frame of reference, which seems like the most solid frame to think in. Imagine there is a landmark such as a large tree at the point where you lost your hat, and call this the origin. When you turn around after one hour rowing upstream, you are 2 km upstream of the origin. Rowing back with a land speed of 4 kph, you get back to the origin in only half an hour. But by that time, the hat has been flowing downstream at 1 kph relative to the land for a total of 1.5 hours, so it has moved 1.5 km further. You can cover that distance in $3/8$ hour at your downstream land speed of 4 kph, but by *that* time the hat will have moved further (another $3/8$ km to be precise). You can cover *that* distance in $3/32$ an hour (for a total time of $1/2 + 3/8 + 3/32 = 31/32 = 0.96875$ hour). The hat will have moved slightly further in that time, so even this answer is not exact. A bit of algebra shows that the exact answer is 1 hour, but instead of an algebra lesson let us practice reframing the problem.

The problem is *much* easier to analyze in the river frame. In this frame the water is always at rest, so the action unfolds as if on a lake with no current. Your hat, after being dropped, is at rest in the river frame; we say that this is the **rest frame** of the hat. Your speed in this frame is always 3 kph, so when you turned around you were 3 water km from the hat. Rowing 3 water km to retrieve your hat

The Elements of Relativity. David M. Wittman, Oxford University Press (2018).
© David M. Wittman 2018. DOI 10.1093/oso/9780199658633.001.0001

will take exactly 1 hour. Solving the problem this way is faster, easier, and yields an exact answer, but there is a deeper benefit as well: it allows us to see that your rowing strength is irrelevant. If you could row much faster you would have gone much further upstream in the initial hour, but the distance would still take 1 hour to row back. In the river frame this story is as simple as leaving your house and walking for an hour; how long does it take to walk back? Framed that way, the problem is easy.

Using the river frame here is not just a shortcut to the numerical answer—the river frame extracts the essence of the problem. This becomes strikingly visible after constructing a **spacetime diagram** of the story. Unlike motion diagrams (Chapter 1), spacetime diagrams depict the march of time explicitly along a time axis, which is traditionally taken to be the vertical axis. The horizontal axis or space axis represents position and is traditionally labeled x; in the hat story we will choose downstream as the direction of increasing x. An event has a specific location and time, and thus appears as a point on the spacetime diagram. Figure 4.1 is a spacetime diagram of the story as seen in the land frame, with the hat-dropping event labeled A, the turning of the boat labeled as event B, and the picking up of the hat event C. You can turn any spacetime diagram into a movie by cutting a horizontal slot in a piece of paper and then sliding the paper from bottom to top across the diagram. The slot will make the story unfold one instant at a time. When you remove the paper again you see the entire story at once. Hence the quotations that open this chapter: spacetime diagrams give you some of the power of The Doctor.

The spacetime diagram in Figure 4.1 has a grid that marks kilometers and hours. (Be aware that these will *not* be convenient units for the high-speed diagrams that appear throughout this book and other relativity texts.) The hat leaves event A and floats downstream at 1 kph, so its path through space and time—called its **worldline**—crosses one horizontal graph-paper square (1 km) for each vertical graph-paper square (1 hour). The boat moves upstream at 2 kph in the land frame—thus covering two squares in x for each square in t—and then reverses course through space (*but not through time!*) after which it moves downstream at 4 kph in the land frame.

Now consider the story in the river frame (Figure 4.2). The hat is at rest in the river frame, so its worldline is vertical: it occupies the same position in space at all times. This removes one unnecessary complication caused by thinking in the land frame. Even more striking is the change in the boat's worldline. We can now see at a glance that it simply reverses its velocity at the turnaround. Regardless of your rowing speed, the time interval between A and C must be double that between A and B. By thinking in a new frame we took the triangle ABC —which was skewed in the land frame—and straightened it out to expose the essential symmetry of the situation. Note that Figures 4.1 and 4.2 have different subscripts on their x coordinates; this is because river-frame positions are determined by a tape measure floating along with the river and thus represent a coordinate system quite different from x_{land}. For convenience I have chosen the origin of

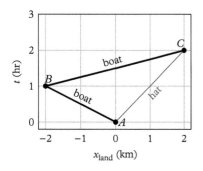

Figure 4.1 *Spacetime diagram of the hat-on-the-river story, as measured in the land frame.*

the river-based tape measure to coincide with the origin of the land-based tape measure at event A.

Note also how the worldlines expose Newton's first law: objects remain at constant velocity unless acted upon by a net force (Section 2.2). Constant velocities appear in spacetime diagrams as straight worldlines. Where a worldline bends or kinks in a spacetime diagram, by definition there is acceleration and so (by Newton's first law) net force has been applied. Conversely, straight worldlines indicate no acceleration and therefore absence of any net force. Diagramming the situation in a different inertial frame will *never* change a straight worldline to a bent one or vice versa. This maps perfectly to our conclusion in Chapter 2 that an application of force manifests itself in *all* frames; the coffee-sloshing test, for example (Section 2.4), would inform all observers that net force was applied at event B and only at event B.

What about the rower frame? The rower does not have a consistent *inertial* frame due to the acceleration at event B. To give the rower a straight worldline we would have to pull apart the graph-paper cells somehow. Spacetime diagrams are therefore not very good tools for analyzing noninertial (i.e., accelerated) frames. In fact, there is no simple way to think about a frame attached to this rower, because relative to the rower everything in the universe outside the boat accelerated at event B! What we *can* do to better understand the rower perspective is to split the rower's journey into separate inertial segments A-B and B-C. We will practice this more later in the book.

Finally, practice visualizing how a spacetime diagram in one frame relates to a diagram of the same story in another frame. Figure 4.3 places two diagrams of the hat story side by side to emphasize their unity. Mentally picture deforming the diagram on the left to become the diagram on the right, and vice versa. Notice that in this process events B and C move horizontally *but not vertically*. This is a consequence of our implicit assumption that time is the same in all frames; for

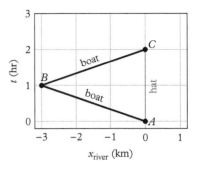

Figure 4.2 *Spacetime diagram of the hat-on-the-river story, as measured in the river frame.*

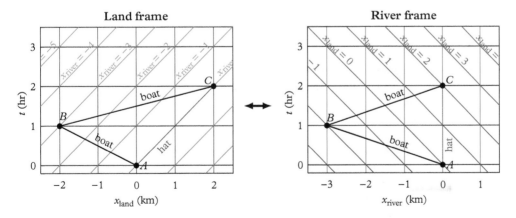

Figure 4.3 *Spacetime diagrams of the same story in two inertial frames. The physical reality underlying the events and worldlines is the same in either frame; only the event coordinates differ.*

example, that clocks at rest on the land tick at the same rate as clocks at rest on the water. Developments in the next few chapters will force us to revisit this assumption.

Check your understanding. Jack the dog is on a walk with his human, who walks at constant velocity and throws a ball that Jack eagerly fetches and returns. Construct a spacetime diagram of this story in *(a)* the human's frame; and *(b)* the land frame.

4.2 Frame-dependent versus frame-independent questions

Thinking in the river frame simplified the hat retrieval problem so much that we may wonder if the river-frame answer is somehow correct only in the river frame. Which aspects of the problem change from frame to frame and which are *frame-independent*?

The existence of an event such as "the hat fell out of the boat" must be frame-independent, because if the hat gets wet in one frame it cannot stay dry in another frame. The hat falling defines an event (labeled A) at which the worldlines of the boat and the hat begin to move apart—in all frames. Similarly, the two worldlines must intersect again at event C in all frames because the hat being plucked from the water cannot be frame-dependent. The physical meaning of the story must be the same in any frame; what changes from frame to frame is merely the coordinate of an event, because each frame defines its own coordinate system. For example, event C occurs at the origin in the river frame, but 2 km from the origin in the land frame.

Because event positions are frame-dependent, so too must be any statement such as *events A and C occurred at the same position*. A good mental picture for this is a car windshield full of splattered insects—they were all splattered at the same position in the car frame, but surely not at the same position in the land frame. Velocities in turn are frame-dependent, because velocity measurements use frame-dependent positions. But if a particle *changes* its velocity (accelerates) in one inertial frame it changes its velocity in *all* inertial frames. Graphically, an acceleration is a kink or a curve in a worldline, whereas velocity is the tilt or slope of the worldline. Viewing a worldline in a different inertial frame changes its slope but does not introduce or remove any kinks or curves; therefore, whether a particle accelerates is a frame-independent question (so long as we restrict ourselves to thinking in inertial frames).

We are not yet ready to make definitive lists of frame-dependent and -independent quantities, but the distinction is absolutely crucial. Distinguishing between frame-dependent and -independent quantities is a key skill in relativity problem-solving, so make sure to practice this skill as you read and solve problems.

Until now, we have described events that are particularly vivid, such as a hat falling in the river, to illustrate that whether an event happened cannot be a

Think about it

When we go beyond Galilean relativity, a key finding will be that not only the distance but also the *time* between any two events is frame-dependent. The frame-dependence of time is a very small effect at everyday speeds, so in this chapter it is a reasonable approximation to say that the time between events is a frame-independent question. Do not become too accustomed to using this approximation!

frame-dependent question. In fact any event, no matter how mundane, happens regardless of frame. Where frames differ is simply on the coordinates they use to label that event. Section 4.3 will develop this notion graphically.

Check your understanding. Alice rows at 3 kph to the north relative to the water while Bob rows at 4 kph to the south, resulting in a head-on collision at 7 km/s. Is this collision speed frame-dependent or frame-independent?

4.3 Coordinate grids of moving frames

Coordinates feel abstract, so let us make them concrete by planting landmarks in the river/hat story. Imagine trees planted every 1 km along the riverbank. The trees are at rest in the land frame so in that frame their worldlines form a series of vertical lines 1 km apart—just like the graph paper grid itself (Figure 4.4). In a sense these landmarks *are* the grid: each landmark traces a given value of x_{land} through time. Thus a spatial grid line such as $x_{land} = 2$ connects all events that ever occur at $x_{land} = 2$: event C as well as an infinitude of events not specifically labeled in Figure 4.4.

Now, in the river frame the trees are moving 1 kph to the north so their worldlines form a skewed grid. The lines are skewed relative to the river-frame graph paper, but are all parallel to each other. *Parallel* in the context of a spacetime diagram simply means that they maintain a fixed distance apart over time; in other words, they are not moving relative to each other. "Not moving relative to each other" is a frame-independent statement, so worldlines parallel in one frame must be parallel in all frames.

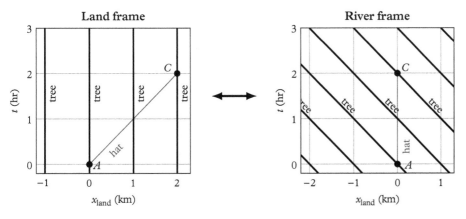

Figure 4.4 *Worldlines of trees in the hat-on-the-river story. Fixed values of x_{land} (represented by trees) form tilted lines in the river frame. The hat worldline shows that fixed values of x_{river} tilt the opposite way in the land frame.*

The same logic applies to worldlines representing fixed x_{river}. If we placed rubber ducks in the river spaced 1 km apart (and the river flowed perfectly uniformly) all their worldlines would be vertical in the river-frame diagram and tilted in the land-frame diagram—by an amount equal and opposite to the tilt of the tree worldlines in the river-frame diagram. Now, if we abstract away from trees and rubber ducks, we can see that these statements are really about the grid lines in the diagrams, which represent fixed locations even without a physical landmark there.

In Galilean relativity, this skewing of the lines of fixed location is the *only* difference between a coordinate grid viewed in its rest frame and viewed in a moving frame. We can thus draw a diagram that represents *both* frames as follows. First, choose a frame in which the grid will be square; that is, in which landmarks are at rest. For the left panel of Figure 4.5 this is the land frame; we say that this panel is "drawn in the land frame" even though we will extend it to represent the river frame as well. Next, we add worldlines representing fixed locations in the *other* frame. In the left panel of Figure 4.5 the tilted red lines represent regularly spaced locations attached to the river. The right panel of Figure 4.5 repeats the entire diagramming process but starting with a square river-frame grid. Either panel is sufficient to represent both frames, but both panels are shown here to emphasize their commonalities.

These two maps of events tell the same story. Each worldline changes its slope from one frame to the other, but all worldlines change their slope in the same way so that many relationships between events and worldlines are preserved. Here are some examples of relationships that are true in either frame: the triangle *ABC* has boundaries formed by the same worldlines and events in the same clockwise order

Think about it

In Figure 4.5 the lines of fixed *time* are drawn in black because they did not change from frame to frame, and the lines of fixed x_{river} remain separated by 1 km as they were in the land frame. These are consequences of the assumption—implicit in Galilean relativity and to be questioned later—that the performance of clocks and rulers is independent of their velocity.

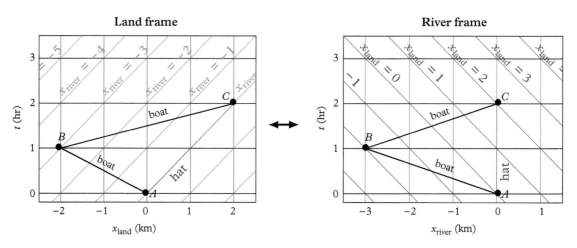

Figure 4.5 Left: *both land (blue) and river (red) grids can be represented on the same diagram if we allow for tilted grid lines in addition to the traditional square grid.* Right: *this idea is repeated in the river frame where the river grid is square and the land grid is skewed.*

ABC; the $x_{\text{land}} = 0$ line runs through event A and bisects the line segment *BC*; and event B occurs before event C (the boat turns around before retrieving the hat). These kinds of relationships are preserved when we consistently change the slopes of all worldlines and grid lines; that is, redraw the diagram in a new frame. With enough practice, you will be able to mentally re-tilt the worldlines and gridlines of a diagram to picture it in a new frame. As you do this, remember that events are carried along with the worldlines or gridlines to which they are attached.

To be clear, the events themselves are not frame-dependent; only their *coordinates* change. There is only one reality, but there are many ways to draw coordinate grids over this reality. A frame is simply a choice of coordinate grid, and choosing which frame to call "at rest" is nothing more than choosing which grid to portray as square in your diagram. In Figure 4.5, picture the intersections of grid lines as hinges that keep the red and blue grids attached to each other even as they allow you to "fold" the grid. Try mentally straightening the red (river) grid in the first panel until it is vertical. This necessarily bends the attached blue grid lines to the left in the second panel.

Check your understanding. Identify three additional frame-independent statements regarding the relationships of worldlines and the events A, B, and C.

4.4 Transverse distances are always frame-independent

Reciprocity is the idea that what frame A measures regarding frame B, frame B must measure regarding frame A. Speed is the most vivid example: when two observers point radar speed guns at each other, they must each measure the same speed. Reciprocity is inherent in the principle of relativity but we will use this specific thinking tool so often that we need a specific name for it.

Reciprocity allows us to prove that distances transverse to the direction of motion must be frame-independent in *any* kind of relativity, not just Galilean. We will rely on this result at several points throughout the book, so follow this argument carefully.

First, I would like to introduce the concept of a *foothold idea*—an idea we tentatively decide to believe so we can see where it leads. The analogy is with a rock climber that tries a foothold to see if it allows him to grasp something else higher up; if it does not, he can always retreat from that foothold and try a different one. Students are often uncomfortable doing this with ideas because they just want the "correct" ideas. But testing ideas is what science is all about! Science has been successful only because scientists have tested—and continue to test—a vast range of ideas. If we study only the results and skip the process, we are not really engaging in science. This section uses a foothold idea that leads to retreat—but the lesson learned from the retreat is valuable. Specifically, we will tentatively assume that transverse distances *are* frame-dependent. When that leads to severe

Think about it

Relativity texts often use "observer" as shorthand for "in a frame attached to this observer." This is confusing because "observer" seems to imply a specific location and limited knowledge, but a frame actually extends everywhere and assigns well-defined coordinates to all events. I will avoid this shorthand, but beware when consulting other texts.

contradictions, we will be assured that no such thing is possible. This is called *proof by contradiction.*

Imagine that in some hypothetical form of relativity the height of a train depends on the speed of the train and that train tunnels are built to accommodate stationary trains. If the train height decreases with speed, the train will easily fit into a standard tunnel at any speed. But by reciprocity, observers on the train should measure the height of the quickly oncoming *tunnel* to decrease with speed. Therefore, the train-frame conclusion is that the train will hit the rock above the tunnel. But whether the train hits the rock cannot be frame-dependent, because a collision would have consequences observable in all frames. Therefore, the height of the train cannot decrease with speed. A similar argument proves that the height of the train cannot increase with speed either; therefore, the train height must be independent of its speed, contradicting our assumption.

The *width* of the train must also be frame-independent by the same type of argument. Therefore, distances in either direction perpendicular to the direction of motion must be frame-independent in any kind of relativity.

Check your understanding. Fill in the details of the proof that the width of the train cannot increase with speed.

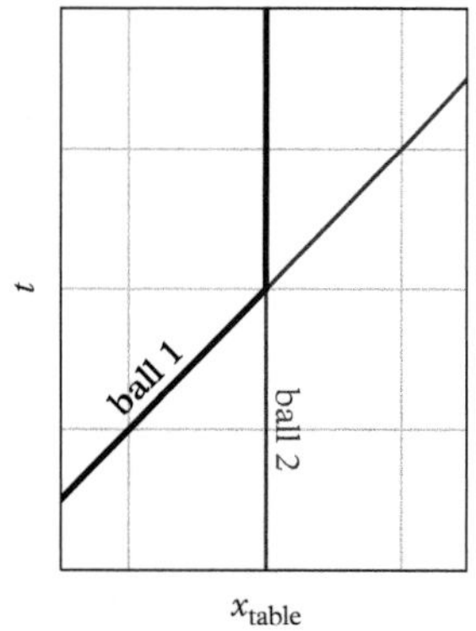

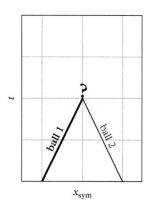

Figure 4.6 *Spacetime diagram of a billiard collision.*

4.5 Billiards

Now that we have established some frame-based thinking skills, we will use those skills to gain new insight into the laws of physics. The case study in this section is inspired by a similar study in N. David Mermin's excellent relativity textbook *It's About Time.*

When a billiard ball rolls across the table and strikes another ball, the first ball tends to stop completely while the second ball takes on the velocity of the first, as in Figure 4.6. *Why does this happen? Why does nature prefer this outcome over all the other outcomes we could imagine?*

The answer is not at all obvious because we tend to think in the frame of the table, but there is another frame in which the answer *is* obvious. Consider the frame in which both balls are moving toward each other at equal speed (Figure 4.7), called the **symmetric frame**. If Ball 1 moves east at 1 m/s in the table frame, the symmetric frame is embodied by a camera moving east at 0.5 m/s. Relative to this camera, Ball 1 would be moving east at only 0.5 m/s before the collision and Ball 2 would be moving an equal 0.5 m/s to the *west*. Make sure you understand the relationship between Figures 4.6 and 4.7 before we move on to consider what happens at the collision.

Now, what must happen at the collision? Because the balls are now identical in all relevant ways—billiard balls have equal mass, and properties such as color cannot matter—the inbound situation is completely symmetric. Therefore, the only possible outcome is also a symmetric one: the balls must exit the collision

Figure 4.7 *The beginning of the story shown in Figure 4.6, but in a symmetric frame. What will happen at the collision?*

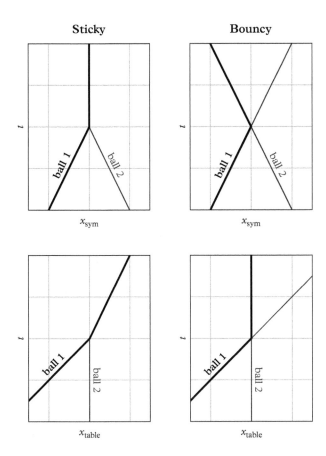

Figure 4.8 *Two possible outcomes of a collision in the symmetric frame. The post-collision worldlines must have equal and opposite velocities by symmetry, but this still leaves a range of possibilities depending on how the particles interact: a sticky collision (left) leaves both particles with zero post-collision velocity while a bouncy collision (right) leaves the sizes of the velocities unchanged.*

Figure 4.9 *As for Figure 4.8 but transformed back into the table frame. The right panel perfectly reproduces our original observation in Figure 4.6 for billiards, while the left panel matches our experience for sticky collisions: the combined object proceeds at a velocity intermediate to the pre-collision velocities.*

with equal speed in opposite directions. (If you have taken a previous physics course you may recognize this result as "conservation of momentum" but focus on the fact that *we can deduce this result from symmetry and frame-based thinking tools alone*.)

But symmetry is not enough to decide everything. Even in the symmetric frame we can imagine different outcomes as shown in Figure 4.8: in the left panel the two objects stick together while in the right panel they reflect off each other. We need to know more about the interaction to determine the *size* of the post-collision velocities, even if symmetry dictates that they must be equal and opposite. Billiard balls are not sticky so we take the right panel as our model. Here we see, in addition to symmetry between billiard balls, symmetry in *time* before and after the collision. In contrast, friction or sticky forces introduce time asymmetry by always reducing relative velocities as time proceeds.

Figure 4.9 shows each of these scenarios transformed back into the table frame. The (nonsticky) diagram on the right indeed reproduces the observed behavior, with Ball 1 motionless relative to the table after the collision. Given that billiard balls are not sticky, *there is no other possible outcome in the symmetric frame, so there is no other possible outcome in the table frame*. In real life, the result is not always exactly as described here because friction is not completely absent and putting spin on Ball 1 can add some interesting effects. But we have explained the basic behavior simply by thinking in a symmetric frame.

Frame-based thinking tools may seem very abstract but you can make them more concrete by imagining a camera mounted on a track alongside the billiard table. You have probably seen views from cameras like this at sporting events. There is nothing special about this camera: it simply records positions over time like any other camera. But changing your frame of reference gives you new insight into the game.

Check your understanding. If Ball 1 had been moving at 3 m/s to the east in the table frame, what is the table-frame velocity of the hypothetical camera recording data in the symmetric frame?

4.6 Accelerated frames

Spacetime diagrams help us understand accelerated frames as well as inertial frames. Recall from Chapter 2 that Newton's first law of motion—objects maintain constant velocity unless acted upon by a net force—is violated in accelerated frames. We can see this vividly with the help of spacetime diagrams.

Imagine an initially stationary car that accelerates relative to the land. Let us put the center of the car at $x_{car} = 0$, the front of the car at $x_{car} = 1$, and the back of the car at $x_{car} = -1$. In the land frame the worldlines of these locations pick up speed over time, so their slopes change as they move toward the top of the spacetime diagram (Figure 4.10, left panel). The worldlines are *curved*. These worldlines define the x_{car} grid, so the entire x_{car} grid is curved.

We can redraw this diagram in the car frame (Figure 4.10, right panel). Now the lines of fixed x_{land} are curved in the opposite direction. There is nothing illegal about drawing the diagram this way. The relationships between events and worldlines will be preserved just as when we compare inertial frames. The new feature is that Newton's first law is not obeyed in the accelerated frame. Consider a ball resting on the land at $x_{land} = 2$; this ball experiences no net force. In the land frame the worldline of the ball is straight, indicating constant velocity, just as Newton's first law predicts for an object that experiences no net force. In the car frame, however, the curved worldline of the ball indicates that the ball is accelerating, in violation of Newton's first law.

This may be easier to see if we place a ball *in* the car. This ball begins at $t = 0$ at the front of the car ($x_{car} = 1$), which coincides with $x_{land} = 1$ at that time. The car then accelerates; what does the ball do in the absence of sticky forces attaching the ball to the car? The ball rolls toward the back of the car while maintaining its *land* position at $x_{land} = 1$. This is a frame-independent statement; imagine the ball striking a passenger as it rolls back. In the land frame, the ball simply follows a straight worldline as demanded by Newton's first law. In the car frame, the curved worldline can be explained only with a fictitious force, which might colloquially be called "the force of inertia." We call this fictitious not only because there is no acceleration in the land frame, but also because the ball itself can tell us there is no force on it. A ball struck with a bat deforms, for example; a smaller force causes a smaller deformation but the principle is the same. The ball accelerates in the car frame without any such interaction, hence the term *fictitious force*.

We conclude this section with a few subtleties regarding Figure 4.10. First, there appears to be a symmetry between the two panels, as if the acceleration could be attributed to either frame. We reiterate that Newton's first law tells us which grid is really accelerating; free particles (such as the ball) have straight worldlines in inertial frames and curved worldlines in accelerating frames. Second, there is a distinction between straight paths through space and straight worldlines in space*time*. You may equate straight worldlines with constant velocity, but you may *not* do so with straight paths through space. In fact, it is common for particles to accelerate (i.e., curve their worldlines) even while following straight paths through space—the car in Figure 4.10 is a perfect example.

> **Think about it**
>
> If we do arrange a straight path for the ball in the car frame—say by keeping it at rest against the seat back as the car accelerates—there *will* be evidence of a force applied, as the seat back deforms. There is no way to satisfy Newton's first law in an accelerating frame.

Check your understanding. (a) Continuing the car example in this section, consider a diagram drawn in the frame of a truck moving at constant velocity relative to the

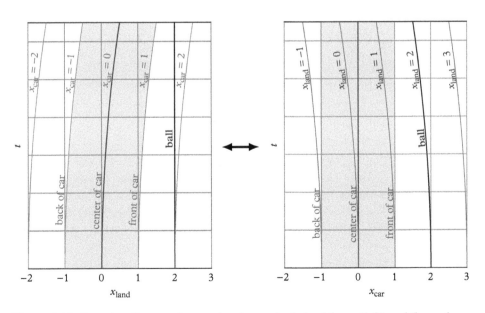

Figure 4.10 *Spacetime diagram of an accelerating car in the land frame* (left) *and the car frame* (right). *In either case a ball is at rest on the land at* $x_{land} = 2$.

land. In this diagram would the car-frame grid be curved or straight? Would the land-frame grid be curved or straight? *(b)* How does a cup of coffee at rest in an accelerating car behave differently than an identical cup at rest on the ground?

4.7 Assumptions

Thinking rigorously requires identifying our assumptions, even if we are satisfied that we arrived at the right answer. In the river/hat and billiard problems we assumed the Galilean velocity addition law. For example, we determined velocities relative to the table using velocities relative to the camera without ever questioning that they simply added. In the river and hat problem, we also assumed that distances were the same in both frames: if you are 2 km from the hat in the river frame, you are also 2 km from the hat in the land frame. *This is a hidden assumption.* This assumption is backed up by everyday experience and measurements, but we must remember that our everyday experience is limited to rather low speeds and is therefore is not definitive. Another hidden assumption is that *time is the same in all frames.* Again, we know this is true at low speeds, but we must be prepared to re-examine this assumption if experiments at very high speeds yield different results. Note that these three assumptions go hand-in-hand. If distance and/or time measurements were different at high speed, then velocity measurements would be affected, because velocity is just distance divided by time. Therefore, the velocity addition law would be affected. So, if experiment disproves any one of these assumptions, the others must be re-examined as well.

CHAPTER SUMMARY

- Thinking in a well-chosen frame can make a problem *much* easier to solve.

- Thinking in a well-chosen frame can yield insight into *why* the solution is the way it is, and therefore how nature works.

- Hints for choosing frames: one problem in this chapter was solved by choosing a frame in which one object was motionless. The other problem was solved by choosing a symmetric frame.

- The answers to many specific questions are frame-dependent, but the really valuable insights are frame-independent. Be ever mindful of the distinction.

- Newton's first law of motion is consistently violated in accelerated frames, making them objectively different from inertial frames.

- The Galilean velocity addition law contains hidden assumptions about space and time.

◢ CHECK YOUR UNDERSTANDING: EXPLANATIONS

4.1 Diagrams may vary depending on the speed and direction of motion assumed for the human and the ball. In the human's frame the human's worldline should be a vertical line, while in the land frame it should be a tilted line. In either case, Jack's worldline should tilt away from and then back toward the human's worldline.

4.2 Frame-independent. Bob observes Alice approaching him at 7 km/s, and vice versa. Even in the frame of an airplane traveling 500 kph to the north, the *difference* between Alice's and Bob's velocity would be unchanged because each of their velocities would be altered by the same 500 kph. Caveat: this assumes the Galilean velocity addition law is correct.

4.3 Some examples: the boat's worldline is kinked at event B; the hat's worldline passes through events A and C but not B; and the hat and boat worldlines diverge from event A but converge on event C.

4.4 If the train width decreases with speed, the train will easily fit into a standard tunnel at any speed. But by reciprocity, observers on the train should measure the width of the quickly oncoming tunnel to decrease with speed and therefore in the train frame the train should collide with the tunnel walls. But whether the train hits the rock cannot be frame-dependent, so the height of the train cannot decrease with speed after all. The same argument applies to a hypothetical *increase* of train width with speed.

4.5 1.5 m/s to the east.

4.6 (a) According to the truck frame, the car-frame grid is curved because it is accelerating in the frame of the truck. Meanwhile, the land-frame grid is straight because it is *not* accelerating in the frame of the truck. (b) The coffee sloshes toward the rear of the car.

Hint: in any exercise or problem with multiple frames, first decide in which frame *you* are most comfortable thinking. Even though you may be asked to describe or draw the solution in some other frame, you may *find* the solution more quickly and confidently in the "easy" frame. Even for problems posed in a single frame, some other frame of your choosing may help you think more clearly.

？ EXERCISES

4.1 A stone may skip across water if it hits the surface of the water at high speed. If you are on a riverbank, is it better to throw the stone upstream, downstream, or is it irrelevant?

4.2 Jack the dog is on a walk with his human, who repeatedly throws a ball that Jack eagerly fetches and returns. To maximize Jack's exercise, should the human throw the ball forward or backward? Draw a spacetime diagram of each option to clarify the answer.

4.3 In Figure 4.11, which worldline(s) are: *(a)* stationary; *(b)* inertial; *(c)* accelerating; *(d)* impossible?

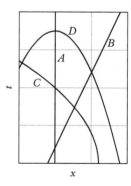

Figure 4.11 *A variety of hypothetical worldlines.*

4.4   Using graph paper to represent a coordinate grid attached to the ground, draw the coordinate grid attached to a train moving to the left.

4.5   Explain why *not moving relative to each other* must be a frame-independent statement.

4.6   Follow the reasoning of Section 4.4 to prove that the *width* of a train cannot increase or decrease at high speed. Does this proof depend on the assumptions of Galilean relativity?

4.7   Modify the billiard story in this chapter so the initially moving ball is the one on the right. Draw the space-time diagrams in: *(a)* the table frame; *(b)* the symmetric frame; and *(c)* the frame in which the left ball is at rest at the end of the story.

4.8   For each of the following questions, will the answer be frame-dependent or frame-independent? *(a)* "How far will this car travel on a full tank of gasoline?" *(b)* "Will this car get me from here to Albuquerque on a full tank of gasoline?"

+ PROBLEMS

4.1   Find the exact answer to the hat story in Section 4.1 by thinking in the land frame and using algebra. Compare the effort to solve the problem in the land frame to that required in the water frame.

4.2   Draw the worldline of a tennis ball dropped from a substantial height, bouncing up and down multiple times and eventually resting on the ground.

4.3   Draw a table-frame spacetime diagram of a collision of billiard balls of *different* masses (draw on your experience to make a qualitatively correct prediction). Can there be a frame in which the velocities before *and* after are symmetric?

4.4   Jack the dog has a ball in his mouth and runs at full speed toward his stationary human companion. Jack stops in front of the human but drops the ball while still at full speed, so the ball rolls far beyond. *(a)* Draw this story (with worldlines for Jack, human, and ball) in the human's frame and the ball's frame. *(b)* Which worldlines are inertial? Are they inertial in

both diagrams? *(c)* Draw the story in Jack's frame. What law of motion is violated in this frame?

4.5   *(a)* Explain why the argument that transverse distances are frame-independent does *not* apply to distances along the direction of motion. *(b)* If we cannot be sure whether distances along the direction of motion are frame-dependent or not, how do you think we can settle the issue? Be specific about the logic or the experiments you would do.

4.6   A spring-loaded device moving at a constant 10 km/s to the north past a space station separates cleanly into two equal-mass halves that move apart at a relative speed of 3 km/s. *(a)* Assume the separation is along the original direction of motion and draw the story in the space station frame and in the rest frame of the original device. *(b)* Assume the separation is *perpendicular* to the original direction of motion and draw the story in the rest frame of the original device. How is your answer in this part similar to that in part (a)?

The Speed of Light

5

You have probably heard that nothing can go faster than the speed of light. In this chapter we will see that the speed of light is even more remarkable: it is the same in all frames. From this surprising fact we will deduce that this speed must also serve as a limit, and that Galilean velocity addition fails to describe how nature works at high speeds. This is our first glimpse of the modern understanding of relativity.

5.1 Observation: the speed of light is frame-independent

"Why can't anything go faster than light?" is one of the most frequently asked questions about relativity. In Section 5.2 we will see that this is a consequence of an even deeper truth: *light travels at the same speed in all frames*. The idea that any speed could be the same in all frames is deeply counterintuitive, so this section aims to make you more comfortable with that premise, before tackling any of the consequences.

First, you may be curious how we can even measure the speed of something as fast as light. Figure 5.1 shows the concept: two "gates" move together such that any projectile entering the first gate must have a very specific speed to exit the second gate. If an entering projectile manages to exit the device, we know its speed must have been the distance between gates divided by the time taken to rotate the exit gate into place. Measuring faster projectiles simply requires faster rotation rates. Physicists have used this concept (with some "gates" made of mirrors rather than holes) to measure the speed of light, denoted as c, with extraordinary precision: $c = 299,792.458$ km/s. This book will round c to 300,000 km/s for convenience.

You might think the invariance of c was discovered by comparing measurements with such devices aboard airplanes, spacecraft, and so on. But physicists discovered this invariance long before such high-speed laboratories were available. As early as the mid-1800s there were compelling reasons to suspect that the speed of light is frame-independent. In 1865 James Clerk Maxwell (1831–1879) worked out the equations of electromagnetism and showed one consequence is that electromagnetic waves—also known as light—must travel at c in any frame. Nineteenth-century astronomers came to the same conclusion empirically by studying stars moving at hundreds of km/s relative to Earth, much faster even

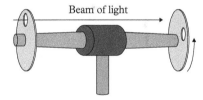

Figure 5.1 *This concept for a speed-of-light measuring machine is inspired by Lewis Carroll Epstein's* Relativity Visualized. *At slow rotation rates light entering the front hits the back wall rather than the exit hole. The rotation rate is increased until light does exit; the speed of light is then the length of the device divided by the time taken to rotate the exit hole into place. In practice physicists use rotating mirrors, but this concept captures the essence of the idea.*

The Elements of Relativity. David M. Wittman, Oxford University Press (2018).
© David M. Wittman 2018. DOI 10.1093/oso/9780199658633.001.0001

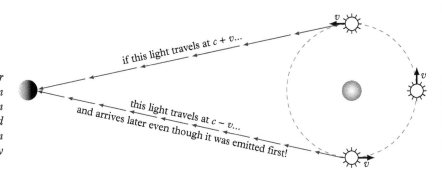

Figure 5.2 *Astronomical argument for the velocity of light not depending on the velocity of the emitter. This diagram is not to scale; the binary star should be drawn billions of times further from Earth. Inspired by a figure in Relativity Visualized by Lewis Carroll Epstein.*

than rockets today. Although astronomers did not directly measure the speed of light from moving stars, they were convinced by the following evidence.

Stars in binary star systems orbit each other and so are constantly changing their velocities. The white star in Figure 5.2 is shown first moving away from and then toward Earth. If the velocity of light relative to us depended on the velocity of the source, then light emitted by the star when it was moving away from us would travel more slowly than light emitted by the star when it was moving toward us. Because the light travels for years before reaching Earth, even a small boost in speed would shave weeks off the travel time. Faster light that was emitted *after* the slower light could then reach Earth first. Astronomers on Earth would then see the star orbit in a mixed-up sequence, with greater mixing for more distant stars. In fact, astronomers never saw anything of the sort, so they routinely assumed—without necessarily thinking of the implications for fundamental physics—that the speed of the light does not depend on the speed of the source.

The binary-star argument does not rule out a model in which the speed of light depends on the speed of the *observer* relative to some preferred frame (Section 3.3). As an analogy, the speed of sound through air is determined by properties of the air alone (such as its temperature) and is unaffected by the speed of the object making the sound. Because the sound speed is fixed in the air frame, observers moving *relative to the air* do measure a different sound speed. A similar model for light was popular in the nineteenth century: a hypothetical medium called the ether was thought to fill space and allow light waves to travel through otherwise empty space. If so, Earth's rotation and motion through this medium would alternately increase and decrease the apparent speed of light in the east-west direction while leaving it unchanged in the north-south direction. This model was discredited when (among other difficulties) experiments showed no such pattern.

So we are stuck with the unsettling fact that the speed of light is always *c*, relative to *any* observer. This is not merely a statement about light: anything that races light to a tie in one frame must do so in all frames, and therefore also has the curious property of having the same speed in all frames. We therefore attribute

Think about it

The 1887 Michelson–Morley experiment comparing north-south to east-west speeds of light is often cited as killing the ether hypothesis, but in fact many physicists tried to save it with a series of complicated tweaks. Einstein stood out by having the courage to take the frame-independence of *c* as a foothold idea and follow the implications wherever they led.

this frame-independence or **invariance** to the speed c itself. The consequences of this invariance are many and profound, and will take us several chapters to work through completely. In this chapter, we will focus on just two.

Check your understanding. A bullet train with headlights on approaches a train station at high speed. Rank the following speeds from slowest to fastest: the speed of the light from the headlights measured by an observer in the station; the speed of the light from the station measured by an observer in the train; the speed of the light from the station measured by an observer on the ground outside the station.

5.2 Implication: nothing can travel faster than *c*

Imagine Carol traveling at a constant 90% of the speed of light—written more compactly as $0.9c$—to the east through Bob's laboratory as in Figure 5.3. If Bob has a laser pointer and points it to the east, he would certainly measure the light to be traveling at c relative to him. What is unsettling is that, due to the invariance of c, Carol must also measure the light to be traveling at c relative to *her*. This directly contradicts the prediction of the Galilean velocity addition law, which is that Carol would measure the light as moving at only $0.1c$ relative to her. (In Carol's frame Bob moves *west* so his velocity is $-0.9c$ and the Galilean calculation is $c - 0.9c = 0.1c$.)

Figure 5.3 *Carol moves through Bob's frame at* $v = 0.9c$.

Which is wrong: the Galilean velocity addition law, or the invariance of c? Each has substantial experimental support, so this seems to be a dilemma. We can wriggle out of this dilemma by noting that experimental tests confirming the Galilean law always take place at everyday speeds, which are *far* lower than c. It could be that the Galilean law is a very good approximation to the true law at low speed, but is a less good approximation at very high speed. So let us adopt invariance of c as a foothold idea (Section 4.4) and see what we can deduce about velocity addition given our two other constraints: the Galilean law must be very close to correct at low speeds, and the same laws of physics apply in the coordinate system attached to Carol as in the one attached to Bob (the principle of relativity).

This approach suggests that while ruler and clock measurements in two frames agree at low relative velocities, those same rulers and clocks will disagree at high relative velocities. By reciprocity, any effect of velocity on rulers and clocks must be such that Carol and Bob each measure the *other's* rulers and clocks to be affected in equal measure. We will work through the details gradually over the next several chapters. Right now, we can deduce a few other qualitative consequences by adding just one more wrinkle to our thought experiment.

Let us zoom out from Figure 5.3 to find that Bob's laboratory is actually a rocket traveling east at $0.9c$ relative to a ship captained by Alice (Figure 5.4). The

Think about it

Light outruns your top driving speed by about the same factor—10 million—that your top driving speed outruns an amoeba. Even rocket speeds are so far below c that we will refer to them as "everyday low speed."

Figure 5.4 *Why velocities must add in a way that leaves the sum always less than c. Carol moves at 0.9c through Bob's ship, which itself moves at 0.9c through Alice's frame, yielding a naive prediction of 1.8c for Carol's speed through Alice's frame. But Alice's ship emits a beam of light, which Carol must observe passing at speed c. Therefore, Carol cannot overtake the beam—and this is a frame-independent statement. Carol's failure to overtake the beam implies that Alice must measure Carol's speed as less than the speed of the beam, c.*

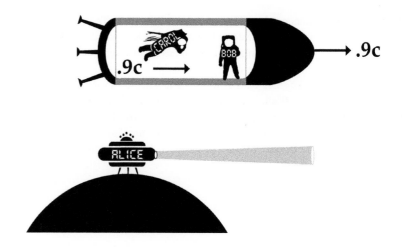

Confusion alert

When we say "Carol's speed relative to Alice" we mean Carol's speed through a coordinate grid attached to Alice, or equivalently what Alice and Carol measure when they point radar speed guns at each other. This is *not* the same as asking how quickly they move apart in a third frame such as Bob's. Given that Bob measures Alice moving west at 0.9c and Carol moving east at 0.9c, for example, in his frame they *must* move apart at a rate of 1.8c. See Box 5.1 for more nuances in interpreting the speed-limit argument.

Think about it

The speed-limit argument presented here only prevents particles from starting out slowly and accelerating to c or faster; it does not prevent particles from being "born" that fast. But there is no evidence that such particles exist, and there is a strong argument that they should *not* exist (Section 10.5).

Galilean model predicts that Alice measures Carol's speed as $1.8c$, the sum of Carol's speed through Bob's frame and the speed of Bob's frame through Alice's frame. But if we take the invariance of c seriously, the following argument proves that $1.8c$ is *not* correct. When Alice turns her headlights on, the front edge of the headlight beam shoots to the east at c relative to Alice, *and* at c relative to Bob, *and* at c relative to Carol. That last fact implies that Carol can never catch up to or overtake the beam of light, and this must be a frame-independent statement: if Carol overtakes the light she could manipulate it in a way that is clear to all observers. Now switch your thinking back to Alice's frame: when Carol fails to overtake the beam it is clear that her speed in this frame is less than c, rather than the $1.8c$ predicted by the Galilean model.

Thus, even without deducing the correct way to add velocities, we must conclude that c is a speed limit, as follows. We can add another observer moving at $0.9c$ eastward relative to Carol, another moving at $0.9c$ eastward relative to that one, and so on ad infinitum, but—because c is the same in all frames—even the fastest observer measures the distance between himself and the beam growing at c. Therefore, no one passes or even keeps pace with the light, and, therefore, no one exceeds or even reaches c in Alice's frame. And Alice's frame was not special, so this applies to all inertial frames: no matter how fast something is moving in a given frame, there can always be something else even faster, yet that something else is still slower than c. Therefore, *no object can reach the speed of light, much less exceed it.* Note that this is a *conclusion* rather than an axiom; the invariance of c is the axiom that enables this conclusion.

There can be only one speed limit, so this conclusion helps us see that c is really special. Speed is a ratio of distance and time, so we will come to see c as *nature's exchange rate between space and time.* Note that c is not special *because* it is the speed of light; rather, light is special because it travels at c. To appreciate the distinction, imagine a universe without light in which scientists spend their time developing extremely fast spaceships and guns that fire extremely fast bullets. The model

150K gun fires a bullet at 150,000 km/s (relative to the gun, of course), but when it is mounted on a spaceship traveling 150,000 km/s relative to Earth, the resulting bullet moves at only 240,000 km/s relative to Earth. After more experiments and analysis, scientists in this universe figure out how to modify the Galilean velocity addition law so that it explains the high-speed data without changing its low-speed behavior. After much experimentation, they find that only one equation works, and that equation involves not only the two speeds to be added, but a third speed—299,792.458 km/s—as well, in a way that ensures that the resulting speed never exceeds 299,792.458 km/s. Because 299,792.458 km/s is indispensable in equations involving velocities, scientists give it a name (c) and call it a *constant of nature*—just as π is a constant that is indispensable in equations involving circles and arcs. These scientists deduce everything about special relativity without ever using the phrase *speed of light*. The only difference with our universe is that for them c remains a bit abstract—an unreachable limit just above the speed of the fastest known particles, a constant of nature, but not exactly the speed of anything in particular—while for us c happens to be the speed of something we deal with every day.

> **Think about it**
>
> An analogy for the hypothetical scenario presented here: just as experiments on *nearly* frictionless surfaces allow us to deduce the laws of motion applying to perfectly frictionless surfaces, scientists in this alternate universe deduce all the laws of special relativity using speeds *near* the critical speed c.

Check your understanding. Estimate Carol's velocity relative to Alice's coordinate system in the scenario in this section. You should be able to give a range of possible velocities; go ahead and take a guess within that range.

5.3 Implications for the velocity addition law

We have seen that successive velocity additions can never add up to a final speed greater than c. How can this possibly work in practice, if we can always give additional boosts to objects that are already traveling at speeds near c? The details will emerge over several chapters; here we start by drawing a useful mental picture. Imagine standing a certain distance d away from a building and thinking about how far you must tilt your head to see the roof. Let us call the angle required for a one-story building a_1 (Figure 5.5).

Now imagine standing the same distance from a two-story building as in Figure 5.6. If the additional angle required to encompass the second story—denoted with a question mark—is equal to a_1 then the total angle is $2a_1$ and doubling the height of the building doubles the angle at which you must crane your neck. But is the additional angle for each additional floor really equal to a_1? We can answer this question nonmathematically by using the thinking tool of taking things to extremes. Consider a much taller building as in Figure 5.7. When we start with a tall building, it is clear that each additional story contributes very little extra angle. Even in the most extreme case of an infinitely tall building the angle will never reach, much less exceed, 90°. But the reduced impact of additional stories was hardly visible when we thought only about adding a second story to a single-story building. It must be that each successive story boosts the angle by successively smaller amounts.

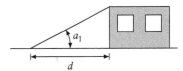

Figure 5.5 *We tilt our heads by an angle a_1 to see the top of a one-story building.*

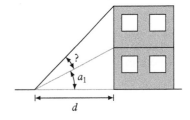

Figure 5.6 *Does the angle double if we double the height of the building?*

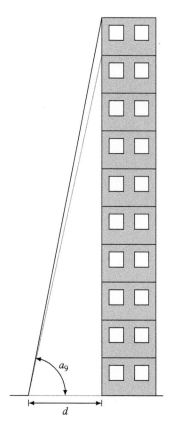

Figure 5.7 *Adding additional floors to the building has little effect on the angle if the building is already tall.*

Figure 5.8 *Diminishing returns: if seeing the top of a one-story building requires tilting your head at 10°, a two-story building requires a bit less than 20°—so close to 20° that you might not notice the difference, but each additional story adds less and less to the angle. No building can quite reach 90° no matter how many stories it adds, just as no object can reach c no matter how much it accelerates.*

This is highly analogous to the addition of velocities in the same direction in special relativity. Just as an arbitrarily large number of additional stories can bring the angle arbitrarily close to—but never quite reach—90°, an arbitrarily large boost or series of boosts can bring the speed arbitrarily close to—but never quite reach—c. There must be some law of diminishing returns that mathematically describes the decreasing effectiveness of successive boosts. In the case of building heights this law is fairly simple to construct if you recall your geometry (Figure 5.8). In the case of special relativity the law we seek—the **Einstein velocity addition law**—will emerge more clearly later in our journey. One thing we know about this law already is that at low speeds it must very closely approximate the Galilean law, because we know that diminishing returns are not noticeable when adding everyday speeds. In our geometric analogy, adding everyday low speeds would be like stacking sheets of paper rather than buildings: the stack is simply not high enough to notice the diminishing-return effect.

You might think the very high speeds required for large departures from the Galilean law would make those departures nearly impossible to detect in practice; even our rockets move at only about $0.00003c$. But we can easily make one of the speeds large by using light. The first experimental indication of the need for a new velocity addition law—although it was not interpreted as such at the time—came about in 1851 due to Hippolyte Fizeau (1819–96), who was studying the motion of light through moving water. Light moves at about $0.75c$ through still water, making this a safe and cost-effective way to obtain a high speed in the laboratory.

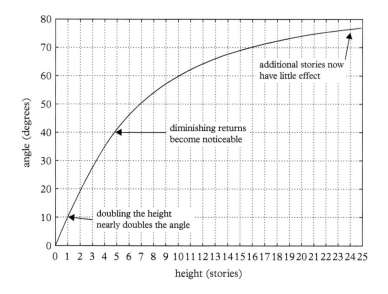

Fizeau then boosted this speed by moving the water through the laboratory and measuring something called the index of refraction, which is related to the speed of the waterborne light relative to the laboratory. He found that the observed change in index of refraction was much smaller than predicted by the ether hypothesis (Section 5.1). After Einstein worked out the correct velocity addition law he showed quite specifically how it explained the results of the Fizeau experiment. Thus, the Einstein velocity addition law was experimentally confirmed with mid-nineteenth-century technology!

Check your understanding. When pillows are stacked they compress, making the height of the stack less than the sum of the heights of the separate pillows. This is a reasonable analogy with velocity addition, but in what way does this analogy fail?

5.4 Graphical interpretation

Spacetime diagrams provide another useful picture of velocity addition. But if we use kilometers and hours to define the grid as in Chapter 4, all speeds approaching c would look like horizontal lines, covering millions of km each hour. So let us define the grid using distances in light-years (about 10 trillion kilometers each) on the x axis and time in years on the t axis. In this case light—which moves one light-year per year by definition—will follow worldlines tilted at 45°. This allows us to easily distinguish speeds below, at, and (hypothetically) above c. Coincidentally, light travels very nearly one foot per nanosecond (one billionth of a second, abbreviated ns), so we can also use units of feet and nanoseconds if we wish to think on a smaller scale. The specific units are less important than the key concept that speeds near c should be easy to identify and distinguish in these diagrams.

Figure 5.9 is a spacetime diagram of the Alice/Bob/Carol story in Section 5.2, drawn in Bob's frame. This frame makes it easy to get started, because it directly relates to every velocity listed in the story. Specifically, it was given that Carol moves east at $0.9c$ through Bob's frame; and that Bob and Alice measure a relative speed of $0.9c$ for each other. Before moving on to the next paragraph, make sure you are comfortable with this representation by convincing yourself of the following points:

- A Bob-frame diagram must show Bob as motionless, so he has a vertical worldline (he remains at one location as time proceeds). The coordinates are labeled x_B and t_B to emphasize that this is Bob's coordinate system.

- In our units light always travels in a 1:1 space:time ratio, or a 45° tilt from vertical. Feel free to draw a hypothetical ray of light on any diagram to

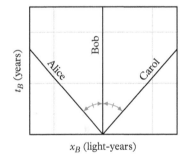

Figure 5.9 *Our spacetime diagrams will henceforth use units of light-years (for distance) and years (for time). Light thus moves one graph-paper square left or right for each square up in time, and the worldlines of our characters will always tilt from vertical by less than this. Here, Bob is motionless (the diagram represents his frame), Alice moves west at 0.9c (0.9 squares left per square up), and Carol moves east at 0.9c.*

provide a convenient reference line. The worldlines of our characters will never tilt this much from vertical.

- Carol's worldline corresponds to $0.9c$ because it covers 0.9 squares of distance in space for each square of time. When you draw your own diagrams on graph paper, you will find it convenient to count nine squares of space for ten squares of time to represent $0.9c$, four squares of space for five squares of time to represent $0.8c$, and so on.

- Alice's worldline corresponds to $0.9c$ in the opposite direction.

- The worldlines intersect at the origin, but this does not necessarily mean that the three characters collided there. A spacetime diagram represents only *one* dimension of space, which we ususaly refer to as the x direction, or east-west. The characters may have a north-south separation we cannot represent on the diagram. However, we will assume any such separation is constant with time because we have *chosen* the x direction to represent the direction of relative motion.

- The term "Bob's frame" should not lead us to think in terms of what Bob *personally* sees. "Bob's frame" really means a coordinate system in which Bob is motionless, and we can record worldlines in this coordinate system without worrying what Bob personally sees.

Now that we have established a graphical language, let us return to the physics. Figure 5.9 shows that in Bob's frame the distance between Alice and Carol grows by 1.8 light-years per year (see the Confusion Alert in Section 5.2 if this surprises you). However, we have already deduced that in Alice's frame the distance between Alice and Carol cannot increase by more than one unit of space per unit of time. These two statements can both be true only if switching coordinate grids—from one attached to Bob to one attached to Alice—is more complicated than Galileo envisioned. We will examine that process in more detail in Chapter 6. To build a foundation for that, let us practice translating everything we do know about the situation from a Bob-frame spacetime diagram (Figure 5.9) to a new Alice-frame diagram (Figure 5.10).

In Figure 5.10 we start by drawing Alice as stationary and labeling the coordinates with A subscripts to make it clear that this is a grid attached to Alice. Next, it was given that Bob moved at $0.9c$ east relative to Alice, so we draw Bob's worldline as moving 0.9 units to the right per unit time. Note that the angle between Alice and Bob worldlines is the same as in Figure 5.9—this reflects the symmetry that $v_{AB} = -v_{BA}$ always (Section 1.3). Next, we ask how to draw Carol's worldline. Carol's velocity through Alice's frame was *not* given, but we deduced that it is somewhere between Bob's velocity ($0.9c$) and c. Carol's worldline in Figure 5.10 is drawn accordingly—you can see it is barely less than c because it just misses the grid intersection marking one unit of space and one of time.

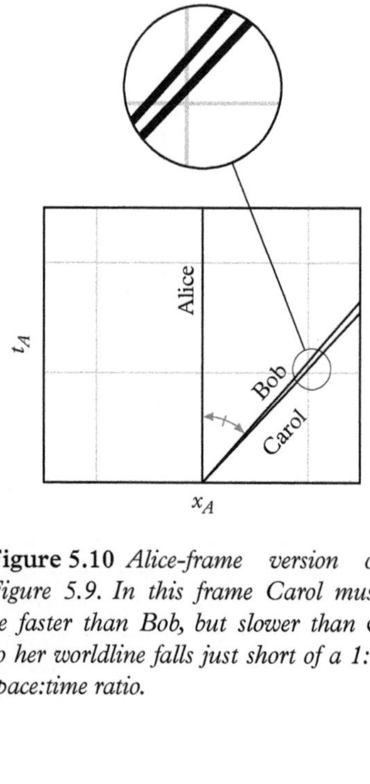

Figure 5.10 *Alice-frame version of Figure 5.9. In this frame Carol must be faster than Bob, but slower than c, so her worldline falls just short of a 1:1 space:time ratio.*

Take a moment to compare Figures 5.9 and 5.10. The relationship between the two frames is something akin to a spacetime rotation: rotating Figure 5.9 clockwise makes Alice's worldline vertical and Bob's worldline tilt right as in Figure 5.10. But the transformation between grids requires some other ingredient as well, so that Carol's worldline does not rotate much. The details will emerge over the next two chapters, but we can already appreciate conceptually why Carol's worldline cannot rotate much in this transformation. The invariance of *c* means that any worldline following a 1:1 space:time ratio (representing *c*) in one frame must do so in *all* frames. Therefore, if we draw a hypothetical ray of light just under Carol's worldline in Figure 5.9, the transformation to Figure 5.10 will leave this worldline unchanged. The 1:1 space:time ratio thus represents a limit that other worldlines may approach but never quite reach, just as buildings in our analogy (Section 5.3) never quite reach 90° viewing angle. Figure 5.11 makes this explicit by showing five hypothetical worldlines in two different frames. Adoption of a new frame leaves unaffected the dashed worldlines representing *c*, so all worldlines must squeeze in toward the dashed worldline.

Check your understanding. Draw the Carol-frame diagram corresponding to Figures 5.9 and 5.10.

5.5 Incomplete versus wrong models

Frequently in science, a model that has been working for us begins to fail when applied to more extreme situations. The Galilean velocity addition law is a good example. In these cases, scientists often say that the old model is not wrong, merely incomplete. The old law can still be used for the same slow velocities it was always used for, but it does not describe *all* velocities.

Sometimes models are just plain wrong, though. The geocentric model of planetary motion (in which planets go around the Earth) was replaced by the heliocentric model (in which they go around the Sun) and we do not say that the geocentric model was merely incomplete. The Galilean velocity addition law qualifies as incomplete rather than wrong because the Einstein velocity addition law is nearly the same as (or *reduces to*) the Galilean law in the limit of very low velocities. Figure 5.12 illustrates the relationships between Galilean relativity, special relativity, and their assumptions and implications. This process of extending everyday physical laws to new regimes will play out in many ways as we work through the consequences of the invariance of *c*.

Check your understanding. Describe another model (not necessarily from physics) that works well enough most of the time but breaks down in extreme conditions.

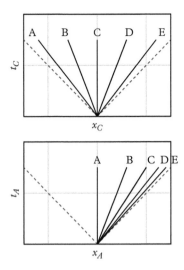

Figure 5.11 *Adopting a different frame by definition retilts worldlines on a spacetime diagram. Yet this must happen in a way that leaves* c *(dashed worldlines) the same in all frames. As a result the change in tilt is smaller for worldlines nearer* c. c *thus represents a limit that other worldlines cannot quite reach.*

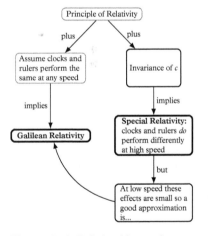

Figure 5.12 *Relationships between Galilean relativity, special relativity, and their assumptions and implications.*

Box 5.1 Speeds apparently faster than c

In Section 5.2, a Confusion Alert briefly explained why Bob measures the distance between Alice and Carol growing at $1.8c$. The speed-limit argument has nothing to say about how to add two displacements (e.g., those of Alice and Carol) that have been *measured in the same coordinate system*. We must add them linearly, just like we measure a single large displacement by adding up the number of meter sticks used. The speed-limit argument is concerned with the much more complicated issue of deducing what Alice would measure regarding Carol when we are *given only measurements in Bob's coordinate system*. This is a subtle question, even if it goes by the simple name of "velocity addition." This box develops additional examples in which the notion of a speed limit must be used with caution.

First, imagine a powerful laser pointer that makes a dot on a screen millions of kilometers away. If you tilt the laser pointer even slightly, the dot will move hundreds of thousands of kilometers across the screen. So shaking the laser pointer rapidly could make the dot move faster than c. This does not contradict the argument presented in this chapter, because light is not moving across the screen at all; it is moving from the laser pointer to the screen (at c) and just happens to intersect the screen in a way that causes the appearance of a moving entity. Our argument in this chapter that *successive boosts from sublight speeds cannot reach or surpass c* really has nothing to do with this situation. A handy criterion for scenarios like this is to ask whether a message could have been transmitted by the motion. Although the laser can carry a message from the pointer to the screen (at speed c) the dot is *not* capable of carrying a message from one point on the screen to another.

Second, to revisit the issue in the first paragraph of this box more concretely, imagine one missile coming at you from the north at 200,000 km/s and another coming at you from the south at 200,000 km/s. The conclusion that you measure the distance between missiles as decreasing by 400,000 km each second is unavoidable given the stated measurements in your frame. The speed-limit argument applies only to *what one missile would measure for the other's speed*, which is an entirely different question. We naturally conflate these two questions when we rely on the now discredited Galilean velocity addition law. Understanding what the missiles *do* measure as their relative velocity will require several more chapters; the important message here is to be on guard against inadvertent use of discredited assumptions. The logic of relativity fits together beautifully given its particular assumptions, but you will find many apparent contradictions if you inadvertently add further assumptions as well. When you find that something in relativity does not make sense, the first step is to search your reasoning for hidden assumptions.

Third, c coincides with the speed of light *in a vacuum*, but light moves more slowly through materials. Even in these materials c remains the constant of nature c relevant to all the effects of relativity. Light moves at $0.75c$ through water, for example, so a particle moving at $0.9c$ through water overtakes light moving through the same water without violating relativity in the least.

In summary, the phrase "nothing moves faster than light" is an oversimplification that often leads students astray. How can we replace this meme with something more accurate but equally simple? I suggest "no *thing* moves faster than c." The italicized word helps you remember what the rule applies to, and replacing "light" with c gives proper respect to this constant of nature.

Box 5.2 Definition of the meter

Behind the statement that c is precisely measured lies a nuance that may be of interest to physics students. In the late twentieth century, measurements of c became more precise than the definition of the meter itself. The most precise way to define the meter came to be "1/299,792,458 of the distance traveled by light (through a vacuum) in one second." Because c is now defined rather than measured, statements such as "c is invariant and precisely measured" should technically be restated as "meter sticks produced according to this definition have identical lengths regardless of the velocity of the laboratory that makes them." This mouthful obscures the central idea of relativity, so we will continue to use the simplified statement.

CHAPTER SUMMARY

- The speed $c = 299,792.458$ km/s is the same in all frames. This is the speed of light in vacuum but the key feature is its invariance rather than its connection to light.

- As a consequence, c is also nature's speed limit.

- The invariance of c violates the Galilean velocity addition law so we must seek a modified law, but there is no contradiction with the principle of relativity.

- At low velocities the modified addition law must still look like the Galilean law. Galilean relativity is an incomplete description of nature but perfectly serviceable at low velocities.

- The angles of worldlines change when we redraw a spacetime diagram in a new frame. The invariance of c implies, however, that light-ray worldlines do *not* retilt—they always tilt at 45° in the usual convention for spacetime diagrams. This in turn limits the extent to which the tilt of slower worldlines can be affected by a frame change. This graphically encodes how c is a speed limit as well as how successive velocity additions must provide diminishing returns.

FURTHER READING

Relativity: The Special and General Theory by Albert Einstein (free online at bartleby.com) is a short book for the general public; Chapter XIII describes how the Einstein velocity addition law explains Fizeau's experimental results.

In *Spacetime Physics*, Edwin F. Taylor and John Archibald Wheeler derive the addition law for the angles of worldlines in spacetime diagrams, in both Galilean and special relativity cases.

◢ CHECK YOUR UNDERSTANDING: EXPLANATIONS

5.1 All these speeds are equal.

5.2 Alice's speed relative to Carol must be very close to c, but not quite c. A good rule of thumb for thinking in Alice's frame is to think of Carol as being 0.9 (90%) of the way between Bob's speed and c, so $0.99c$ is a good estimate. (For comparison, $0.994475c$ is the answer given by quantitative laws we will deduce later.) Alice's velocity in Carol's frame is then $0.99c$ to the west.

5.3 The analogy is good at first glance, but it eventually fails because with enough pillows we can still make a stack as tall as we please. In contrast, the ultimate height of any "stack" of velocities is always less than c.

5.4 Carol's worldline must be vertical, Bob's must slope nine squares left for every ten squares up, and Alice's worldline slopes a bit more than that (but less than ten squares left for every ten squares up).

5.5 Creativity is encouraged, so answers may vary widely.

? EXERCISES

5.1 If you are traveling at 99.99% of the speed of light while holding a mirror and you look in the mirror, do you see anything unusual? Does the light from your face take a long time to reach the mirror because the mirror is moving away from the light so rapidly? Explain your reasoning.

5.2 Imagine that you eat lunch and dinner in the same restaurant. In the frame of your city, there is zero distance between the lunch and dinner events. Is this distance frame-dependent?

5.3 Is velocity frame-dependent?

5.4 Why can no worldline tilt more than 45° from vertical on a spacetime diagram?

5.5 If you saw a horizontal worldline on a spacetime diagram, what velocity would it represent?

5.6 Invent another analogy for velocity addition, even if it is a very rough analogy. In what way(s) does your analogy work, and in what way(s) does it not work? If you do not know how to start, think about the phrase *diminishing returns*.

5.7 (See Box 5.1.) *(a)* One missile approaches you from the north at 200,000 km/s and another approaches you from the south at 200,000 km/s. Explain why you can state that the distance between missiles decreases by 400,000 km/s without violating the speed-limit argument. *(b)* Can the missiles send messages to each other at 400,000 km/s? Explain how this relates to part (a).

+ PROBLEMS

5.1 A strobe light emits eastbound and westbound flashes of light simultaneously. *(a)* In the strobe frame, what is the rate at which the distance between the flashes increases with time? *(b)* Is the rate the same in all frames? Justify your reasoning.

5.2 A Klingon warship approaches the planet Vulcan. It can fire laser pulses that at the speed of light, "phasers" that travel at $0.9c$, or torpedoes that travel at $0.1c$ (these numbers are of course relative to the ship that fires them). *(a)* The Klingons fire while approaching Vulcan at 0.9c. *Estimate* the speed the Vulcans measure for each of the three incoming weapons. *(b)* Repeat for a Klingon approach speed of 0.1c.

5.3 Draw a Klingon-frame spacetime diagram of the situation in Problem 5.2.

5.4 Draw a Vulcan-frame spacetime diagram of the situation in Problem 5.2. Explain why you must estimate the phaser and torpedo velocities while you can draw tha laser velocity exactly.

5.5 For the mathematically inclined: indentify any additive function with the required properties, or at least *some* of the properties. State clearly which properties it satisfies and which, if any, it does not. *Hint:* does subtracting a speed from itself yield zero speed, or something else?

5.6 *Return to Bender:* revisit the problem in Chapter 3 regarding Bender, the torpedoed robot. Given our new velocity addition law, how would your answers to that question change? Assume the spacecraft has a very high speed.

5.7 *(a)* What is the maximum worldline tilt from vertical allowed in *any* model of relativity? *Hint:* Consider the meaning of a worldline tilted close to horizontal. *(b)* Given your result for part (a), can worldline angles add linearly in any model of relativity? Explain your reasoning.

6

Time Skew

Ship frame

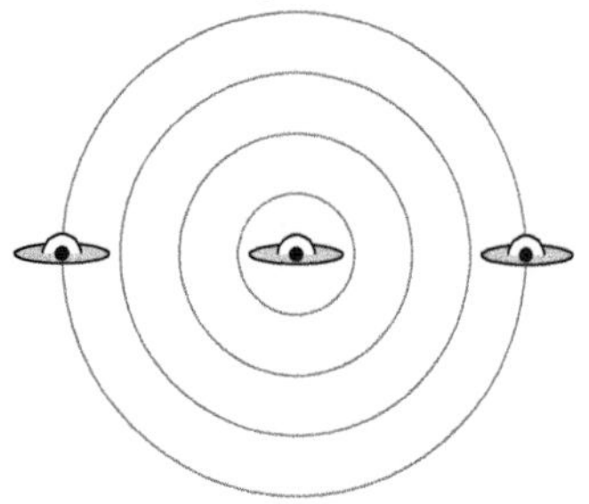

Figure 6.1 *Three ships at rest. The flanking ships are equidistant from the central ship, so flashes of light (shown as rings) emitted by the central ship reach the flanking ships simultaneously.*

In this chapter, we will discover another consequence of the invariance of c: events that are simultaneous in one frame are not necessarily simultaneous in other frames. Investigating this in more detail, we will find that time is surprisingly flexible: the time coordinates of events are just as frame-dependent as their positions. This is no accident, but a symmetry between space and time.

6.1 Simultaneity is frame-dependent

Two events may occur at the same place in one frame, but not in another; for example, two beeps of a car horn occur at the same place in the car frame but not in the street frame. The invariance of c will force us to recognize that *occurring at the same time* is just as frame-dependent a statement as *occurring at the same place*. We will prove this by setting up two events that are guaranteed to occur simultaneously in one frame, and then analyzing the same setup in a different frame. The following is inspired by a story and figures in Lewis Carroll Epstein's *Relativity Visualized*.

Imagine three stationary spaceships: a central command ship plus flanking ships at equal distances to the east and west (Figure 6.1). The central ship emits regular flashes of light in all directions, represented by the rings. Because the flanking ships are equidistant from the source, any given flash reaches each flanking ship simultaneously.

Now think in another frame—call it the planet frame—in which the ships are moving east (Figure 6.2). The faded ships show the positions when the flash was emitted, and the full ships show the current positions. For clarity, only the outer flash is now highlighted. At this instant in the planet frame, the flash has already passed the trailing (western) ship, but has yet to catch the leading (eastern) ship. Therefore, the ships do not receive the flash simultaneously in this frame. In general, then, statements such as "these two events are simultaneous" must be frame-dependent.

Take a moment to see *why* we are forced to this conclusion. In the planet frame I drew the outer flash just as in the ship frame, rather than having it move along with the ships. Why? Because light travels strictly at speed c in *this* frame, not only in the ship frame. If the speed of the light had been augmented by the speed of the emitting ship, our conclusion would have been quite different. So, the frame-dependence of simultaneity is a direct consequence of the invariance of c.

The Elements of Relativity. David M. Wittman, Oxford University Press (2018).
© David M. Wittman 2018. DOI 10.1093/oso/9780199658633.001.0001

Let us see how this works on a spacetime diagram. First, note that the procedure shown in Figure 6.1 is a way of *defining* the time coordinates of events. Let us call the event where the flash hits the east ship E, and the event where the flash hits the west ship W. Without the light flashes from the central ship, how could we be sure that events E and W were simultaneous? The easy answer is to have each ship note the time on its clock when the flash hits—but how do we know the clocks are working and properly synchronized? The way to check for sure is just what we did: send a flash of light from the midpoint between the clocks. If each flanking clock is one light-second (300,000 km) from the central ship, the central clock can emit a flash at 11:59:59, and then we will be sure that the flanking ships will receive it at 12:00:00. This idea is expressed in the form of a spacetime diagram in the left panel of Figure 6.3 (but with larger time differences to make the clocks more readable). Repeating this process at different times and places, we can build a full coordinate grid as shown in the right panel of Figure 6.3. Pencil in flashes of light starting from random grid intersections in that panel until you are convinced this procedure works to define a spacetime grid. And keep this clock grid in mind whenever you look at a coordinate grid—we may not bother to draw the clocks but they are still conceptually there.

Next, we ask what this grid looks like in the planet frame. We start drawing the left panel of Figure 6.4 with a square planet-frame grid, the worldlines of the three eastbound clocks, and regular clock ticks for the central clock. Where and how should we draw the time readings on the other clocks? Let us use the light rays fired from the central clock at the tick before noon. The left panel of Figure 6.4 shows the situation after waiting a while *in the planet frame*. As in the

Planet frame

Figure 6.2 *As for Figure 6.1, but in a "planet" frame where the ships are moving east. At this instant in the planet frame, the trailing (western) ship has already intercepted the flash, while the leading (eastern) ship has not; the flashes are not received simultaneously.*

Think about it

It may help to think of the central clock as emitting flashes encoding the message "the current time is . . ." Each receiving clock then adds the known light travel time and displays the result.

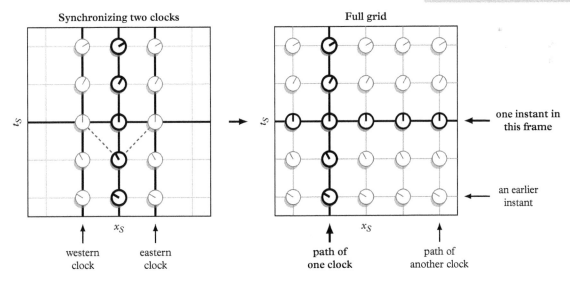

Figure 6.3 Left: *spacetime diagram representation of the clock synchronization principle shown in Figure 6.1.* Right: *the same principle can be extended to build and verify a full spacetime coordinate grid. Draw a flash of light from any grid intersection to verify that the clocks flanking it are synchronized.*

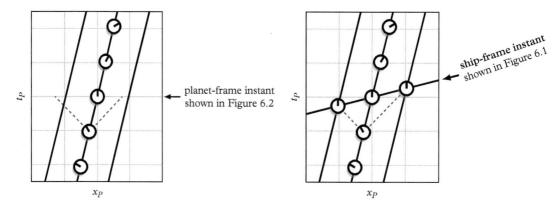

Figure 6.4 Left: *spacetime diagram representation of Figure 6.2.* Right: *the events at which the flanking ship clocks strike noon are identified at the intersections of the light rays with the clock worldlines. These events are* not *simultaneous in the planet frame.*

cartoon version of this situation, Figure 6.2, the westbound light has already hit the west ship, but the eastbound light has yet to catch the east ship. Make sure you understand how Figure 6.2 and the left panel of Figure 6.4 show the same situation before moving on!

We can already see where the westbound light hits the west ship. Because that *defines* noon on the west ship, we draw a clock displaying noon at that intersection in the right panel of Figure 6.4. Next, we locate noon on the east ship simply by extending the eastbound light ray a bit further until it hits the ship worldline. Connecting the clocks reading noon on the three ships, we get a line of simultaneous events—simultaneous in the *ship* frame, that is. The set of events simultaneous in the ship frame forms a *skewed* line in the planet-frame diagram.

To appreciate this further, we can build out the ship grid (as measured in the planet frame, remember) by firing additional light rays from any clock shown in Figure 6.4 and following the same logic. The skewed-line pattern simply repeats over and over as in Figure 6.5. Again, I encourage you to pencil in flashes from a few random clocks in this figure until you are completely comfortable with the reasoning.

Figure 6.5 contains nearly everything you need to understand special relativity, so study it well. This diagram contains *two* grids—the faint square grid representing the planet frame (hence the P subscripts on the coordinate labels along the sides of the large square) and the skewed grid representing the ship frame. Any given event can be specified in terms of its planet-frame coordinates *or* its ship-frame coordinates. Different frames are simply different ways of laying out grid lines, and there is no one right way to do so. Nevertheless, there are clear patterns in the relationships between the two grids. First, for any grid that moves (relative to the grid we draw as square) the lines marking fixed *locations* are tilted from vertical. (This is simply what "moving" means, and we already saw several examples of this in Chapter 4.) Second, grid lines marking *time* in the moving

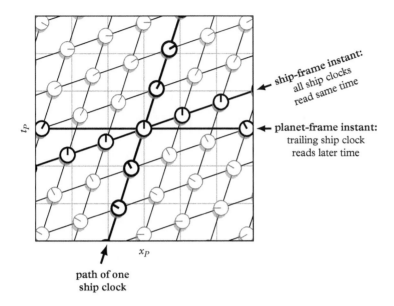

ship-frame instant:
all ship clocks
read same time

planet-frame instant:
trailing ship clock
reads later time

t_P

x_P

path of one
ship clock

Figure 6.5 *The principle illustrated in the right panel of Figure 6.4 can be extended to build and verify the full ship-based coordinate grid* as measured in the planet frame. *The thought process developed here will apply to any two grids in relative motion and will always yield similar patterns.*

frame are tilted from horizontal by the same amount so the grid appears to "fold up" symmetrically.

Had we been clever, we could have predicted this result quite directly from the invariance of c as follows. Light moves one light-year per year, so on a square grid its worldline appears rotated halfway from a vertical grid line toward a horizontal grid line; see the dashed light ray in the left panel of Figure 6.6. The invariance of c then dictates that in *any* frame light travels midway between the grid lines marking space and those marking time. If the spatial grid lines tilt right of vertical (which, by definition, is a grid moving to the right) then keeping light midway between the grid lines requires the temporal grid lines to tilt equally up from horizontal. This symmetry between space and time explains *all* the skewed grid lines in these figures. Indeed, **spacetime symmetry** is a key idea to keep in mind throughout your study of relativity.

The right panel of Figure 6.6 introduces a stripped-down version of a spacetime diagram, typically drawn when graph paper is not available. Instead of a full square grid, we simply draw the left and bottom sides of a representative cell. The left side is labeled t (with P subscript for planet frame in this case) and has an arrow pointing in the direction of increasing time. We call this the "t axis" or "time axis" because it points in that direction. Similarly, the bottom side is labeled x, has an arrow pointing in the direction of increasing x, and is called the x axis or space axis. "Time axis" and "space axis" sound mysterious but they are nothing more than representative grid lines to stand in for the full grid when you have no graph paper.

The axis representation applies just as well to moving grids. Copy the left side of a skewed cell in the left panel of Figure 6.6, and you have the t_S axis in the right panel; copy the bottom side of that cell and you have the x_S axis. When you see

> **Confusion alert**
>
> Students may be tempted to rotate the grid lines marking time to maintain perpendicularity with the grid lines marking space. This does *not* keep c (a 45°-tilted worldline) midway between space and time grid lines. Practice drawing spacetime diagrams until this point is perfectly clear; think spacetime *symmetry* rather than spacetime perpendicularity.

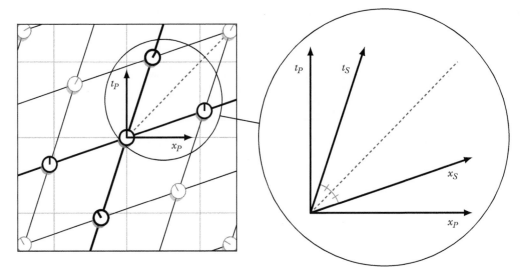

Figure 6.6 Left: *the dashed light ray travels one light-year per year, or midway between horizontal and vertical on a square spacetime grid. On a moving grid, which by definition has "vertical" lines tilted from vertical, this worldline can remain at the same speed only if the "horizontal" grid lines tilt up from horizontal by the same amount. Thus, the entire skewed grid follows simply from the invariance of* c. Right: *if you have no graph paper, you can still draw a spacetime diagram by using the left and bottom sides (the "axes") of a representative grid cell to stand in for that grid. The resulting diagram is less cluttered and more clearly shows how the axes of any grid are equidistant from the light ray.*

axes without grids such as in the right panel of Figure 6.6, practice filling in the rest of the grid mentally until this becomes second nature. And remember that, even without graph paper, you can always check your axes by making sure that light travels midway between them.

Check your understanding. (a) In a frame in which the ships move to the *west*, which ship's clock displays 12:00:00 first? *(b)* Do you think this effect will be more pronounced, less pronounced, or remain the same at higher speed?

6.2 Practice with skewed grids

This section helps you absorb some of the details of the skewed coordinate grids developed in Section 6.1. First, we should clarify that although we refer to the skewed grid as "the moving grid" we merely mean that it is moving relative to the grid drawn as square. It would be wrong to think that one is "really" moving and the other is "really" stationary. In fact, you will eventually master the skill of reframing any diagram so that the formerly skewed grid looks square and the formerly square grid looks skewed.

Second, think of the grid skew as tracing how time and space mix differently in different frames. In your own frame time and space appear orthogonal, but we can now recognize that as merely a convenient choice of coordinates rather than an ironclad law of physics. And while a generic inertial grid looks more complicated than the orthogonal one you choose to call stationary, it actually mixes space and time in a fairly simple, orderly way. A set of randomly jumbled grid lines could not serve as an inertial coordinate system because it would not trace inertial landmarks and clock readings.

Now, let us zoom in to this small section of Figure 6.5:

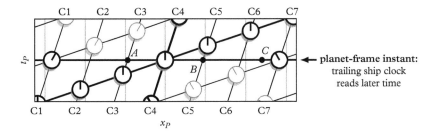

I have added labels for each clock worldline so we can refer to specific clocks more easily. We can visually compare Clocks 1, 4, and 7 at the planet-frame instant $t_P = 0$ and see Clock 1 reading a later time. The clock faces, however, are training wheels that help you learn to read the grid; they will be omitted as we add more details to the grid, so we must practice reading grid lines rather than clock faces. Clock 3, for example, passes through $t_P = 0$ at event A. We can infer its reading at A by interpolating between the grid intersections just before and after A; it must read just a bit later than noon. Similarly, by looking at the grid intersections where Clock 5 brackets event B, we can see that B occurred just *before* noon in the ship frame. Practice this interpolation procedure yourself on clocks C2 and C6.

Once you understand how to interpolate in the time direction, try interpolating in the space direction as well. To find the ship-frame time at event C, we can pencil in the worldline of a hypothetical ship-frame clock passing through that event; make it parallel to the worldlines of its neighbors Clocks 6 and 7. Our new clock reads noon where its worldline intersects the heavy black line above C, and one tick before noon where it intersects the lighter grid line below C. Event C is about 0.15 of the way from one reading to the other along the penciled-in worldline, so the ship-frame time at C is 0.85 ticks before noon.

A common task in interpreting these diagrams is identifying simultaneous events. Figure 6.7 shows how to find events simultaneous to B: draw a line through B, parallel to the grid lines marking time. A horizontal line through B identifies events simultaneous to B *in the frame represented by the square grid*—the planet frame in this case (note the P subscripts on the square grid). Events simultaneous to B in the *ship* frame fall on a skewed line parallel to the other skewed lines marking time in that frame. Use a straightedge to draw such a line through B in

Figure 6.7 *To identify events simultaneous to B, draw a line through B parallel to the grid lines marking time. For simultaneity in the planet frame (square grid here) draw a horizontal line; for simultaneity in the ship frame (skewed grid here) draw the dashed line. This is shown both in the full-grid style (left) and the the spare style with axes standing in for the full grid (right). The shaded area under the latter helps you see how it matches the footprint of the former.*

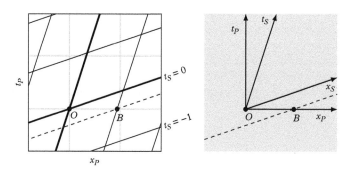

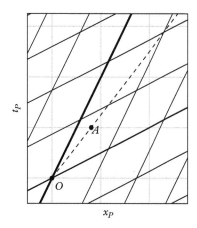

Figure 6.8 *Add velocities graphically by defining a worldline relative to one grid and then measuring it relative to another. Here, the dashed worldline moves at $\frac{1}{2}$c relative to a skewed grid that already moves at $\frac{1}{2}$c relative to the square grid; this yields 0.8c in the square grid (e.g., event A is 0.8 squares of space and 1 square of time from the origin).*

Think about it

A more exact velocity addition can be obtained by extending the grid further and looking for an exact intersection of the dashed line with the square grid; you will find one at $x_P = 4$ and $t_P = 5$, indicating $v = \frac{4}{5}c = 0.8c$ exactly.

Figure 6.7; it should match the dashed line I drew. The right panel of Figure 6.7 repeats this idea, but with the spare style of spacetime diagram you would use in the absence of graph paper. Notice the spacetime symmetry in each panel: if we were to identify events at the same *location* as B in the ship frame, we would draw a line tilted from vertical by the same amount as the dashed line is tilted from horizontal.

Another common task is to read the coordinates of an event in either grid. Let us read the coordinates of event B in the left panel of Figure 6.7, assuming that O marks the origin ($x = 0$ and $t = 0$) in either grid. We begin in the planet frame, which is drawn here as the square grid. Event B is on the $t_P = 0$ line, and about 95% of the way from $x_P = 0$ to $x_P = 1$. Its planet-frame coordinates are therefore $x_P \approx 0.95$ and $t_P = 0$ (the $\approx$ symbol means approximately equal to). In the ship frame, event B is right on the $x_S = 1$ worldline, and about 1/3 of the way from $t_S = 0$ to $t_S = -1$. Its ship-frame coordinates are therefore $x_S = 1$ and $t_S \approx -0.33$. Now, practice locating coordinates yourself: where on Figure 6.7 would you mark an event that occurs at $x_S = 0.5$ and $t_S = 0.5$?

Now that you know how to read coordinates in the skewed grid, you can see how velocity addition really works. Figure 6.8 shows a skewed grid moving east at $\frac{1}{2}c$ relative to a square grid; please verify this by tracing the heavy black worldline up and right from the origin O until you find a dot marking exactly one unit of distance traveled in two units of time. Furthermore, the dashed worldline moves east at $\frac{1}{2}c$ relative to the skewed grid; verify this by tracing the dashed line up and right from the origin O until you find a dot marking exactly one skewed cell of distance traveled in two skewed cells of time. We have not yet learned the algebraic form of the velocity addition law, but we can add $\frac{1}{2}c + \frac{1}{2}c$ graphically by tabulating the velocity of the dashed line *through the square grid*. Tracing the dashed line up and right from the origin, we find that in one square cell of time (at event A) it has crossed about 0.8 square cells of distance; its velocity is thus $0.8c$. The skewness of the grid forces velocities to add to less than Galileo would predict, and thus never pass c.

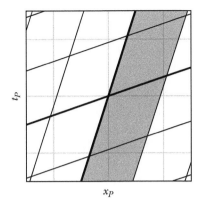

Figure 6.9 *The shaded area is the set of events between* $x_S = 0$ *and* $x_S = 1$. *This worldsheet could represent a large space station that is stationary in the ship frame.*

Another task we will encounter is marking objects too large to be represented by a single worldline. If a large space station extends from $x_S = 0$ to $x_S = 1$ in the ship frame, then at some time or another it touches all events between $x_S = 0$ and $x_S = 1$. This area is shaded in Figure 6.9 and is called a **worldsheet**. In principle, all macroscopic objects should be represented by worldsheets rather than worldlines. For example, you are a worldsheet bounded by the worldlines of your left and right hands. But on a diagram representing a large area, the distance between your left and right hands may be so small that for practical purposes you are a single worldline.

To recap the skewed grid, we will construct the planet-frame coordinate grid as it appears in the ship frame. In the ship frame (where the t_S and x_S axes are perpendicular by definition) the planet moves *west*, so the grid lines marking planet-frame positions tilt west (left) from vertical (Figure 6.10). If you have graph paper, make sure the tilt is one square to the left for every three squares "up" in time to accurately portray the $\frac{1}{3}c$ speed used in the previous ship figures. This simple procedure yields half the skewed grid—and we have used no physics yet, only the definitions of velocity and coordinate system (see Section 4.3).

How do we add the remaining grid lines? Start with the $t_P = 0$ line because we know one thing about this line already: it goes through the origin. Now add the physics: the invariance of c dictates that light travels one light-year per year, or midway between the axes, *in the planet grid just as in the ship grid*. Figure 6.11, shows how to use this fact to draw the $t_P = 0$ line. First, draw a light ray from the origin by making sure that it travels midway between the existing t_S and x_S axes. Then double the angle between $x_P = 0$ (the original heavy black line) and the light ray to draw $t_P = 0$ (the new heavy black line). This is the *only* way to make sure that the light ray travels midway between the axes.

Finally, we add grid lines marking $t_P = 1$, $t_P = 2$, and so on. These must be parallel to the line marking $t_P = 0$; the only question is how far apart to draw them.

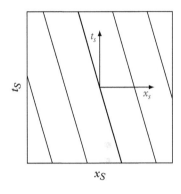

Figure 6.10 *Constructing the planet-frame coordinate grid as measured in the ship frame.* Step 1: *lines marking* x_P *are essentially worldlines of planet landmarks, so their tilt follows directly from the definition of velocity. The heavy line marks* $x_P = 0$ *(the* t_P *axis).*

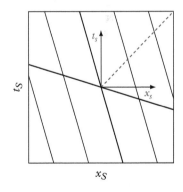

Figure 6.11 Step 2: *a light ray from the origin (dashed) must travel midway between the* t_P *and* x_P *axes. This dictates the orientation of the* x_P *axis (additional heavy line).*

Again, we must respect the principle that light travels one light year per year in the planet frame as well as in the ship frame. The only way to do this is to place the grid lines marking t_P exactly as far apart as the grid lines marking x_P, as in Figure 6.12. (Readers following along on their own graph paper may wonder how many square cells should separate each pair of skewed lines; that issue is deferred to Chapter 7.)

With the full grid in place it is clear that the origin is not special. A light ray emitted in either direction from *any* grid junction will travel through the skewed grid in a 1:1 space:time ratio. In fact, this is true of light emitted from *any event*; the grid junctions merely make this pattern easier to visualize.

Comparing Figure 6.12 to, say, Figure 6.9 we see that when measured in the ship frame the planet grid appears "folded up" in the direction of motion, and vice versa. This is an example of reciprocity (Section 4.4): effects observed by frame A regarding frame B must also be observed by frame B regarding frame A. Neither frame is special.

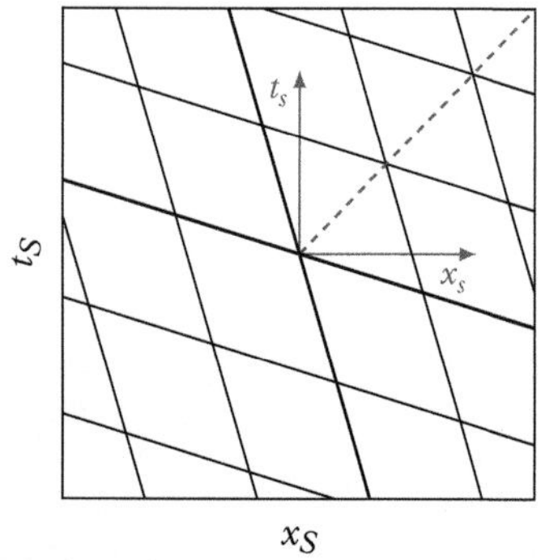

Figure 6.13 shows how reciprocity applies to the simultaneity of events in the ship and planet frames. Events E and W are simultaneous in the ship frame, while events A and B are simultaneous in the planet frame. In either panel the ship frame is the black grid and the planet frame is the red grid; choosing a frame in which to measure amounts to choosing which grid should be considered square, and you should visualize how squaring up one grid necessarily skews the other. Events are shown in black, but be mindful that *events do not belong to any specific frame*. Events have an independent existence, and a frame merely corresponds to a specific way of laying grid lines over the map of events. We can always redraw the map so that an originally skewed grid becomes square, but this does not change anything that happened.

With that in mind, study the events in Figure 6.13. Events E and W mark simultaneous ship-frame events (the stroke of noon aboard two different ships) so they are connected by a black (ship-frame) grid line *no matter which grid we draw as square*. But W is lower down on the red (planet-frame) grid, so the planet frame measures the west clock as ticking over to noon *before* the east clock does;

Figure 6.12 Step 3: *addition of lines marking* t_P, *parallel to* $t_P = 0$. *To ensure that light travels one light year per year in the new coordinate system, place these lines exactly as far apart as the lines marking* x_P.

Figure 6.13 *Ship-frame (black) and planet-frame (red) coordinate grids as measured in the ship frame* (left) *and in the planet frame* (right). *The relative speed has been reduced to 0.1c.*

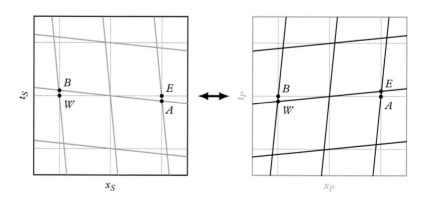

the trailing (west) clock reads a later time. Again, this is true regardless of the frame we choose to draw as square, even if it is easier to spot relationships such as "before" and "trailing" relative to a square grid. The ship frame can make similar statements about the nonsimultaneity of events A and B, which are simultanous in the planet frame. There is no right or wrong answer regarding which pair is *really* simultaneous; the answer is frame-dependent.

You may be concerned that if the order of events is frame-dependent then the order of cause and effect could be frame-dependent. We will address this in Section 6.4 but for now, note that both frames agree that A occurred before E, and that W occurred before B. Therefore, not *all* questions about the order of events need be frame-dependent.

Check your understanding. (a) Can you think of a frame in which event A is simultaneous with W? Describe its velocity relative to the ships and to the planet. *(b)* In the right panel of Figure 6.13, identify events simultaneous with A in the ship frame.

6.3 Time skew

We have seen that events that are simultaneous in one frame are not necessarily simultaneous in other frames. We will call this effect **time skew** because it corresponds to skewing the grid lines marking time in a spacetime diagram. This section explores the factors affecting time skew and lays the groundwork for answering (in Section 6.4) the question about cause and effect posed at the end of Section 6.2.

First, are there any event pairs that are simultaneous in *both* frames? Yes—remember that the spacetime diagram shows only *one* spatial direction: the direction of motion, which we call x here. This direction highlights time skew because it highlights ships moving directly toward or away from the command ship's light flash in Figure 6.14. In contrast, the top and bottom ships in Figure 6.14 are hit by the flash of light at the same time, so there is no time skew perpendicular to the direction of motion. (The top and bottom ships in Figure 6.14 do seem to outrun the light by an *equal*—and tiny—amount; we will return to this in Section 7.1.) For this reason, physicists do not bother to represent directions perpendicular to the direction of motion (call them the y and z directions) in spacetime diagrams. However, it may be instructive to imagine the y or z direction as popping straight out of the page in Figure 6.13; the grids do extend in this direction but without skew. Time skew thus affects only those event pairs with some separation along the direction of grid motion.

Second, larger separations along the direction of motion yield greater differentials in clock readings. Look again at Figure 6.13: if the spatial separation between east and west ships had been tiny, the planet-frame time difference between events E and W would also be tiny. In the limiting case of zero separation along the

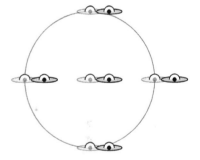

Figure 6.14 *An elaboration on Figure 6.2: ships separated perpendicular to the direction of motion do receive the flash of light simultaneously.*

direction of motion, there can be no effect on clock readings. This implies that two events that happen at the same time *and* the same place can never be separated by clock skew in *any* frame; they are, for all intents and purposes, the same event.

Third, time skew must increase with speed, because speed determines the tilt of the grid lines. This explains why time skew is negligible at everyday speeds, which are tiny compared to c. Furthermore, because speeds have a definite upper limit (c), the amount of time skew has a definite upper limit. Section 6.4 shows how this upper limit is exactly what nature needs to keep cause and effect from ever reversing order.

Fourth, trailing clocks always read a later time. Of course, "trailing" is a frame-dependent adjective: given a set of clocks there are frames through which they move west (so the eastern ones trail), other frames through which they move east (so the western ones trail), and yet other frames in which they are stationary so none trail and all are synchronized. With that in mind, the trailing-clock rule is best seen in Figure 6.5, where western ship clocks trail through the planet frame and any horizontal slice through the diagram yields a planet-frame snapshot. Most of our diagrams will not have clock readings specifically drawn in; the readings are implicit in the grids in diagrams such as Figure 6.13. If bare grids are too abstract for you at first, practice penciling in clock readings wherever it helps. Furthermore, you should be able to take a *ship-frame* instant in Figure 6.5 and use the square grid to confirm that trailing (eastern) *planet* clocks read a later time. The skewed grid pattern is really the fundamental idea here, but "trailing clocks read a later time" is a useful mnemonic.

To reinforce this concept, Figure 6.15 shows a moving array of clocks as it would be seen in actual snapshots at two different times. Trailing clocks read later times, and there is no desynchronization in the direction perpendicular to the motion. Furthermore, the desynchronization *pattern* is fixed—if we wait until one clock has advanced 15 seconds, then all clocks have advanced 15 seconds

Confusion alert

The "trailing clock reads a later time" rule means the trailing clock strikes noon *before* the leading clock. Avoid confusion by using specific statements such as "clock A reads a later time" rather than, say, "clock A is ahead" (which fails to specify a coordinate system, and whether you mean ahead in time or in space). Use caution also when relating terms like leading/trailing to terms like eastern/western because the former is frame-dependent while the latter is not.

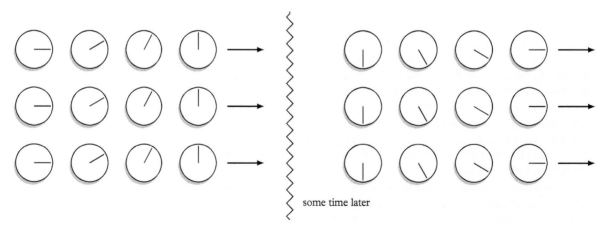

some time later

Figure 6.15 *A parade of clocks (synchronized in their rest frame) marching down a street, recorded at a single instant in the street frame* (left) *and again at a later instant* (right). *This illustrates the pattern of time skew while highly exaggerating the size of the effect.*

(this is guaranteed by the fact that they remain synchronized in their rest frame). To test your mastery of this concept, make sure you can relate the snapshots in Figure 6.15 to the spacetime diagram view of desynchronization; for example, in Figure 6.5.

All these patterns stem from the basic idea illustrated in Figure 6.2, which traditionally goes by the name "desynchronization of moving clocks" or "relativity of simultaneity." But the desynchronization pattern is so specific that it deserves a more specific name, so I am coining the term *time skew*. This term evokes the skewed spacetime grid and thereby helps you remember these patterns.

Time skew describes a pattern in the readings of *different clocks at a single instant*. In Chapter 7 we will tackle the distinct question of how much time elapses on a *single* clock from one event to another.

Check your understanding. *(a)* If a train travels west relative to the land and each car has a large clock display, sketch how the clock readings would appear at one instant in the land frame. Assume a near-lightspeed train so the effect is apparent. *(b)* Explain how this relates to Figure 6.13. Choose a panel and identify which color represents the train frame and which the land frame; then verify that the grid shows the same effect as your drawing.

6.4 Causality

Could time skew ever change the order of events so much that an effect could *precede* its cause in some frame? Such a violation of **causality** would be truly astonishing, so this question is worth exploring carefully.

We first review causality in one frame before extending it to multiple frames. Given some event A, let us try to identify all the subsequent events that A could cause or affect in any way. Messages launched from, or consequences of, A could fan out east or west at a range of speeds, up to and including c. On a spacetime diagram then, an eastbound light ray represents the eastern limit of events that could be affected by A, and a westbound light ray represents the western limit. Figure 6.16 shows this area shaded in; it forms a cone and is called the **future light cone** of event A. Each event has its own future light cone, as illustrated by the different cone attached to event B.

Practice interpreting the causal relations between labeled events in Figure 6.16. First, a light ray from A clearly could affect D. Second, A cannot emit anything that travels fast enough to affect B (or vice versa). Third, A can affect C, even without anything traveling through space. If A is the setting of an alarm clock, C is when the alarm sounds at the same position some time later. Event B can also affect C and D; for example a rocket could be launched from B that reaches C or D if its average speed is $0.5c$ (1.5 units of space over 3 units of time).

We can also identify all events capable of *causing* event A by drawing light rays *arriving* at A and shading the region of spacetime between them; this is called the **past light cone** of A and is labeled explicitly in Figure 6.17 (left panel). Regions

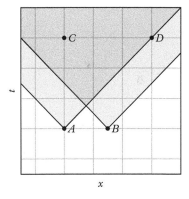

Figure 6.16 *Future light cones of events* A *and* B.

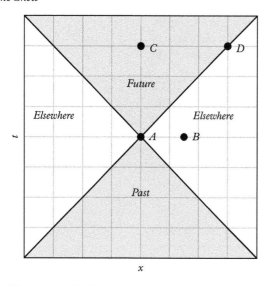
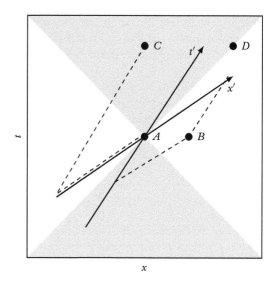

Figure 6.17 Left: *causal structure of the spacetime centered on event* A. Right: *in another frame, the causal relationships among these events are preserved. For example,* A *and* B *still have an elsewhere relationship because they are still separated by more space than time. The light cones are the same in all frames.*

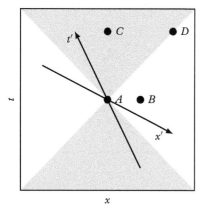

Figure 6.18 *On the scenario of Figure 6.17 we plot the axes of a west-moving frame. Use the skewed axes to estimate the time and space displacements in this frame between event pairs* AB, AC, *and* AD. *From these displacements, verify that the causal relationships we saw in other frames still hold in this frame.*

of spacetime outside the past and future light cones of an event are considered to be **elsewhere**. These events cannot cause or be caused by event A.

Figure 6.17 (right panel) illustrates how these causal relationships are preserved in another frame, moving east relative to our original frame. To avoid clutter, axes are drawn in place of a full grid; these axes are labeled with primes to indicate a generic other frame. We find $\Delta x'$ and $\Delta t'$ between A and B by drawing the skewed dashed lines from B to the axes. Although B occurs *before* A in this frame, it cannot *cause* A: B and A are separated by more space than time, so any signal from B would have to exceed c to reach A. Thus, the "elsewhere" causal relationship between B and A is preserved in the new frame. Similarly, although C is now separated from A by some amount of space (the dashed line is placed along the x' axis to avoid sending it off the top of the diagram), the separation in time is greater so an alarm clock can still travel from A to C. The same alarm clock that was stationary in the original frame is, in this frame, moving west at appreciable but sublight speed. Finally, A can still cause D only via light ray because D is still midway between the space and time axes. In summary, the new frame preserves *all* the causal relationships we had noted. Figure 6.18 allows you to practice this reasoning in a new frame, westbound relative to our original frame.

The deeper idea behind all these examples is that *the light cone is the same in all inertial frames.* If an event is on the edge of the light cone in one frame, it is so in all frames, so no choice of frame can make any event jump from inside to outside the light cone or vice versa. Therefore, the entire causal structure (past light cone, future light cone, and elsewhere) is frame-independent.

You may think that the region between past and future should be called "now" or "simultaneous to A" but this would not be accurate. The set of events simultaneous to A is highly frame-dependent. Each event in the elsewhere region is simultaneous to A in some frame, before A in some other frames, and after A in yet other frames. Thus, we cannot categorize elsewhere events as occurring before or after A in any frame-independent sense. If this bothers you, ask yourself what is physically meaningful about the before/after distinction; you will probably conclude that it has something to do with cause and effect. But we have already shown that cause and effect are preserved by the past/future distinction. This relegates the before/after distinction to the narrow sense of "having an earlier/later time coordinate," which is clearly a frame-dependent issue. Because elsewhere events can neither affect nor be affected by event A, we need not worry that the before/after question has no definitive answer.

Prior to special relativity, we thought of *before*, *after*, and *simultaneous* as three separate and nonoverlapping categories. Time skew forces us to conclude that this categorization is frame-dependent. *Past*, *future*, and *elsewhere* is a more meaningful categorization because it is preserved in all frames.

Check your understanding. (a) In the westbound frame in Figure 6.18 event B occurs after A. Prove that A cannot cause B in this frame. *(b)* Again referring to that frame: events B and C now appear to be in the same position in this frame. Does that affect their causal relationship?

CHAPTER SUMMARY

- Think about simultaneous events by thinking about synchronized clocks striking a certain time together. A set of clocks synchronized in their rest frame will always be measured as desynchronized (in a highly specific pattern) in other frames. Thus, events that are simultaneous in one frame are not simultaneous in others.

- The desynchronization pattern seen in a set of comoving clocks is: trailing clocks read a later time; the more they trail in space, the later the time they read; clocks separated perpendicular to the direction of motion remain synchronized; and the faster the motion, the greater the effect (we do not notice it at everyday speeds).

- On a spacetime diagram, this effect skews the grid lines marking time. This exposes a symmetry between space and time: if the latter are tilted clockwise from vertical, the former must be tilted counterclockwise from vertical by the same amount (and vice versa).

- This counter-rotation or folding of the spacetime grid ensures that c is invariant: a light-ray worldline stays midway between space and time axes (crossing one unit of space per unit of time) in all frames. "Occurring at the same time" is just as frame-dependent a statement as "occurring at the same place."

- In a skewed grid, identify simultaneous events by drawing a skewed line through any event of interest.

- Relationships such as *before*, *after*, and *simultaneous* are frame-dependent, and should be seen as artifacts of the chosen coordinate system rather than physically meaningful. But causal relationships between events are nevertheless frame-independent. Light cones highlight the causal relationships *past*, *future*, and *elsewhere*.

FURTHER READING

Lewis Carroll Epstein's *Relativity Visualized,* which inspired Figures 6.1, 6.2, and 6.14, is full of great visualizations of many of the effects of relativity.

The *Minute Physics* video series has a good visualization of why simultaneous events must form a tilted line when drawn in a moving frame. The two-minute video titled *Einstein and The Special Theory of Relativity* is available at https://www.youtube.com/watch?v=ajhFNcUTJI0.

CHECK YOUR UNDERSTANDING: EXPLANATIONS

6.1   (a) the eastern ship; (b) more pronounced.

6.2   (a) Yes, a frame at intermediate velocity. In this frame the ships are eastbound at lower speed and the planet is westbound at lower speed, so the black grid in Figure 6.13 tilts to the right but less than in the right panel, and the red grid tilts to the left but less than in the left panel. This will make A and W simultaneous. (b) Draw a line through A, parallel to the other not-quite-horizontal black lines.

6.3   (a) Cars further to the east should read progressively later times, like this:

$$\longleftarrow \boxed{10:00}\ \boxed{10:01}\ \boxed{10:02}\ \boxed{10:03}\ \boxed{10:04}$$

In practice, the effect would be much smaller: a maximum of about 100 nanoseconds per car even for a near-c train because train cars are about 100 feet long and light travels one foot per nanosecond.

(b) Choosing the left panel of Figure 6.13, black must represent the land frame and red the train frame. According to the grid, at any instant in the black (land) frame, a red (train) clock that is further east (more trailing) reads a later time. This matches the sketch.

6.4   (a) There are two ways to prove this. The detailed way is to draw skewed lines to show the space and time displacements between A and B as shown below. The displacement in space is larger than the displacement in time, so not even a light ray could travel from A to B.

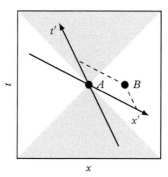

The more conceptual answer is to note that c is the same in all frames, so if a light ray cannot travel from A to B in the original frame, it cannot do so in any frame. (b) No; a sub-lightspeed particle can travel from B to C in the other frames as well. The frame shown happens to be one in which such a particle has $v = 0$ but there is no physical relevance to that fact.

❓ EXERCISES

6.1 Consider the story in Section 6.1, using the Galilean model of relativity. Explain how Figure 6.2 would be different, and why the Galilean model does not predict time skew. *Hint:* think about the Galilean velocity addition law.

6.2 *(a)* In Figure 6.6, what is the velocity of the ship frame relative to the planet frame? *(b)* If the ships moved at a higher speed in the same direction, how would the right panel of that figure look? Sketch it. *(c)* If the ships moved in the opposite direction, how would the right panel of that figure look? Sketch it.

6.3 A fleet of flying saucers flies in a circular formation at constant velocity as in Figure 6.19. Relative to you, they are all moving east at uniform velocity. The central ship emits a flash of light in all directions. *(a)* In the ship frame, in what order do the ships receive the light? *(b)* In your frame, in what order do the ships receive the light?

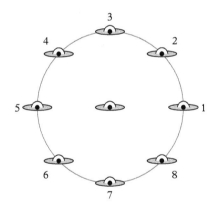

Figure 6.19 *A fleet of alien ships with serial numbers.*

6.4 In Figure 6.8, what are the coordinates of events A and B in *(a)* the square grid; *(b)* the skewed grid?

6.5 On Figure 6.8, identify events that are simultaneous to event A in *(a)* the square grid; *(b)* the skewed grid.

6.6 On Figure 6.8, identify events that are at the same location as event A in *(a)* the square grid; *(b)* the skewed grid.

6.7 In Figure 6.13, what are the coordinates of events A, B, E, and W in *(a)* the planet frame; *(b)* the ship frame? Assume the central intersection is the origin for both grids.

6.8 Bob moves east at $\frac{2}{5}c$ relative to Alice. *(a)* Think in Alice's frame. On graph paper, draw spacetime axes for Alice's frame and label them with A subscripts. Now add spacetime axes for Bob's frame and label them with B subscripts. *(b)* Now think in Bob's frame. On a new area of graph paper, repeat the steps in part (a).

6.9 Draw a row of five alien ships heading north and label the times their clocks would read at one instant in your frame. Aim for conceptual, rather than quantitative, correctness.

6.10 Draw a train with five cars heading east. Each car has a vertical stack of three clocks. Label the time each clock would read at one instant in your frame. Aim for conceptual, rather than quantitative, correctness.

6.11 Consider Figure 6.16. *(a)* Describe a frame in which event A occurs before D, or if this is impossible, state why it is impossible. *(b)* Describe a frame in which event D occurs before A, or if this is impossible, state why it is impossible. *(c)* Describe a frame in which event A occurs before C, or if this is impossible, state why it is impossible.

6.12 Consider Figure 6.16. *(a)* Could event D cause or affect event B? If so, describe how (e.g., with a light ray); if not, explain why not. *(b)* Could event D cause or affect event C? Describe how, or why not. *(c)* Could event D cause or affect event A? Describe how, or why not. *(d)* Are your conclusions frame-dependent? Justify your reasoning.

+ PROBLEMS

6.1   Newcomers to relativity, after understanding how the axes of a right-moving grid fold up around a right-moving light ray, often propose to draw the axes of a *left*-moving grid as in Figure 6.20. (As Figure 6.18 suggests, physicists have *not* adopted this proposal.) *(a)* In what way does this proposal reflect a good understanding of relativity? *(b)* What is the weakness of this proposal? *Hint:* identify an unnecessary complication in the relationship between frames A and B in Figure 6.20 when they have no relative motion.

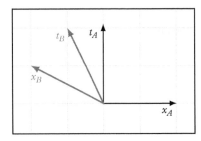

Figure 6.20 *Unconventional proposal for labeling the x axis of a moving frame.*

6.2   For each part in this problem, copy the events of Figure 6.16 onto a clean area of graph paper. *(a)* Shade in the set of events that can be affected by events A and D. *(b)* Shade in the set of events that can be affected by both events A and B. *(c)* Shade in the set of events that can be affected by both events A and C.

6.3   Estimate the number of seconds of desynchronization observed in the land frame between clocks at the front and back of a 1 km long train moving at *(a)* 100 kph; *(b)* 100,000 kph. Assume, of course that the train clocks are synchronized in the train frame. You will have to develop a quantitative model of time skew from the conceptual model and grids in the text.

6.4   Why do we not imagine synchronizing clocks in a given frame by moving a master clock from clock to clock and making sure each clock displays the same time as the master clock when they are side by side? You need not predict exactly what *does* happen in this scenario, but do use the concepts in this chapter to explain why this is not guaranteed to work.

The remaining problems are inspired by a scenario in The Elegant Universe *by Brian Greene. Two warring countries agree to sign a peace treaty, but neither wants to sign first. They ensure simultaneous signing by sitting equidistant from a central strobe light. A neutral official such as the Secretary-General of the United Nations pushes a button to make the light flash, and each president will sign their copy when they receive the light flash. The question is then whether they sign simultaneously in some other frame. Keep in mind that we are* not *concerned with when a country* sees *the signing, but only with the time coordinates of the signing events themselves (in various frames).*

The first problem is essentially identical to Greene's original scenario, and the others provide additional levels of complication.

6.5   The presidents agree to sign a peace treaty on a moving train. They synchronize their watches perfectly so they can sign the treaty at the same time. They sign as the train rolls over the border between the two countries, as shown in Figure 6.21. *(a)* Do Country 1 and Country 2 measure the presidents as signing simultaneously? Why or why not? If not, which president signed first? *(b)* Do Country 1 and Country 2 agree on what happened? Why or why not? If there is a disagreement, explain what it is about and what each country thinks.

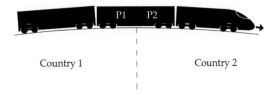

Figure 6.21 *Treaty-signing scenario for Problem 6.5. P1 and P2 represent the presidents.*

6.6   The presidents agree to have the Secretary-General stand on the border of the two countries, while the presidents sit in their respective countries an equal distance from the strobe light as in Figure 6.22. With this setup, they reason that they are guaranteed to sign at the same time. *(a)* The CNN helicopter covering the event is flying to the east at constant velocity. Who, if anyone, signed first in the CNN coordinate system? Explain your reasoning. *(b)* The Fox News helicopter covering the event is flying to the west at constant velocity. Who, if anyone, signed first in the Fox coordinate system? Explain your reasoning. *(c)* Who is correct about who signed first: CNN, Fox, or the Secretary-General? *Warning:* the question is about what is measured in a *hypothetical coordinate system attached to* each helicopter, rather than what any one character actually saw.

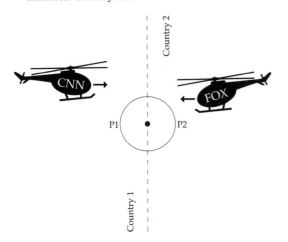

Figure 6.22 *Treaty-signing scenario for Problem 6.6. P1 and P2 represent the presidents.*

6.7 Countries 1 and 2 are at war yet again. This time, their presidents agree to sign a peace treaty on a boat moving down the river that forms the border between the two countries, as shown in Figure 6.23. Country 1 and Country 2 agree to share all video footage taken from cameras set up all along the riverbank. Pretend that the boat moves fast enough to measure relativistic effects. *(a)* Do Country 1 and Country 2 measure the presidents as signing simultaneously? Why or why not? If not, which president signed first? *(b)* Do Country 1 and Country 2 agree on what happened? Why or why not? If there is a disagreement, explain what it is about and what each country thinks. *(c)* The official photographer of Country 2 was late and missed the boat. He steals a speedboat from the

Country 2 side of the river and heads straight for the main boat as shown, as fast as he can. Do the president sign simultaneously in a coordinate system attached to the photographer? Why or why not? If not, which president signed first?

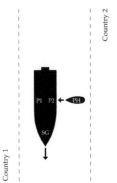

Figure 6.23 *Treaty-signing scenario for Problem 6.7. P1 and P2 represent the presidents, SG the Secretary-General, and PH the photographer.*

6.8 *Four* countries are now at war. The presidents of four agree to sign a treaty simultaneously. They board an airplane, synchronize their watches, and sign the treaty at an agreed-upon time just as the plane flies over the border of the four countries, as shown in Figure 6.24. *(a)* In which order did the presidents sign in Country 1's frame? *(b)* In which order did the presidents sign in Country 2's frame? *(c)* In which order did the presidents sign in Country 3's frame? *(d)* In which order did the presidents sign in Country 4's frame?

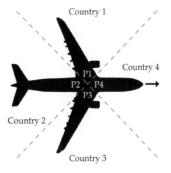

Figure 6.24 *Treaty-signing scenario for Problem 6.8. P1 represents the president of Country 1, and so on.*

Time Dilation and Length Contraction

In Chapter 6, we discovered that different frames mix time and space differently. The high-speed universe has a lot more in store for us: in this chapter we will discover that speed causes time to run slowly and space to contract. Of course, the truth is a bit more subtle than that, because when two frames are in relative motion each frame measures the *other* as the high-speed frame where time runs slowly and space contracts. By the end of this chapter, you will understand how all this fits together with time skew in a fully consistent model that beautifully fits a wide range of evidence.

7.1 Time dilation

The effect of speed on a clock will become apparent if we follow up on something we noted but did not explore in the ship-and-planet time skew story in Chapter 6. But first, let us be absolutely clear that "the effect of speed" is really about how *coordinates* in one frame relate to another. There is no actual effect on the "moving" clock, which marks time perfectly in its own frame. Either frame can be considered stationary or moving, depending on how you choose to view the situation.

With that in mind, let us return to Figure 6.14 from Section 6.3; it is repeated here as Figure 7.1. The point made in Section 6.3 was that ships separated perpendicular to the direction of motion—the top and bottom ships in this case—are equally placed relative to the one-light-second circle and thus read the same time. Here, we follow up on the fact that those ships' centers are nevertheless slightly *outside* the circle. Recall that each flanking ship clock completes a one-second tick exactly when it receives the light. Yet the circle demonstrates that those flanking clocks still have not ticked after one full second in the planet frame (the frame of the page). Therefore, the planet frame observes each ship clock as ticking too slowly. The ship clocks work as expected when measured in their own frame, so this is not a malfunction—it is a frame-dependent effect called **time dilation**.

The Elements of Relativity. David M. Wittman, Oxford University Press (2018).
© David M. Wittman 2018. DOI 10.1093/oso/9780199658633.001.0001

Time dilation is distinct from time skew in several important ways:

- Time dilation involves time *intervals*, that is, separate readings of (in principle) a *single* moving clock, whereas time skew involves *instantaneous readings of separate clocks.*

- Time dilation pertains equally to all clocks in a given frame; it is not a position- or direction-dependent effect. This is not necessarily clear from Figure 7.1 because the east and west flanking clocks exhibit time skew *as well as* time dilation, making it difficult to see the latter, smaller effect. Because time skew dominates along the direction of motion, we were able to neglect time dilation throughout Chapter 6 where we focused only on that direction. Conversely, we focus now on perpendicular directions to study time dilation without complicating factors—but to see how everything fits together we will eventually return to the direction of motion as well.

- Time dilation determines the clock tick rate and therefore affects a space-time diagram by determining *how far apart* we draw the grid lines marking time. Time skew, in contrast, determines (and takes its name from) the *tilt* of these lines.

Both effects increase with speed, but in different ways. Look again at Figure 7.1: the top and bottom dots are *barely* outside the circle despite having a speed of 0.4c (between snapshots the ships have traveled a distance equal to 0.4 times the radius of the circle). At this speed—120,000 km/s or 270 million mph—time skew is already substantial, but time dilation is just starting to become apparent. We will come to understand this more quantitatively in the sections that follow.

Check your understanding. How does the time dilation effect depend on *(a)* direction of motion; *(b)* speed?

7.2 Light clocks and γ

To quantify the relation between speed and time dilation, the scenario in the previous section can be reduced to the bare essentials of one light ray traveling between two locations in a self-contained "light clock" (Figure 7.2). In this hypothetical device, a flash of light bounces back and forth between two opposing mirrors. Light travels from one mirror to the other in a well-defined, repeatable amount of time (a "tick"), and after each tick the next tick starts promptly as the light reflects off the mirror. For concreteness we will assume that each tick is one nanosecond, which puts the mirrors about one foot apart. We can put the mirrors at either end of a tube and outfit the tube with a big bright time display readable by any observer. This last feature is not necessary but reinforces the point

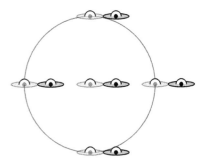

Figure 7.1 *In the ship frame all ships maintain their faded positions and one second passes between emission of the flash and its reception. In the planet frame the ships move to the eastern positions within one second. The one-second circle has not quite reached the top and bottom ships, so less than one second has passed in the ship frame.*

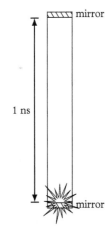

Figure 7.2 *Snapshot of a light clock. One nanosecond (ns) after this snapshot, the flash of light will arrive at the top; another nanosecond, back at the bottom, and so on. We can attach a device that counts the nanosecond ticks and displays the time prominently.*

that the reading displayed by a particular clock at a particular event will never be in dispute and is not frame-dependent; what is frame-dependent is the way that *different* events relate to each other.

The light clock will surely seem like a swindle to beginners. How can we be sure this hypothetical light clock will behave the same way as a real clock? Let us try a proof by contradiction (Section 4.4): assume the light clock *does* behave differently and see if that leads to a dead end. So take this hypothetically ill-behaved light clock aboard a train, along with your trusty watch. Before the train starts moving, adjust the distance between light-clock mirrors until the light clock ticks at exactly the same rate as your watch. Now, let the train move at some constant high velocity. If the light clock differs from your watch at this new velocity, you have just discovered a way to determine your state of motion without reference to anything outside the train! This would violate the principle of relativity, so we conclude that the light clock in fact must agree with other types of clocks regardless of the laboratory's state of motion. The light in a light clock is thus merely an *indicator*—rather than a cause—of how nature relates time and space. By the same argument, the ticking of the light clock cannot depend on whether the light clock is oriented along the direction of train motion, perpendicular to it, or somewhere in between.

The following light clock discussion and some of the drawings in this chapter are inspired by and adapted from the elegant presentation by Lewis Carroll Epstein in *Relativity Visualized*. We will put the light clock in a train moving eastward at constant velocity v, and analyze the clock in both train and ground frames. Furthermore, we initially orient the light clock perpendicular to the direction of motion ("vertically") because perpendicular lengths are guaranteed to remain frame-independent (Section 4.4). This orientation is convenient for the following argument, but remember that the clock keeps time regardless of orientation. Therefore, any conclusions we reach about the timekeeping function in this orientation will automatically transfer to other orientations.

We start the experiment when the light is at the bottom of the clock, and we let one nanosecond elapse on our identically constructed ground-frame clocks. We first do a dry run of the experiment with the train parked to ensure that light does indeed reach the top after one nanosecond. Then, we start the experiment again with the train moving. The "target" at the top of the clock is now moving, so the light—traveling at the fixed speed c—fails to reach the top as shown in Figure 7.3. The circular arc in the figure shows all locations one foot from the starting location (as measured in the ground frame); because c is fixed the light *must* be somewhere on this arc after one ground-frame nanosecond. The light must also be in the tube, because the clock must continue to function normally for train passengers. After one ground-frame nanosecond the light must therefore be where the arc passes through the new location of the clock. Knowing this location, we can draw the path of the light throughout the ground-frame nanosecond—this is the straight line in Figure 7.3.

The light has only reached partway up the clock in one ground-frame nanosecond, so only a fraction of a nanosecond has passed aboard the train. By the same

Think about it

If nature is ever found to violate the principle of relativity, we will have to revisit the logic in this paragraph. But this seems unlikely, because the logic here leads to many predictions that have already been confirmed by experiment.

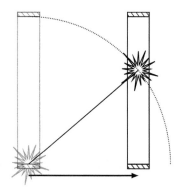

Figure 7.3 *Time dilation: snapshots of an east-moving clock are taken one tick apart as measured by an identical clock at rest. The arc shows the distance light can travel in that time: it cannot reach the top, so the moving clock completes less than one tick per stationary-clock tick.*

token, we must wait *longer* than one ground-frame nanosecond for the train clock to finish its tick as shown in Figure 7.4. If the light was two-thirds of the way up the clock after one ground-frame nanosecond then the ratio a/b in Figure 7.4 is 2/3 and the ratio b/a is 3/2. Thus, we must wait 3/2 of a ground-frame nanosecond before a train-frame nanosecond is completed.

Regardless of the specific number, think in terms of the *ratio* of elapsed times rather than a difference in times. How do we define a ratio that is generally useful, and not tied to the specifics of train and ground? The essence of the comparison here is time elapsed *in the coordinate system of our choice* versus time elapsed *in the clock's rest frame*. **This is a fundamental distinction that students must always remember.** To keep track of these different versions of time, we have a special name for time in the clock's rest frame—**proper time**—as well as a special symbol, the Greek letter tau (τ). The term for time elapsed in the coordinate system of our choice is simple—**coordinate time**—and its symbol is familiar: t. For either kind of time measurement, we generally think in terms of the time *elapsed between two events*: $\Delta\tau$ for proper time or Δt for coordinate time. The crucial distinction between the two is that $\Delta\tau$ belongs to a specific clock that may be free to move in complicated ways, while Δt is always just the change in the time coordinate of a coordinate system. That said, if a moving clock is consistently embedded in a coordinate system of its own, the elapsed time on that clock may stand in for the elapsed time in its own coordinate system.

Figures 7.3 and 7.4 show that *any* motion of the light clock through our coordinate system will increase the coordinate time between proper-time ticks (which are completed when the light hits a mirror). The *minimum* amount of coordinate time per unit proper time occurs if we happen to choose a coordinate system in which the clock is at rest; then the coordinate time is equal to the proper time. Therefore, the ratio of coordinate to proper time can range from one (for stationary clocks) up to much larger numbers (for rapidly moving clocks). We use this ratio so often in relativity that we always denote it with a specific Greek letter, gamma (γ). Gamma is defined as the ratio of coordinate time to proper time: $\gamma \equiv \frac{\Delta t}{\Delta\tau}$. This quantifies, for a moving clock (or particle or observer), the *displacement in the time coordinate per unit proper time*, so it is a kind of "speed through time." Practice viewing t as just another coordinate in a coordinate system, whereas τ is the time experienced by the clock—and anyone or anything comoving with the clock.

As long as we stick to inertial (constant-velocity) coordinate systems, the mathematical relationship between $\Delta\tau$ and Δt is simpler than you might think. In Figure 7.4, a right-angled triangle is formed by the path of the light (the hypotenuse), the ground-frame distance moved by the train (the horizontal leg of the triangle), and the height of the light clock (the vertical leg of the triangle). This triangle is reproduced in Figure 7.5, along with mathematical expressions for their lengths:

- The height of the light clock, by design, is $c(\Delta\tau)$: the distance light travels between ticks of proper time.

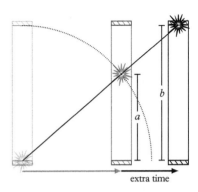

Figure 7.4 *Figure 7.3 is extended until the moving clock reaches one full tick. The elapsed ground-frame time is longer than the clock-frame time by the ratio* b/a, *which is also the ratio of the full diagonal line to the radius of the arc. This ratio is called* γ.

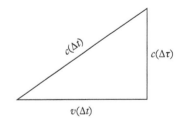

Figure 7.5 *The essential geometry of Figure 7.4.*

- The length of the hypotenuse is the distance moved by light in the elapsed coordinate time: $c(\Delta t)$.

- The length of the horizontal leg is the coordinate distance moved by the clock during the story: $v(\Delta t)$.

You may recall the Pythagorean theorem for right-angled triangles: the square of the hypotenuse is the square of the horizontal leg plus the square of the vertical leg. For our triangle this would be written

$$c^2(\Delta t)^2 = c^2(\Delta \tau)^2 + v^2(\Delta t)^2 \tag{7.1}$$

Separating the $\Delta \tau$ and Δt terms, we find:

$$c^2(\Delta \tau)^2 = c^2(\Delta t)^2 - v^2(\Delta t)^2 \tag{7.2}$$
$$= (c^2 - v^2)(\Delta t)^2.$$

Dividing through by c^2 yields

$$(\Delta \tau)^2 = (1 - v^2/c^2)(\Delta t)^2 \tag{7.3}$$

and then taking the square root yields

$$\Delta \tau = \sqrt{1 - v^2/c^2}(\Delta t). \tag{7.4}$$

This shows mathematically how the elapsed proper time $\Delta \tau$ is always smaller than (or at most, equal to) the elapsed coordinate time Δt: the $1 - \frac{v^2}{c^2}$ factor can never be greater than one, and is less than one if the clock moves at all.

We can rearrange Equation 7.4 to find the ratio of coordinate time to proper time:

$$\gamma \equiv \frac{\Delta t}{\Delta \tau} = \frac{1}{\sqrt{1 - v^2/c^2}} \tag{7.5}$$

Figure 7.6 plots γ as a function of v. Students should become intimately familiar with the behavior graphed in Figure 7.6:

- $\gamma = 1.0$ corresponds to the low-speed Galilean world of our intuition.

- γ increases only slowly as v climbs to a substantial fraction of c. For example, at $v = 0.5c$ (also written $\frac{v}{c} = 0.5$) γ is still only 1.15.

- γ spikes up and increases without bound as v approaches c. "Increases without bound" or "becomes arbitrarily large" means that the closer we push v to c, the larger γ becomes, so that γ would become *infinite* if v could actually reach c.

- γ is the same regardless of the sign (direction) of v, because even if a velocity is negative, its square is positive.

Think about it

The height $c(\Delta \tau)$ is invariant by the argument of Section 4.4. Thus, by keeping $\Delta \tau$ and Δt terms on opposite sides of the equation we separate invariant and frame-dependent quantities. This will become important in Chapter 11.

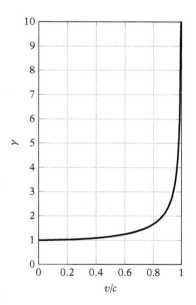

Figure 7.6 γ *as a function of* v. γ *is very close to 1.0 for everyday speeds, but increases without bound as* v *approaches* c *(in other words, as* $\frac{v}{c}$ *approaches 1).*

There are two ways you might use Figure 7.6. To find the speed required to achieve a given time dilation ratio, say $\gamma = 2$, find that number on the left, look across until you hit the curve and then look down to read the speed: a bit shy of $0.9c$ in this case. To find γ for a given speed, say $v = 0.8c$, find 0.8 on the horizontal axis, look up until you hit the curve, then look left to identify γ: about 1.7 in this case. This means that if a clock moves at $0.8c$ through some coordinate system, 1.7 seconds will elapse in that coordinate system for each second elapsing on the clock.

The moving clock is just a stand-in for *any* time-dependent process in the moving frame, because all such processes stay in sync with each other. If Alice and Bob live aboard a rocket that moves through Earth's coordinate system at $0.8c$, they will age only 1 year for each 1.7 years marked by Earth-frame clocks. They will not *feel* themselves aging slowly, because their metabolisms proceed according to the same time measured by their wristwatches. Clocks measure time; therefore, it is often stated that "time itself runs slowly in the moving frame." This is somewhat misleading, however, because time unfolds at a rate of one second per second in the rocket frame just as it does in Earth's frame. The situation is better described as a speed-dependent "exchange rate" between units of time in different coordinate systems. Physicists see this every day with fast-moving subatomic particles. These particles have known lifetimes at rest, and they live *much* longer (as measured by Earth-frame clocks) when moving near c through Earth's coordinate system (Section 7.4).

To cement our understanding, let us translate some of the points on the curve in Figure 7.6 back to the geometric view that led us here. Figure 7.7 shows a range of triangles comparable to Figure 7.5, but representing a range of velocities. The invariant vertical leg of the triangle is now placed at left to catch your eye first. To obtain the labels on the large triangle, I have divided each label in Figure 7.5 by $c(\Delta\tau)$; for example, the hypotenuse label is $\frac{c(\Delta t)}{c(\Delta\tau)} = \frac{(\Delta t)}{(\Delta\tau)} = \gamma$. By the definition of γ, the hypotenuse must be γ times longer than the invariant (vertical) leg. With the hypotenuse representing γ units of coordinate time, the horizontal leg is $\frac{v}{c}$ times this length because it represents the distance moved in this time. Because $\frac{v}{c}$ can never quite reach one, the length of the horizontal leg ($\frac{v}{c}\gamma$) can never quite match that of the hypotenuse (γ)—but it can come arbitrarily close, making an arbitrarily long triangle and therefore arbitrarily large γ. At the other extreme, low speeds imply that the horizontal leg is short compared to the hypotenuse. Given the invariant vertical leg, this forces the hypotenuse to be barely longer than the vertical leg, or γ barely more than one. There is, of course, a range of possibilities in between, but these extremes best illustrate the behavior of the curve in Figure 7.6.

Check your understanding. Estimate γ for each of the speeds in Figure 7.7 by comparing lengths of line segments. Check your estimates by comparing them with Figure 7.6. Explain why the triangle method works to estimate γ.

Think about it

The exchange rate analogy is imperfect because (by reciprocity) the exchange rate must be the same in *both* directions. We will figure out how this works in Section 7.5.

Confusion alert

The stretching triangles will appear throughout the book, but may be flipped vertically or horizontally to give you practice recognizing the geometry in a variety of situations. Note that either type of flip leaves the invariant leg as the vertical one.

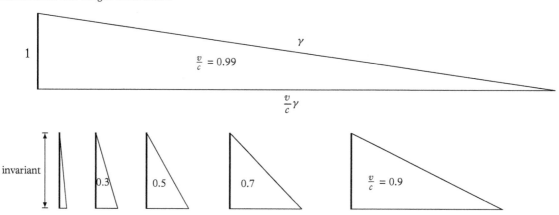

Figure 7.7 Top: *a version of Figure 7.5 at higher speed, with the invariant leg on the left and with the labels divided through by* $c(\Delta\tau)$. Bottom: *the triangle is repeated for various values of* v/c. *In all cases, the horizontal leg is* v/c *times the length of the hypotenuse, which is itself* γ *times the length of the invariant leg. Near light speed,* v/c *is close to one so the horizontal leg is nearly as long as the hypotenuse. This in turn forces the hypotenuse to be much longer than the invariant leg (* γ *is much larger than one). At low speed, the horizontal leg nearly disappears, making the hypotenuse (* γ *) barely more than one.*

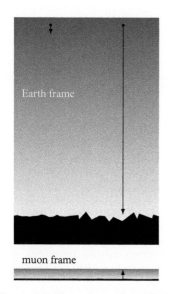

Figure 7.8 Top: *absent time dilation, muons should die before traveling far (arrow at left). Time dilation explains why they live long enough to hit the ground (arrow at right).* Bottom: *in the muon frame, the atmosphere is contracted so the ground is reached even in a short lifetime.*

7.3 Length contraction and reciprocity

From time dilation we can immediately deduce the final major counterintuitive effect in special relativity. Say you run along a track, and track-frame measurements show that you complete a 100 m dash in a certain time. Your watch recorded an elapsed time smaller than the track-frame time, by a factor of γ. But to satisfy reciprocity you must measure the track passing by you at the same speed that the track measures you passing by it. The speed can be preserved only if you measure a track *length* that is smaller than 100 m, by the same factor of γ that affected your time measurement. As with time dilation, this **length contraction** is surprising because we have no intuition for situations where γ is not one. But the logic is ironclad: if we have time dilation we must also have length contraction. Time dilation and length contraction are two sides of the same coin; you can't have one without the other. There is one important difference in their application, though: time dilation has no directionality while length contraction happens *only* along the direction of motion (Section 4.4 proved that perpendicular distances are always preserved).

To better appreciate the power of the reciprocity argument, we turn to an example where γ is quite large. In laboratories, physicists have found that subatomic particles called **muons** have a typical lifetime of 2.2 microseconds (millionths of a second, abbreviated to μs) before undergoing radioactive decay. Therefore, the maximum distance you may expect a muon could ever travel is its maximum speed (c) times its lifetime of 2.2 μs; this works out to less than a kilometer. It turns out that muons can also be created at the top of our atmosphere when energetic particles from space hit the air. These muons are often found to hit the ground *tens of kilometers* below. How is this possible if the maximum distance they can ever travel is less than a kilometer?

One way to view this is time dilation: these muons have speeds close to c so γ is large, say $\gamma \approx 100$. Thus they live about 100 times longer in our frame, enough to travel about 100 km before decaying, rather than a mere 1 km (Figure 7.8). But this does not explain what happens *in the muon frame*, where muons live a mere 2.2 μs yet still objectively reach the ground! In this frame the explanation is length contraction: the roughly 100 km Earth-frame thickness of the atmosphere is contracted by a factor of $\gamma \approx 100$ to a mere 1 km. The choice of frame affects how we word the explanation, but in either case the observable result is that muons routinely hit the ground while living only 2.2 microseconds in their rest frame.

We now look at length contraction in the context of spacetime diagrams. We start by reintroducing a thinking tool from Section 4.5: the **symmetric frame**. This tool is valuable because it makes the symmetry between inertial frames explicit. When we think about, say, a ground frame and a train frame, we tend to put ourselves in one frame or the other and thereby become unable to see the reciprocity or symmetry between frames. Reciprocity is much more apparent if we *begin* by thinking in a frame in which the ground and train have equal and opposite velocities. The symmetric frame also makes it easier to deal with effects that are independent of the direction of motion. For example, train and ground clocks, despite moving in opposite directions, have the same speed in the symmetric frame and therefore the same time dilation factor γ. This frees us from the burden of taking time dilation into account when comparing train and ground clocks. And if we are not sure how speed affects rulers, we can at least be sure that in the symmetric frame it will affect ground and train rulers equally.

That said, viewing three grids (symmetric, train, and ground frames) in one diagram can be overwhelming, so we start in Figure 7.9 with the symmetric (square) and train (eastbound) frames. The shaded area is the worldsheet of a meter stick aboard the train. To recap the worldsheet concept (Section 6.2), the west boundary of the sheet is the worldline of the west end of the ruler, the east west boundary of the sheet is the worldline of the west end of the ruler, and all events between them are touched by some part of the ruler. Grid lines marking *time* in the train frame cross this worldsheet periodically. The black line segments crossing the shaded region thus constitute snapshots of the ruler taken at specific instants *in the rest frame of this ruler*.

Figure 7.10 adds an an equally speedy ground-frame ruler moving in the opposite direction. The full red and black grids are omitted for clarity, but representative grid cell sides (i.e., axes) are drawn for each frame—thin line segments for one unit of time in each frame and thicker segments for one unit of space. Each thick line segment is a snapshot of that color ruler extending from

Think about it

We will not use the symmetric frame for detailed calculations, but if you do, note that a train/ground relative speed of v yields symmetric-frame speeds greater than $\frac{v}{2}$ thanks to the velocity addition law.

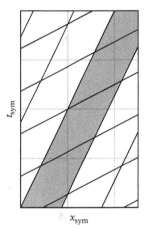

Figure 7.9 *Worldsheet of a meter stick moving east. Each line segment crossing the shaded area is a snapshot of the ruler at some instant in its rest frame.*

Think about it

The worldsheets in Figure 7.10 overlap, but the rulers are not necessarily passing through each other because they may be separated in some direction other than the direction of motion.

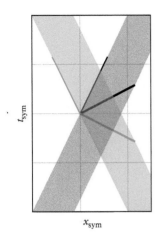

t_{sym}

x_{sym}

Figure 7.10 *To Figure 7.9 we add an equally speedy westbound red meter stick. Line segments stand in for full grids to reduce clutter. The heavy red (black) line segment is a snapshot of the red (black) ruler at one instant in its rest frame.*

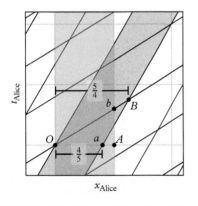

t_{Alice}

$\frac{5}{4}$

b B

O a A

$\frac{4}{5}$

x_{Alice}

Figure 7.11 *The skewed grid spacing satisfies reciprocity if and only if the ratio $\frac{OB}{Ob}$ is the same as the ratio $\frac{OA}{Oa}$. In terms of Alice's x coordinate, this simplifies to $OB = \frac{1}{Oa}$, and indeed event B occurs at $x_{Alice} = \frac{5}{4}$ while a occurs at $x_{Alice} = \frac{4}{5}$.*

$x = 0$ to $x = 1$ in its rest frame. Trace your finger from the west end of the black ruler to the east end, along the heavy black axis. As you did so, you also happened to start at the west end of the *red* ruler—but your finger ran out of red ruler long before it reached the end of the black ruler. In other words, *the red ruler is shorter than the black ruler if we measure at one instant in the black frame*. Now, repeat this process at one instant in the *red* frame by tracing your finger across the heavy red axis—you will find that you run out of *black* ruler before you run out of red ruler. In other words, *the black ruler is shorter than the red ruler if we measure at one instant in the red frame*. But the rulers are physically identical, as is clear in the symmetric frame. Spacetime diagrams thus prove that the length of a moving object (along the direction of motion) is shorter than its rest length—in other words, length contraction. The moving object does not *feel* squeezed because this is an artifact of the way different coordinate systems mix space and time differently—but the effect on *measurements* of moving objects is very real, as we will see in a fully worked example in Chapter 8.

Take a moment to appreciate that we have deduced the existence of length contraction using *only* time skew (which tilts the red and black line segments in Figure 7.10) and symmetry. Time skew is what makes it possible for each ruler to measure the *other* as contracted. Similarly, in Section 7.5 we will deduce time dilation using only time skew and symmetry, and see that time skew is what makes it possible for each set of clocks to measure the *other* as ticking slowly. Neither time dilation nor length contraction would make for a logically consistent universe without time skew. These three effects form a tightly woven, logically consistent whole, which is expressed graphically by the skewed spacetime grid. Time skew determines the tilt of the lines in this grid but not their spacing. Time dilation and length contraction tell us that the spacing is γ.

"The spacing is γ" has a very specific meaning, illustrated by Figure 7.11. There are two grids, named Alice (the square grid) and Bob, coinciding at their origins (event O). The spacing rule is that, at a given Alice instant, Alice's grid columns and rows are γ times as wide as Bob's. By reciprocity this must also work out so that at a given Bob instant, Bob's grid columns and rows are γ times as wide as Alice's. The next three paragraphs help you unpack this statement in more detail.

First, measure the relative speed of the grids in Figure 7.11 using any tilted worldline; you should find $v = 0.6c$ or $\frac{v}{c} = 0.6$. Plugging this into $\gamma \equiv 1/\sqrt{1 - (v/c)^2}$ we find that $\gamma = 1.25 = \frac{5}{4}$. Now consider one instant in Alice's frame: the width of her grid column is the distance from O to A, which we abbreviate as OA. At the same Alice-instant, the width of Bob's skewed column is only Oa; measure this with a ruler and you will find that Oa is $\frac{4}{5}$ of OA. Thus Alice's grid column is $\frac{5}{4}$ or γ times as wide as Bob's. A key point here is that we measured this at one Alice-instant. When you start on a fresh sheet of square graph paper, there will be no ambiguity: you will be staring at a set of Alice instants. Simply space Bob's lines $\frac{1}{\gamma}$ (which can also be expressed as $\sqrt{1 - v^2/c^2}$) squares apart.

A skeptic could ask: how can we be sure this is the way nature works, rather than just a choice I made when drawing the grid? The answer: this is the only spacing that makes reciprocity work. To confirm this, focus now on the (skewed) Bob-instant that goes through O. The width of his grid column is now OB, which is $\frac{5}{4}$ the width of Alice's column, Ob. Reciprocity is satisfied because *both* ratios $\frac{OB}{Ob}$ and $\frac{OA}{Oa}$ are equal to γ. We can simplify this further using Alice's x coordinate because $x_{\text{Alice}} = 1$ for both Ob and OA. The general statement of reciprocity, $\frac{OB}{Ob} = \frac{OA}{Oa}$, then becomes $OB = \frac{1}{Oa}$; the x_{Alice} coordinate of event B is literally the reciprocal of the x_{Alice} coordinate of event a.

We can even show that this is the *only* spacing that satisfies reciprocity. If we were to space Bob's grid more widely as in Figure 7.12, we would see that the ratio $\frac{OB}{Ob}$ is now much larger than the ratio $\frac{OA}{Oa}$. Now, look back at Figure 7.11 and visualize spacing Bob's grid a bit more finely; this would make the ratio $\frac{OB}{Ob}$ *smaller* than the ratio $\frac{OA}{Oa}$. These counterfactual scenarios strongly suggest that at any given speed there is just one spacing (γ) that allows the skewed grid to satisfy reciprocity (this can be proven rigorously with a bit more geometry). Thus, if we had never thought about the light clock we still would have discovered γ with this reciprocity argument.

Hidden in plain sight in Figure 7.11 is yet another proof of γ: the cells of each grid have the same *area*. The area of a cell is indicative of the number of spacetime events that can be packed inside, and this number cannot be frame-dependent—if it were, there would be an objective difference between inertial frames. Now, the cell area depends sensitively on the spacing between grid lines, and it turns out that spacing by γ is the *only* spacing that preserves the cell area in the face of velocity-dependent skew. Thus, our spacing rule survives a test of self-consistency we had not even imagined when we devised the rule. Scientists value this quite highly— had special relativity yielded even one contradiction scientists would long ago have moved on in search of a better model.

To help recap this section, let us define **rest length**, L_R, as the length of an object when measured in its rest frame. This section outlined multiple ways to prove that in other inertial coordinate systems the measured length is smaller: $L = \frac{L_R}{\gamma} = L_R\sqrt{1 - (v/c)^2}$. A useful mental picture is that moving meter sticks contract, and therefore all distances in the moving frame contract even as observers in that frame are unable to *measure* any contraction of their objects. In fact, nothing *really* contracts; the differing measurements reflect the use of differing spacetime grids to assign coordinates to the same underlying reality.

If it is unclear how all this works in practice, rest assured that Chapter 8 works through time dilation and length contraction examples in detail. This chapter focuses on the deductive reasoning relating time dilation, length contraction, and time skew. Each effect can be deduced from the principle of relativity and the invariance of c, but none of them makes sense without the others. For example, Figure 7.10 demonstrates how time skew necessarily leads to length contraction, and we will see in Section 7.5 how time skew necessarily leads to time dilation. Similarly, neither length contraction nor time dilation could satisfy reciprocity if

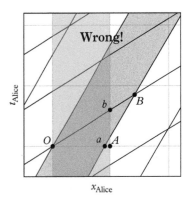

Figure 7.12 *A hypothetical scenario showing how an incorrect γ (i.e., incorrect skewed-grid spacing) leads to violations of reciprocity. Compared to Figure 7.11 $\frac{OB}{Ob}$ has increased while $\frac{OA}{Oa}$ has decreased; they are clearly not equal.*

Think about it

L is smaller than L_R in the same way that the proper time between events along the worldline of a clock, $\Delta\tau$, is smaller than the coordinate time Δt. This is spacetime symmetry at work.

the grid lines marking time were not skewed. Practice thinking of time dilation, length contraction, and time skew as inseparable pieces of a single model that consistently satisfies the principle of relativity and the invariance of *c*. This model is completely embodied by the skewed grid with correct spacing.

We have now completed the logic at the heart of **special relativity**, which describes how inertial frames relate to each other. The ideas fit together so tightly that I have introduced them in quick succession, so you may feel overwhelmed at this point. The following sections and chapters apply these ideas in more detail so you can understand how they play out on the spacetime stage. The skewed coordinate grids will serve as a tool throughout, so understanding the grids should be your primary goal in this chapter.

Check your understanding. A rocket is 100 m long in its rest frame and points east. Roughly what is its length in *(a)* a frame in which the rocket moves east at 0.6*c*; *(b)* a frame in which the rocket moves west at 0.99*c*? You may find Figure 7.6 useful. *(c)* How, if at all, does the size of the rocket differ in frames where it moves north or south (but still points east)?

7.4 Experimental proof

At everyday speeds γ is very close to 1.0 so we are not be able to detect time dilation with low-precision devices like stopwatches. At jetliner speeds ($v/c = 0.0000008$) clocks tick every 1.0000000000003 seconds according to identically constructed "stationary" clocks. This difference *has* been measured with ultraprecise atomic clocks flown aboard jetliners, but time dilation is more salient for faster particles. Every day, there are many atom-smashing accelerators around the world that each move millions of particles at very large γ factors. Particle lifetimes when moving are always observed to be γ times longer than when stationary, and this makes time dilation one of the most well-tested effects in all of physics. Nature can also be very effective at accelerating particles; the muons described in Section 7.3 provided early experimental confirmation of time dilation.

A less obvious application of time dilation is in astronomy, where we routinely use atoms as clocks. For example, hydrogen at rest commonly emits light waves such that a wavecrest leaves the hydrogen atom every 2.18918×10^{-15} s; think of this as a clock that ticks every 2.18918×10^{-15} s. But astronomers observe the wavecrests to arrive much less frequently when the light is emitted from high-speed particles; for example, in a stream of particles—called a jet—leaving an active galactic nucleus. This is a natural consequence of moving clocks ticking slowly in our frame. Astronomers use this effect to infer the speed of the jet, which can reach more than 0.9999*c* (inferred from the fact that γ is around 100). For jets pointed toward or away from us, time dilation is not the only factor that affects the observed frequency (as we will see in Chapter 9), but for jets perpendicular

Scientific notation

Scientists express very small and very large numbers with **scientific notation**, which prints a count of the decimal places; 0.0000000000003 is written 3×10^{-13} because it contains a three in the thirteenth place after the decimal point. This is pronounced "three times ten to the minus thirteenth" and -13 is called the **exponent**. Large numbers are written with positive exponents, so 3,000,000,000 is written as 3×10^9. The positive exponent is the number of digits between the first digit and the decimal point.

to our line of sight time dilation *is* the only factor. This *transverse Doppler effect* is therefore one of the hallmarks of special relativity.

Length contraction is also tested in accelerators, if a bit more indirectly. Physicists use accelerators to collide a vast range of particles, ranging from protons to lead atoms 200 times more massive than a proton. All these particles are roughly spherical in their rest frame, but when moving near *c* length contraction should turn them into pancakes. Physicists analyzing the results of these collisions find that the data make sense only if they account for the pancake effect.

A completely different context for length contraction is in electricity and magnetism. Wires contain equal numbers of fixed positive charges and freely moving negative charges called electrons; collective motion of the electrons is called an *electric current*. Imagine you are an electron in one of two parallel wires carring parallel currents; the electrons in the other wire are comoving with you, so you do not see them contracted. You do, however, see the *positive* charges in the other wire as contracted, so you see the other wire as packing in a higher density of positive charges than negative charges. This in turn is attractive to you (a negative charge) because opposite charges attract. The net effect is that both wires attract each other. The attraction of parallel currents is a classic demonstration in introductory physics, where it is explained in terms of the magnetic field generated by each current. However, relativity shows that magnetic effects are simply electric effects viewed in a different frame. See the articles listed in Further Reading for more details.

Finally, millions of people unknowingly use relativity every day through their Global Positioning System (GPS) devices. Figure 7.13 provides a cartoon of the concept on a two-dimensional map: if you know you are 2 km from Alice *and* 1.5 km from Bob, *and* 1 km from Carol, you know you are at the one point where all three circles intersect. The difficulty in accurately locating yourself is really in measuring each of these distances accurately. The invariance of *c* suggests that an accurate method is to measure the travel time of a flash of light (or a radio pulse, which is just another form of light) emitted by each character. So Alice, Bob, and Carol (representing GPS satellites) broadcast streams of messages stating their current time; you note your time when you receive any message, and the difference between reception and emission time tells you the distance to each satellite. But for this to work, you all need synchronized clocks, and the clocks on the GPS satellites experience time dilation due to their speedy orbits (as well as other relativistic effects; Box 13.2). If the GPS system did not correct for the effect of motion on the satellite clocks, your device would give you very much the wrong position. Each time you get a correct position, then, remember that relativity has just passed another experimental test.

Check your understanding. Has time dilation been confirmed directly with clocks, or only with indirect astronomical arguments?

Think about it

Technically, no amount of evidence can *prove* the correctness of a model, because some future experiment could reveal a need to refine the model. But the predictions of special relativity are so diverse and so precisely matched by experiment that "proof" is an appropriate word here.

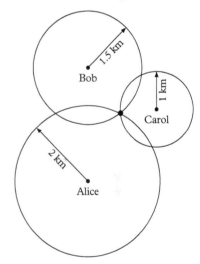

Figure 7.13 *How to find your location: if you know your distance from each of three landmarks, you can only be at the one point where the three circles intersect.*

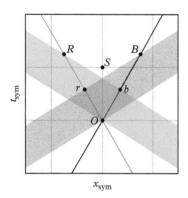

Figure 7.14 *Time dilation from time skew and symmetry. The black frame measures the red clock as time-dilated because the red worldline is only about halfway from O to R when the black second (shaded area) expires. Similarly, the red frame measures the* black *clock as time-dilated. Time dilation is reciprocal, and necessary, because the slices of time are skewed.*

Confusion alert

A common trap is thinking that if the black frame measures the red clock as ticking slowly, the red frame must measure the black clock as ticking rapidly. This is not true because the two measurements are done separately in different coordinate systems. When you approach a friend from a distance, you *each* see the other as gradually looming larger. Your friend appearing larger to you does *not* imply that you appear smaller to her, because you have different coordinate systems.

7.5 Time dilation with spacetime diagrams

This section introduces no new concepts, but reviews time dilation in the context of spacetime diagrams. The motivation here is that after studying time dilation with a light clock and length contraction with spacetime diagrams, you may solidify your understanding by studying time dilation with spacetime diagrams as well. Use this section as a quick review if you have a good grasp of the diagrams and reasoning in Section 7.3, or as a chance to start over in a new context if you found those diagrams befuddling.

Pretend for a moment that you know nothing about time dilation; we will see that it can be deduced independently from time skew in the spacetime diagram alone. Figure 7.14 shows a symmetric-frame diagram with an eastbound black clock and an equally speedy westbound red clock. The black clock reads $t_{\text{black}} = 0$ at the origin O and $t_{\text{black}} = 1$ at event B; likewise, the red clock reads $t_{\text{red}} = 0$ at the origin O and $t_{\text{red}} = 1$ at event R. The entire slice of time from $t_{\text{black}} = 0$ to $t_{\text{black}} = 1$ (the set of events with t_{black} coordinates between 0 and 1) is shaded in black, and likewise for red.

The key feature is that when the black frame completes its first second the red clock is at event r, only halfway from O to R, and thus has completed only half a tick. Similarly, when the red frame completes its first second, the black clock is at event b, only halfway from O to B and thus has completed only half a tick. Each coordinate system measures the other as time-dilated by the same factor, and this is a direct consequence of each slice of time being skewed. This argument works even in the absence of any previous knowledge of time dilation, because shaded regions of *any* height would have led to the same conclusion—as long as the black and red heights were equal. The symmetric frame makes it clear that time skew leads directly to time dilation.

To practice our skills, let us compare the black frame to the "stationary" symmetric frame rather than the equal-and-opposite red frame. A symmetric-frame clock passing through O in Figure 7.14 first ticks at event S, which looks to be at $t_{\text{black}} = 5/4$ based on the fact that a line from O to S runs out of shaded area (the black second) about 4/5 of the way up. Event B then satisfies reciprocity by occurring at $t_{\text{sym}} = 5/4$. Had I not known the formula for γ, I would not have known exactly where to place B along the black worldline when constructing the diagram—but I would have found that only one place satisfies this reciprocity. This would have led to yet another way to discover time dilation. As with the analogous length contraction argument in Section 7.3, it is possible to work through the geometry with a general v rather than the fixed $v = 0.6c$ shown in Figure 7.14, and the result is, of course, that $\gamma = 1/\sqrt{1 - (v/c)^2}$.

Our analysis of Figure 7.14 showed that that $\gamma = 5/4$ when comparing the symmetric and black frames, but $\gamma \approx 2$ when comparing black and red frames. This is because the black and red frames have a higher relative velocity. We thus

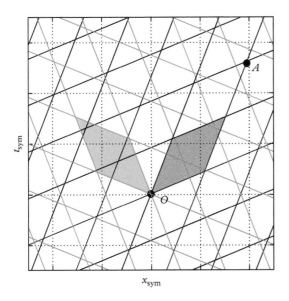

t_{sym}

x_{sym}

Figure 7.15 *Coordinate grids moving at* v = −0.4c *(red) and* v = +0.4c *(solid black lines). One cell in each moving grid is highlighted to make each grid pattern more distinct. An object moving from O to A has* $\frac{v}{c} = \frac{1}{2}$ *in the black frame and approximately* $\frac{1.9}{2.6} \approx 0.75$ *in the square frame.*

conclude this section by revisiting velocity addition with skewed grids, which was only briefly introduced in Section 6.2.

Figure 7.15 overlays two grids, moving at $v = \pm 0.4c$, in addition to a faint square grid at rest. To help make the two grids distinct, one cell in each grid is highlighted. Think of the side walls of each cell as worldlines of clocks at rest in the given frame, and the cell tops and bottoms as delimiting one unit of time elapsing on these clocks. Think of the cell width as corresponding to a ruler extending from one clock to the other. To see how velocities add, let us find the square-grid velocity of a particle that moves at $0.5c$ in the black grid. An example of such a particle would be one that moves from event O to A, because starting from O we count two skewed black cells up and one to the right before arriving at A. This is a space-to-time ratio of 1:2 in the black frame, or $\frac{v}{c} = 0.5$. How fast is this particle moving through the square grid? From O to A is about 1.9 squares of space and 2.6 squares of time (verify this by counting squares in the dotted grid), so $\frac{v}{c} \approx \frac{1.9}{2.6} \approx 0.75$. We can also draw a worldline from O to A with a straightedge; this worldline crosses about three quarters of a square of space in the first square of time, so it has $v \approx 0.75c$ in the square frame. Again, we see graphically why velocities add nonlinearly and why no sum of velocities can reach or exceed c; we will work out the algebraic formula in Section 9.4.

Figure 7.15 also exposes a practical aspect of how each frame measures the *other's* clocks as ticking slowly. Think of each tilted-from-vertical line as the worldline of a clock. A single black clock crosses paths with multiple red clocks, which then collectively judge the black clock to be ticking slowly. Likewise, a single *red* clock crosses paths with multiple black clocks, which then collectively judge the red clock to be ticking slowly. A popular myth about relativity is that "each clock measures the other as ticking slowly," which gives the impression of a head-to-head clock battle with a self-contradictory result. The truth is that each *coordinate system* measures the other's clocks as ticking slowly, and this is easily accommodated by the differing skews of the grid lines. Time coordinates, like space coordinates, depend on the frame you choose.

Check your understanding. In Figure 7.15, what is the velocity of the black frame relative to the red frame?

7.6 Light clock along the direction of motion*

Section 7.2 used a moving light clock oriented perpendicular to the direction of motion to deduce time dilation. In this section, we see what more we can deduce when we orient the clock along the direction of motion. We will call this the "horizontal" clock because we picture it stretched out along the floor of a train.

We start with some easy math describing a "vertical" clock. Let us call the height of this clock in its rest frame (the train frame) L_R for "rest length." In the train frame the round-trip time for the light is the round-trip distance ($2L_R$) divided by the speed (c), or $\frac{2L_R}{c}$. In the ground frame, the round-trip time is γ times longer (Section 7.2), or $\frac{2L_R\gamma}{c}$. Because all clocks in a frame keep the same time regardless of orientation (Section 7.2), we can trust that the ground frame also measures $\frac{2L_R\gamma}{c}$ for the round-trip time on the *horizontal* clock. From this elapsed ground-frame time and the speed of light, we will be able to deduce the ground-frame *length* of the horizontal clock.

Notice that I used L_R to describe the height of the vertical clock in *both* frames, because distances transverse to the direction of motion cannot be affected by length contraction (Section 4.4). When we orient this clock along the direction of motion, its length is *still* L_R in the train frame regardless of the train's velocity—otherwise, passengers would be able to use its varying length to determine when the train is *really* moving (thus violating the principle of relativity). Only the length of the clock in the ground frame is in question, so let us call it L (to distinguish it from L_R) and try to determine if L differs from L_R.

Figure 7.16 shows ground-frame snapshots of a horizontally oriented light clock aboard a train moving to the right; they are drawn with small vertical offsets for clarity but the displacements in our thought experiment are purely horizontal. The three snapshots depict the light leaving the left end, reflecting off the right end, and returning to the left end. The snapshots are drawn evenly displaced, but we should not assume that these events are evenly spaced in time.

In the first part of the round trip, the right end of the clock moves away from the light, so light closes the gap at a rate of $c - v$. The time required to close a gap of length L at this rate is $\frac{L}{c-v}$. In the return leg the left end of the clock moves

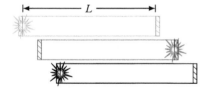

Figure 7.16 *Light making a round trip in a clock moving to the right; snapshots are also separated vertically here for clarity. The length of the clock is* L *in this frame.*

Confusion alert

The "speeds" $c - v$ and $c + v$ are *not* the speed of the light as measured by any observer, so this does not violate invariance of c; see Box 5.1 to review this issue.

toward the light, so the gap is closed at a rate of $c+v$, hence in a time $\frac{L}{c+v}$. Equating the round-time for vertical and horizontal clocks, we get:

$$\frac{2L_R\gamma}{c} = \frac{L}{c-v} + \frac{L}{c+v}.$$

To solve for a simple relationship between L_R and L, eliminate the denominators by multiplying through by $(c-v)(c+v)$:

$$\frac{2L_R\gamma}{c}(c+v)(c-v) = L(c-v) + L(c+v)$$

Then simplify each side a bit:

$$\frac{2L_R\gamma}{c}(c^2 - v^2) = 2Lc$$

$$L_R\gamma \frac{(c^2 - v^2)}{c^2} = L$$

$$L_R\gamma(1 - v^2/c^2) = L$$

We know $\gamma = \frac{1}{1-v^2/c^2}$ so plug that in:

$$\frac{L_R(1 - v^2/c^2)}{\sqrt{1 - v^2/c^2}} = L$$

$$L_R\sqrt{1 - v^2/c^2} = L \qquad\qquad (7.6)$$

This shows that the length of the light clock in a coordinate system of our choice, L, is smaller than its rest length L_R (or at most equal, if we happen to choose a coordinate system in which the clock is at rest). To complete the parallel with time dilation, we can rewrite Equation 7.6 as

$$\frac{L_R}{L} = \gamma.$$

> **Confusion alert**
> _____
>
> Rest length is in the numerator here, while rest time (or proper time) is in the *denominator* of the expression for time dilation. As explained in Section 7.3, this is because two frames must agree on their relative speed.

This effect is not limited to light clocks! Train-frame observers can use the light clock as a ruler to measure the length of *anything*, and if the ruler contracts along the direction of motion then *all* distances along this direction contract. As with time dilation, train-frame observers do not notice or feel this contraction because nothing in the train changes its length in terms of meter sticks. Length contraction is very real in the sense that "stationary" observers *measure* a rapidly moving sphere to be more like a pancake, but it is an artifact of the way different coordinate systems mix space and time differently rather than an intrinsic change in the moving object.

Check your understanding. Could we have analyzed the horizontal clock before the vertical clock?

CHAPTER SUMMARY

- Time dilation: moving clocks run slower than stationary clocks by a factor of $\gamma = \frac{1}{\sqrt{1-v^2/c^2}}$.

- Length contraction: moving rulers are measured as shorter by the same factor due to symmetry of space and time.

- Time skew, time dilation, and length contraction work together to ensure that reciprocal experiments in different frames yield identical results. Considering any one or two of these without the others leads to logical contradictions.

- We deduced how to construct moving-frame coordinate grids that capture all this behavior. Spacetime does not really differ from frame to frame; the coordinate grid is simply drawn differently on the same map of spacetime.

- Understanding these grids will be immensely useful in solving the problems of the next few chapters, so do not be satisfied with memorizing the points listed here. Use them as a *starting point* for practicing your grid skills.

FURTHER READING

Relativity Visualized by Lewis Carroll Epstein has many intuitive illustrations to help you visualize time dilation and other effects.

An Illustrated Guide to Relativity by Tatsu Takeuchi is a rich resource for practice with spacetime diagrams, and for thinking in the symmetric frame. This book made me realize the power of the symmetric frame.

Physics students familiar with electricity and magnetism may appreciate *Special Relativity and Magnetism in an Intro-* *ductory Physics Course*. This article by R. G. Piccioni in *The Physics Teacher* (vol. 45, p. 152, 2007) shows how the magnetic field of a current in a wire can be deduced from length contraction of the set of moving charge carriers. The explanation is at an introductory level (with algebra) and provides references to deeper examinations of the same topic.

CHECK YOUR UNDERSTANDING: EXPLANATIONS

7.1   (a) Direction of motion does not matter, because the effect is the same for *all* clocks in the moving frame. (b) Increases with speed.

7.2   The γ values (to more precision than you can estimate graphically) are 1.005, 1.05, 1.15, 1.4, 2.3, and

7.1. The ratio of the hypotenuse to the vertical leg is the ratio of coordinate time to proper time, which is the *definition* of γ.

7.3 (a) 80 m; (b) 14 m. (c) In frames where the rocket moves perpendicular to its 100 m rest length, its width or height is contracted.

7.4 Yes, precise atomic clocks aboard aircraft have been used to directly measure time dilation.

7.5 Start from event O and trace the black worldline up until it crosses a red grid line. This intersection occurs about 0.7 red cells to the right of the red worldline through O and exactly one unit of red time since O, so the space-to-time ratio is 0.7. In other words, $\frac{v}{c} = 0.7$. Compare this to the Galilean prediction of 0.8.

7.6 No. The horizontal clock is affected by *both* time dilation and length contraction, and the one equation yielded by the horizontal clock would not have allowed us to solve for *both* of these effects. The vertical clock isolated the effect of time dilation for us so that we could solve for the length contraction. If, however, we recognized that any time dilation and length contraction would have to be described by the *same* ratio, we could set up and solve the equation for this single ratio.

? EXERCISES

Exercises 7.1–7.4 refer to a hypothetical express train that travels at a substantial fraction of c.

7.1 The train's dining car serves meals every six hours according to train time. How often are meals served as measured in the ground frame: more often than every six hours, every six hours, or less often than every six hours?

7.2 Meals in the train stations are also served every six hours according to station time. What do passengers on the train measure as the time between meals served in a given station: less than six hours, six hours, or more than six hours?

7.3 Train cars are 30 m long as measure by rulers aboard the train. Do ground-frame observers measure the cars as less than 30 m, 30 m, or more than 30 m?

7.4 The distance between stations is 100 km as measured by meter sticks at rest on the ground. How long is the distance between stations as measured in the train frame: less than 100 km, 100 km, or more than 100 km?

7.5 *(a)* Using graph paper, lay out triangles like those in Figure 7.7, but for $v = 0.2c, 0.4c, 0.6c$, and $0.8c$. *(b)* Explain why the "triangle" for $v = 0$ is simply a vertical line segment. *(c)* Explain why you cannot draw the shape representing $v = c$; use geometry rather than simply referring to the speed limit.

7.6 A new company advertises a way of staying forever young. For $100,000, they will launch you aboard a rocket at 0.999c, so that you will live much longer than you would have lived had you stayed on Earth. Do you get what you pay for; that is, will you find yourself living a longer life? Explain your reasoning.

7.7 A new company advertises a way of appearing thinner. For $100,000, they will launch you aboard a rocket at 0.999c, so that you will be very thin in Earth's frame. Do you feel thinner? Explain your reasoning. If you accept the offer, what will you notice about people on Earth?

7.8 Refer back to Figure 7.11 and explain how it satisfies reciprocity, using only coordinates in Bob's frame.

7.9 Physicists often send subatomic particles around and around a circular track at high speed. Do you expect these particles to "age" more slowly than similar particles kept stationary in the lab, or would a hypothetical observer traveling with them say that *we* appear to be aging more slowly? Explain your reasoning.

7.10   In Chapter 6 I wanted to illustrate time skew without bringing in the additional effect of time dilation. Is there a direction (along the direction of motion or perpendicular to it) in which time dilation is absent? If not, how did I minimize the effect of time dilation in the diagrams in Chapter 6? *Hint:* study Figure 7.6.

✛   PROBLEMS

7.1   Explain why we would *not* deduce time dilation if we analyzed the story in Section 7.1 using Galilean relativity.

7.2   A strobe light in the center of a train car emits a flash of light at $t = 0$ that reflects off mirrors at either end at $t = 1$ and returns to the center of the car. *(a)* Draw a spacetime diagram of the story in the train frame. *(b)* Redraw this diagram in another frame. *(c)* Explain how this diagram shows time dilation.

7.3   Explain how the spacetime diagram for Problem 7.2 relates to the "horizontal" light clock algebra in Section 7.6.

7.4   Consider two events, A and B, that in some frame F happen at the same position but different times. It is reasonable to call F a rest frame here because in this frame no motion is required to get from event A to event B. When we measure in another frame, the distance between A and B is *not* contracted—it must actually be *longer*. Explain why the length contraction concept should not be applied to these events. Make your answer into a general rule about what types of event pairs qualify for the length contraction concept. *Hint:* consider event pairs with small and large spatial displacements compared to their time displacements.

7.5   Lewis Carroll Epstein devised the "cosmic speedometer" shown in Figure 7.17. This is an abstraction of the light clock situation in Figure 7.3: the arc again shows the distance light can travel from the bottom of the clock, and the needle of the speedometer is the path of the light over one unit of coordinate time. At low speed the needle points almost vertically, and at speeds near c the needle points almost horizontally. *(a)* The ratio of the needle length to the height of the needle tip corresponds to

what familiar ratio? (*Hint:* try some extreme cases.) *(b)* The distance from the left edge of the speedometer to the needle tip, divided by the needle length, forms what familiar ratio? *Comment:* The speedometer compactly represents many of the ideas represented by the triangles in Figure 7.7 but is not used more widely here because it has no *invariant* property such as the fixed height of the vertical legs in Figure 7.7.

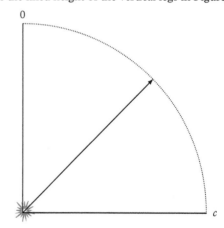

Figure 7.17 *Epstein's cosmic speedometer: compare to Figure 7.3.*

7.6   Bob measures Alice as moving 0.8 feet in an experiment lasting one Bob-nanosecond ($v = 0.8c$). Because $\gamma = \frac{5}{3}$ at this speed, Alice measures only 3/5 or 0.6 nanosecond as elapsing during this experiment. But if she took only 0.6 nanosecond to travel 0.8 feet, she is moving faster than c! What is wrong with this picture?

7.7   Why do rulers contract while time intervals dilate (expand)? Taken literally, these words seems to imply an asymmetry between time and space. Starting with a spacetime diagram showing symmetrically rotated axes, explain how the asymmetry really lies in how we use and talk about moving clocks vs. moving rulers.

7.8 In the spacetime diagrams in Section 7.3, we started with a symmetric frame to show the conceptual need for length contraction. Why did we not continue with the symmetric frame for the next goal, showing *how much* contraction was needed at a given speed?

8

Special Relativity: Putting it All Together

The last few chapters have introduced many ideas in quick succession, because none of the ideas can really be understood in isolation. This chapter works through examples in some detail so you can practice applying the ideas. Reading this chapter is only the beginning; you still must practice! A good approach is to close the book after reading each example and do your best to work through it yourself before proceeding to the next example.

8.1 Solving problems with spacetime diagrams

Learning to solve problems with spacetime diagrams is a bit like learning to ride a bike: you will need to practice a lot before being able to do it smoothly. In preparation for working through detailed examples in the following sections, this section reviews the thinking tools at your disposal. If these thinking tools are fresh in your mind, consider skipping ahead and using this section as a reference later as needed.

The advice here is organized into a suggested sequence of steps, but use this sequence as a guide rather than a recipe. Depending on the problem, steps will vary in importance and sequence.

Read the problem carefully and sort the given information according to frame. Identify the frame in which you wish to begin drawing.
Why: the most common cause of confusion in relativity is forgetting that most statements are frame-dependent. If, for example, two events are described as simultaneous or two objects moving relative to each other are described as having the same length, it is crucial to identify the frame in which this is true.
Best practices: Make a table of what you know about each frame and what is frame-independent. This will prevent many mistakes later. Beginners should start drawing the frame with the most given information. This will make it easier to fill in the gaps, even if you later have to translate the whole diagram into a second frame to best answer the question that is posed. As you become more expert, you may find that you can diagram both frames in parallel, or skip directly to the "aha" frame.

The Elements of Relativity. David M. Wittman, Oxford University Press (2018).
© David M. Wittman 2018. DOI 10.1093/oso/9780199658633.001.0001

In the next several steps, we focus only on your chosen frame. If you later draw another frame, revisit these steps in that frame.

Draw worldlines of stationary objects.

Why: this is the easiest place to start—stationary worldlines are vertical and there is no length contraction between given locations.

Best practices: This step is often where you choose a scale. For example, if there is a 1 km tunnel, you must draw separate worldlines for the east and west ends of the tunnel, which means committing to a certain number of squares per km. Choose wisely so the drawing is spacious without eventually running off the page (diagrams often grow quite a bit when you add tilted worldlines). Label all worldlines very specifically; for example, "east end of tunnel" to avoid confusion later. Forcing yourself to label clearly helps you think clearly as well. If you have separate worldlines for east and west ends of an object, you may lightly shade in between to identify all events in the object's worldsheet.

Draw worldlines of moving objects.

How: Use graph-paper squares to help you represent velocities accurately. For example, if $v = 0.5c$, this equals $\frac{1}{2}c$, so you should draw the line as moving one square of space for every two squares of time (Figure 8.1). If the moving object has a size that is relevant to the story, make parallel worldlines for the front and back. Before you commit to a spacing between moving worldlines, think about length contraction. For example, if you are given a train's rest length but you are drawing in the ground frame, the distance between the tilted worldlines will be smaller than the rest length by γ—and you may have to calculate γ from v.

Best practices: shade in the worldsheet of a moving object; this the moving object pop out from all the other lines on your diagram. Label each worldline with descriptive labels.

Checkpoint: until now you have been focusing on the worldlines mostly one at a time, so this is a good time to check that all the worldlines make sense *in relation to each other.*

Locate events, part one: using information given in the frame of the drawing. Note that events are not "in" any particular frame, but information about the *relationship* between two events is given in some frame. A common task runs along the lines of "find the event at the rear of the train that is simultaneous (in the ground frame) with event F."

Why: we prevent mistakes by starting with events that are easily located in this frame. Furthermore, events that are simultaneous in this frame are easily identified (they can be connected with a horizontal line).

How: you will usually start with some important event that you know occurs at the intersection of two worldlines. Mark this—call it F for now—then find the horizontal graph paper line through F (or pencil one in yourself) to visualize

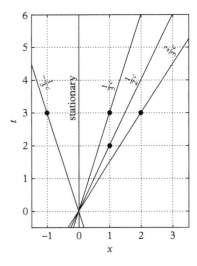

Figure 8.1 *Placing worldlines through the origin:* $v = \frac{1}{2}c$ *is a line through a point 1 square to the right and 2 squares up, and so on. Worldlines need not go through the origin; parallel worldlines indicate objects at rest relative to each other.*

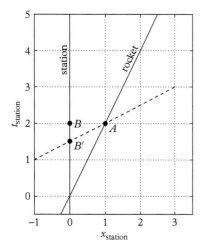

Figure 8.2 *If the rocket moves one square of space per two squares of time, draw a line that crosses two squares of time per one square of space to locate simultaneous events in the rocket frame. Thus B′ (rather than B) is simultaneous with A in the rocket frame.*

Think about it

If you draw the story first in frame A and then frame B, the A worldlines in the B diagram must have a tilt equal and opposite to that of the B worldlines in the A diagram.

---

the set of events that are simultaneous to F in this frame. Amongst this infinite set of events, there is usually one specific event of interest, such as the one that occurs at the rear of the train in the example task here. That means it occurs along the worldline of the rear of the train. The desired event is therefore at the intersection of this worldline with your horizontal line.

Best practices: label each event clearly. Any horizontal lines you draw should be drawn lightly; keep them distinct from worldlines.

Locate events, part two: using information given in the other frame.

How: this is like the previous step, but now the set of simultaneous events is a line tilted from horizontal. If the moving worldlines tilt clockwise from vertical, then the lines of simultaneity tilt counterclockwise from horizontal, by the same amount. As with the worldlines, use graph paper squares to calibrate the tilt, then use a straightedge to draw the line through your reference event (Figure 8.2).

Checkpoint: once you have placed all events, double-check your event locations and labels to make sure they are consistent with the story.

Draw the story in a second frame if necessary.

Why: some problems specify a situation in one frame, then ask you how it can be physically consistent in the other frame. In other cases, you need to see the second frame to enhance your own understanding.

How: To a large extent, repeat the steps you performed to build the diagram in the first frame. But you can also refer to your first diagram to speed up the process. Make sure the event labels are consistent between the two frames! For example, if F labels a train-front worldline crossing a tunnel-exit worldline in the first frame, you must also use F to label the same intersection in the second frame.

Checkpoint: make sure the two frames are consistent with each other. Three events form a triangle, and the triangle should be recognizable in both frames (but rotated and stretched when frames are changed). A rectangle in one frame should become a parallelogram in another frame with the same events at its corners. These are just examples; you must find the appropriate consistency checks for your particular problem.

Interpret your diagram in words. For example, if a torpedo launched in frame A misses its target because the target was contracted, what is the explanation in the target frame where contraction is not an issue? This explanation usually involves, at least in part, the fact that events simultaneous in one frame are not simultaneous in the other.

8.2 Measuring the length of a moving object

Exercise: A meter stick travels eastward at $0.5c$ through your laboratory. You and your assistant choose an instant to simultaneously make chalk marks on the wall at the locations of the front and back of the meter stick. Diagram the story in both frames, and for each chalk-marking event identify events that are simultaneous *in the stick frame* and located at each stick end.

Diagram: The story is told in the lab frame, so let us start with a diagram in that frame. On a piece of graph paper draw the worldline of the front of the stick: use a straightedge to draw a line that moves one square over for every two squares up. (Refer to Figure 8.3 repeatedly throughout this paragraph.) Now, where is the rear of the stick? We need to calculate γ for $v = 0.5c$ to know how much the stick contracted. Figure 7.6 is a good resource, or just enter $1/\sqrt{1 - 0.5^2}$ into your calculator to find $\gamma = 1.15$. Next, decide on a scale: how many squares of graph paper will equal one meter? I will use five squares per meter here, so the rear worldline will be $\frac{5}{1.15} \approx 4.3$ squares behind the front worldline in the lab frame. With both worldlines marked, it is easy to identify simultaneous (in the *lab* frame) chalk-marking events F and R (for front and rear) along these worldlines. You may choose any value of t_{lab} for these events, because there are no other given events or locations to relate them to. So far, we have drawn everything in Figure 8.3. Note that the rear worldline must be to the *left* of the front worldline if the direction of travel is to the right.

Now, to find events that are simultaneous to the chalk-marking events *in the stick frame*, we need to remember that the grid lines marking time in the stick frame will be tilted up one square for every two squares to the right; this is the inverse of the ratio we used for the worldlines. Use a straightedge to draw such a line through event F; the corresponding dashed line in Figure 8.4 marks the set of events simultaneous to F in the stick frame. These events cover a range of locations, but the exercise particularly asks us to identify events occurring at each end of the stick. The rear worldline defines events at the rear of the stick, so the event that is on *both* the rear worldline *and* on the dashed line through F is the event that happens at the trailing end of the stick *and* is simultaneous with F in the stick frame. This event is marked R_S in Figure 8.4. In the same way, events simultaneous to R in the stick frame fall on the dashed line through R. I leave it to you to identify which of these events occurs at the front of the stick.

We can now see why you and your assistant found the length of the stick to be less than 1 m. According to the stick frame, you and your assistant did not mark the front and rear simultaneously at all! Given that you marked the front at event F, the stick coordinate system insists that your assistant should have marked the rear at event R_S. But your assistant actually marked the rear at the later event R. The rear of the stick moved forward in the intervening time, so your assistant's mark is—again, according to the stick coordinate system—too close to your mark,

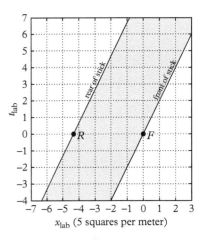

Figure 8.3 *Front and rear ends of a moving meter stick are marked simultaneously in the lab frame.*

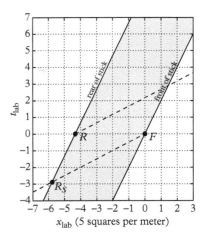

Figure 8.4 *Event R_S is located at the rear of the stick, and simultaneous with F in the stick frame. Identify the event located at the front of the stick and simultaneous with R in the stick frame.*

thus underestimating the length of the stick. But this is not a mistake on the part of your assistant; it is inherent in the lab-frame coordinate system. The stick really is shorter at a given instant in the lab frame because "at a given instant" has a different meaning in the lab frame than in the stick frame.

8.3 Train in tunnel paradox

Problem: If the stick really is shorter in the lab frame, let us do an experiment to prove it. Instead of a stick, take a 1 km long (rest length) train and send it through a 1 km long (rest length) tunnel at $0.4c$. If the train really is shorter than the tunnel, we can drop gates at each end of the tunnel to trap the train and provide frame-independent proof that the train fits inside the tunnel. But wait: in the train frame the *tunnel* is contracted so the train will stick out of the tunnel and the gates will not be able to close. What will happen?

Solution: this "paradox" is very closely related to our discussion of how to measure the length of a moving object in Section 8.2, so we will move more quickly through the basics. Figure 8.5 is drawn with a scale of five squares per km in the tunnel frame, where the train is contracted. To find the length of the train in this frame, we need to compute γ: $1/\sqrt{1 - 0.4^2} \approx 1.09$. The length of the train in the tunnel frame is therefore $\frac{1}{\gamma} \approx 0.92$ km. (*Exact* coordinates are unnecessary when drawing by hand, but conceptually we do need to depict the train as a bit shorter than 1 km.) The train worldlines tilt two squares over for five squares up in order to represent $v = 0.4c$. The only substantive difference from Figures 8.3 and 8.4 is that a 1-km long stationary tunnel worldsheet has been added. I have added this before identifying any events because it will help us organize our thinking about the events.

In this frame, the train fits in the tunnel, as the problem statement describes. To make this clear, the heavy lines in Figure 8.5 represent gates at the tunnel ends. The exit gate is closed at the start of the experiment and is lifted just in time to prevent a collision with the front of the train, while the entrance gate is initally up and then slammed shut as soon as the rear of the train enters the tunnel. From $t_{\text{tunnel}} \approx 0$ to $t_{\text{tunnel}} \approx 1$ the gates are closed with the train completely inside the tunnel. All frames must agree on this because a collision between train and gate would be clear to all frames.

So how do we respond to the objection that in the train frame it is the *tunnel* that is 0.9 km long so the train must protrude from the tunnel? We will eventually diagram the situation in the train frame, but a complete answer should show how this is understandable in the tunnel frame as well. Figure 8.5 makes it clear that "the train fits in the tunnel" is equivalent to saying that the rear of the train enters the tunnel before the front of the train exits, so let us add an event X for the front of the train exiting the tunnel and identify events

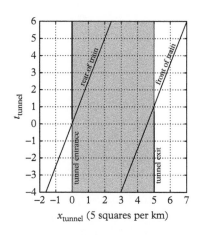

Figure 8.5 *Tunnel-frame diagram of the train "paradox." The shaded area is the tunnel worldsheet.*

simultaneous to X in the train frame by drawing a dashed line with slope $2\!/\!5$ through this event.

The result is Figure 8.6. The dashed line meets the rear of the train *well outside* the tunnel, indicating that in the train frame the rear has not yet entered the tunnel when the front of the train exits the tunnel at event X. In this frame, the train does *not* fit in the tunnel, but a collision with the gates is avoided because the tunnel entrance closes *after* event X rather than simultaneous with it. It turns out that the statement "the train fits in the tunnel" *is* frame-dependent once it is fully broken down into a set of events. With most relativity "paradoxes" the first step toward understanding is to translate the loose language in the problem statement into a very specific statement about events such as "the exit gate is lifted after the entrance gate is closed." This enables you to diagram the situation, and that usually reveals the key to the "paradox."

To double-check our understanding, let us redraw everything in the train frame and compare the frames side by side (Figure 8.7). The skewed parallelogram in the left panel becomes a rectangle in the right panel, because lines of fixed train time become horizontal and the train worldlines become vertical in the train frame. Furthermore, in the train frame the tunnel worldsheet is contracted and skewed toward the rear of the train. Compare the two panels of Figure 8.7 in detail to make sure you understand all the relationships.

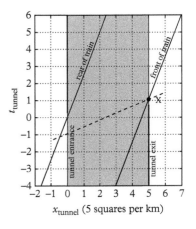

Figure 8.6 *The dashed line shows that at the train-frame instant the front of the train exits the tunnel (at event* X*), the rear of the train is still well outside the tunnel.*

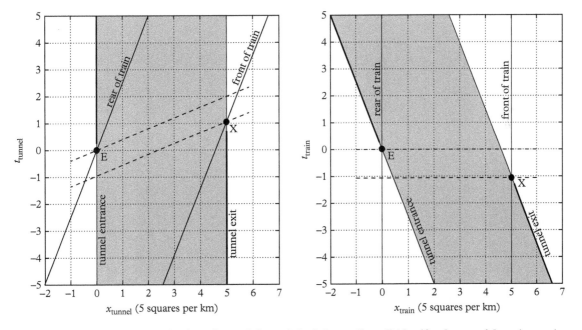

Figure 8.7 *The complete story of train and tunnel, drawn in both frames. Event* E *identifies the rear of the train entering the tunnel.*

8.4 Velocity addition

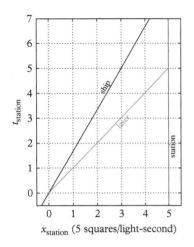

Figure 8.8 *Station-frame diagram of the basics in the torpedo problem.*

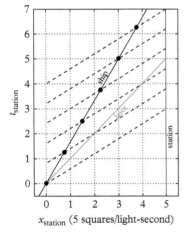

Figure 8.9 *Adding dashed grid lines marking time in the ship frame.*

Problem: a starship approaches a battle station at high speed ($v = 0.6c$). When the ship is one light-second away from the station (as measured by the station), the ship simultaneously fires a laser as well as a torpedo that travels at $v = 0.5c$ *relative to the starship*. Diagram the situation and *graphically* find the speed of the torpedo in the station frame. How much time elapses between the laser hitting the station and the torpedo hitting the station in the station frame? In the ship frame?

Solution: The basic idea here is to use the information given to construct a worldline for the torpedo through a coordinate grid attached to the station; the torpedo speed is the ratio of space to time displacements along this worldline, which we can read off the graph (Section 7.5). The coordinates will be easier to read if the station-frame grid is square, so let us draw it that way. Figure 8.8 shows the worldlines described directly in the problem: the ship, the station, and the laser (which is light and therefore travels at c in any frame).

Representing the torpedo in the station frame requires more thought. We know only that the torpedo moves one grid cell in space per two grid cells in time *in the ship frame*, so we need to build a dashed ship-frame coordinate grid on top of Figure 8.8 and see how much the resulting worldline tilts. In the station frame, the ship moves at $\frac{3}{5}c$ so the grid lines marking ship-time are tilted three squares up per five squares over. Let us put the launching of the torpedo at the origin, and draw the $t_{\text{ship}} = 0$ line through it—this is the lowest dashed line in Figure 8.9. The next ship-time grid line should go through the event where the ship clock completes one tick; for this discussion a tick will be 1/5 second because one square is 1/5 light-second. Time dilation tells us that this event occurs after γ squares in the station frame. At the ship speed $\gamma = 1.25$ so place the ticks along the ship's worldline at every 1.25 squares of station time, and finally draw lines through these events that are parallel to the dashed line through the origin. Look at the resulting set of lines in Figure 8.9: they intersect each vertical grid line with a spacing of 1/γ squares due to reciprocity. "Space the lines 1/γ squares apart in the station frame" is a concise way of stating the recipe for constructing the lines.

The grid lines marking ship-frame positions are marked in a similar way: the lines are spaced 1/γ squares apart, and they slope parallel to the ship worldline (Figure 8.10). Finally, use a straightedge to draw a torpedo worldline that moves one cell of space per two cells of time in the ship frame, and extend this worldline until it hits the station. The final result is clear in Figure 8.10: by event T the torpedo traveled about 5.9 squares of time for five squares of space, for a station-frame speed of about 5/5.9 or $0.85c$. This is much less than the Galilean prediction that $0.6c + 0.5c = 1.1c$.

Figure 8.10 also marks event L where the laser hit. Events L and T are separated by about 0.9 squares of time, or 0.18 seconds in the station frame. Because these events happen *at the same location in the station frame*, we can view this as 0.18 seconds elapsing on a single station clock; because of time dilation we know this will be measured as $0.18\gamma \approx 0.23$ seconds in the ship frame.

The final section in this chapter demonstrates why *at the same location in the station frame* was a key attribute in this argument.

8.5 Clocks

Problem: Bob flies a rocket through Alice's laboratory. If Alice measure's Bob's clock as ticking slowly, how can Bob *also* measure Alice's clock as ticking slowly? Draw pictures of clock displays to show how this is possible.

Solution: time skew desynchronizes the Bob clocks in Alice's frame, and vice versa, so that they each obtain a reciprocal result. Let us draw each observer as having as line of five clocks so we can see the pattern robustly. In Alice's frame, all her clocks read :00 at the start of the experiment, but due to time skew Bob's clocks must display a *range* of times at a given instant in Alice's frame. We know the trend—Bob's trailing clocks must read later times—so penciling in the time on *one* of Bob's clocks is sufficient to determine the readings on all the others.

The problem statement does not specify any more details, so we are free to choose the time on Bob's clocks as long as we respect the time skew pattern. The top panel of Figure 8.11 shows the two sets of clocks at the start of the experiment, after choosing Bob's leading clock to match the Alice-frame clock at the same position. This pattern of Bob-clock readings at an instant in Alice's frame is a pictorial representation of the pattern seen on the spacetime diagram in Figure 6.5, and a digital version of the pattern seen in each row in Figure 6.15.

At the end of the experiment (bottom panel of Figure 8.11), each of Bob's clocks has ticked one second since the start of the experiment while each of Alice's clocks have ticked two, so Bob's time is indeed running slowly by a ratio of

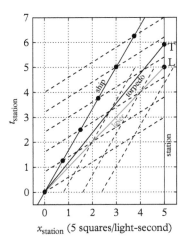

Figure 8.10 *After adding dashed lines marking position in the ship frame, we can draw the torpedo's worldline because we know its velocity in the ship frame.*

Think about it

In this conceptual problem we are not concerned with the size of the time skew but, for reference, the skew can be nearly one second between clocks only if the clocks are one light-second (300,000 km) apart and traveling at a speed near c.

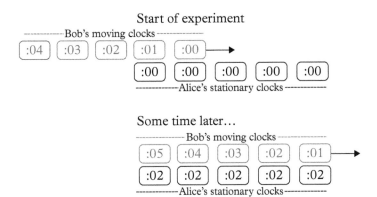

Figure 8.11 *Alice's (black) and Bob's (gray) clocks at one instant in Alice's frame (top) and two seconds later in Alice's frame (bottom). Bob's clocks add only one second for Alice's two, indicating time dilation by a ratio $\gamma = 2$. Nevertheless, Alice also measures Bob's clocks to be running slowly; for example, the Bob-clock reading adjacent to Alice's leftmost clock jumped* four *seconds compared to Alice's two.*

$\gamma = 2$. But notice that we had to compare Bob's leading clock with *two* of Alice's clocks, at different positions. If instead we track a *single* Alice clock throughout the experiment, we get a different story. Focus on Alice's west clock: the adjacent Bob clock reads :01 at the start of the experiment and :05 at the end. In other words, Bob finds that only two seconds elapsed on this Alice clock while four seconds elapsed on his clocks. Bob concludes that *Alice's* clock ticks slowly by a ratio of $\gamma = 2$!

Alice's conclusion that Bob's clock ticks slowly does not contradict Bob's conclusion that Alice's clock ticks slowly, because they are doing different experiments: Alice's experiment tracks one Bob clock as it passes two different Alice clocks, and Bob does the reciprocal experiment. Can we compare one Alice clock and one Bob clock side by side at the start *and* at the finish of the experiment to see which one is "really" ticking slowly? This requires accelerating at least one clock out of its inertial coordinate system, so we defer this experiment to Chapter 10.

 FURTHER READING

Readers seeking more practice with spacetime diagrams may benefit from *An Illustrated Guide to Relativity* by Tatsu Takeuchi. The book focuses very much on spacetime diagrams and offers many problems and solutions.

Very Special Relativity by Sander Bais is an elegant small book built entirely around spacetime diagrams.

 PROBLEMS

8.1   Two rockets of identical rest length speed rapidly toward a near head-on collision, with a very small distance separating the ships perpendicular to the direction of motion. At a certain moment, as shown in the left side of Figure 8.12, the ships are side-by-side, with the nose of the upward-moving darker ship alongside the tail of the downward-moving lighter ship. At this instant the darker ship fires a weapon mounted in its tail. Because the lighter ship is contracted, the weapon must miss. However, in the frame of the lighter ship, the *darker* ship appears contracted as in the right side of Figure 8.12, so the weapon should hit the lighter ship. Does the weapon in fact hit or miss, and how do you know? Explain in words, but also support your explanation with a spacetime diagram or other appropriate tool. Assume the ships are so close to each other that the weapon takes negligible time to cross the gap between ships.

8.2   Illustrate the answer to Problem 8.1 by drawing spacetime diagrams in *each* frame. Use graph paper and make sure all events, worldlines, and axes on the spacetime diagrams are clearly labeled. The exact relative velocity and γ you use are not important as long as they are consistent with each other and with the story.

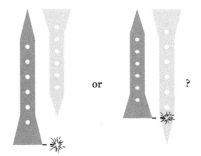

Figure 8.12 *Rockets of identical rest length traveling in opposite directions are side by side for an instant. Does the weapon hit or miss its target? Assume the perpendicular distance traversed by the weapon is so small as to be negligible.*

8.3 A high-speed train moves to the right. Two daredevils stationary with respect to the ground agree to jump across the tracks simultaneously, one just in front of the train and the other just behind. They know the length of the train in their frame and they stand this distance apart, ready to jump simultaneously. *(a)* Draw a spacetime diagram in the ground frame. The exact velocity of the train is not important as long as the graph is consistent with the story and with itself. Let the origin be the event marking the rear jump. Clearly label the worldlines of the front and back of the train, and the events that are the jumps (label F for front jump and R for rear jump). Also label the event that, in the train frame, is simultaneous with the rear jump and at the location of the front of the train (label this R′). *(b)* Draw a spacetime diagram in the train frame. Clearly mark the same three events as in part (a) and label all relevant worldlines. *(c)* In the train frame, the distance between the jumpers is less than the length of the train, so one of the jumpers should hit the train. Does this actually happen in the train frame? If so, which one (or both) jumper hits? If not, why not? Refer to your spacetime diagrams for the explanation.

8.4 A tortoise and a hare agree to a ten-foot race. You are the race official and you have assistants with synchronized clocks recording the locations and times of events in your frame, so all of the numbers below are

in your frame. The hare finishes in 20 nanoseconds (light travels at roughly 1 foot per nanosecond). The hare then stops and waits at the finish line for the tortoise, who takes 100 nanoseconds to get there. *(a)* Draw a spacetime diagram in your frame, including everything through the reunion of the tortoise and the hare at the finish line. Be sure to label the worldlines and important events. Label H for the hare crossing the finish line and T for the tortoise crossing the finish line. *(b)* How much time elapses on the hare's watch from the origin to H? From H to T? And from the origin to T? *(c)* How much time elapses on the tortoise's watch from the origin to T? *(d)* Redraw everything in the tortoise's frame (guesstimate rather than compute how fast the hare travels in this frame). Find and label the event H′ that (in the tortoise's frame) is where the tortoise is when the hare crosses the finish line. *(e)* Find and label H′ in the original diagram. In your frame, does H′ occur before, after, or simultaneous to H? *(f)* In the hare's frame, how much distance was there between the start and finish lines?

8.5 Alice and Bob decide to have an unusual sort of race: they will start at rest in the same spot but run in opposite directions until they reach separate finish lines 100 m apart. You will be the judge and will remain in the "stationary" frame while they each accelerate instantly to their maximum speed in opposite directions. The first one to reach a distance of 100 m in your frame wins. When the race happens, you see that they have the same speed (half the speed of light) in your frame and therefore the race is a tie. *(a)* draw a spacetime diagram in your frame. Label the space and time axes of all participants. *(b)* In Alice's frame, who finished first? Explain in words after marking on the diagram an event (call it A) that marks Alice crossing the finish line, as well as an event (call it F′) that, in Alice's frame, marks Bob's position at the instant Alice crossed the line. *(c)* Does Bob agree with Alice? Explain why or why not, and mark the diagram as necessary to support your explanation.

8.6 Repeat Problem 8.5 but for a race in which Alice and Bob start 200 m apart and run *toward* each other, finishing in the same spot halfway between their start lines. Add a part (d): why are some of the conclusions different compared to those of Problem 8.5?

8.7 Write a computer program to generate skewed graph paper for any input velocity v. You need only two loops: one to draw skewed "vertical" lines and one to draw skewed "horizontal" lines (black belts may consider combining these into one loop). Pay close attention to the spacing between lines. You may wish to include a comparison square grid.

Doppler Effect and Velocity Addition Law

We now look at the effect of motion on communications between observers. This will help us look at the twin paradox in the next chapter, and will prove crucial to understanding the effects of gravity on time.

9.1 Doppler effect basics

Imagine Alice and Bob wearing digital clocks that flash the time once per second. (The thought experiments in this chapter are inspired by N. David Mermin's book *It's About Time*.) Figure 9.1 shows Alice and Bob approaching each other at $\frac{v}{c} = \frac{3}{5}$. Because $\gamma = \frac{5}{4}$ at this speed, Alice's worldline has dots indicating clock ticks every 5/4 second in Bob's frame. The light flashes from Alice and Bob are shown as solid and dashed gray respectively.

Now we can observe how often the flashes are *received*. Bob receives Alice's flashes when they cross his worldline, and that happens *twice* per second (i.e., twice per grid cell in Figure 9.1). Likewise, from the dots marking Alice's time we see that she receives Bob's flashes twice in each of her seconds. (We will always use the emitter's clock to measure emission rates, and the receiver's clock to measure reception rates.) Messages are consistently received more frequently than they are emitted, because motion toward each other reduces the time each successive flash spends "in flight." This is one aspect of the **Doppler effect**, named after Christian Doppler (1803–53) who discovered a similar effect with sound waves. We will analyze the Doppler effect quantitatively in Section 9.3; this section focuses on conceptual understanding.

Because each flash corresponds to a clock tick, the Doppler effect determines the rate at which Alice *sees* Bob's clock ticking, and vice versa. *This marks an important addition to our thinking tools.* Until now, we always asked what was measured in Bob's frame rather than what Bob personally sees. The distinction is that Bob's frame is a *coordinate grid* while Bob himself is fixed to one grid line and cannot personally record any events away from that worldline. Recording Bob-frame coordinates for all events in practice requires a network of minions or webcams on parallel worldlines (at rest relative to Bob) with synchronized clocks, continuously recording timestamped data. With our focus on constructing coordinate grids and understanding the relationships between them, we naturally

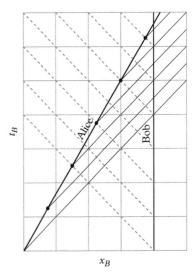

Figure 9.1 *Doppler effect: as Alice and Bob approach each other, each intercepts the other's flashes more frequently than they emit flashes, indicating that each sees the other's clock as ticking too quickly.*

The Elements of Relativity. David M. Wittman, Oxford University Press (2018).
© David M. Wittman 2018. DOI 10.1093/oso/9780199658633.001.0001

assumed this omniscient point of view. But now we turn to what each character personally sees, which is limited to events they directly pass through and light rays crossing their worldline. To get this information from a spacetime diagram, simply run your finger along a character's worldline and ignore everything else. Use this method now to confirm that in Figure 9.1 Bob *sees* Alice's clock ticking twice as frequently as his even though the *coordinate system* attached to him measures her clock as time-dilated.

Alice's clock marks time aboard Alice's rocket, so Bob must see *everything* aboard her rocket elapsing quickly as she approaches him. We study the Doppler effect in detail because it is *the* thinking tool for understanding the rate at which an observer sees time pass in other locations. This, in turn, will become important for understanding the relationship between gravity and time.

We start by defining some terms. The **frequency** of some type of event is the number of times it happens per second. Frequency is the inverse of elapsed time: $f = \frac{1}{\Delta t}$. (Test-drive this statement: if a strobe light flashes 10 times per second, how many seconds elapse between flashes?) The frequency of emission, f_{emit}, is always measured in the emitter's rest frame. We will use subscripts to label the receiver and emitter; for example, Alice receives Bob's flashes with frequency f_{AB}. In Figure 9.1 Alice receives Bob's flashes twice as frequently as f_{emit}, so $\frac{f_{AB}}{f_{\text{emit}}} = 2$; she sees Bob aging twice as fast as she would if there were no relative motion between them. The ratio of reception frequency to emission frequency is called the **Doppler ratio** and quantifies how one observer sees the other as living quickly (if this ratio is greater than one) or slowly (if this ratio is less than one). This section focuses on three fundamental properties of Doppler ratios.

Reciprocity of Doppler ratios. Whatever Doppler ratio Alice observes for Bob, Bob must also observe for Alice: $\frac{f_{AB}}{f_{\text{emit}}} = \frac{f_{BA}}{f_{\text{emit}}}$. For example, we saw in Figure 9.1 that Alice receives Bob's flashes twice per second *and vice versa*. This is no coincidence; it reflects the symmetry of two inertial observers in a universe where only relative motion matters. In fact, radar speed guns work by measuring the Doppler effect, so we can visualize this reciprocity quite easily: if Alice and Bob point speed guns at each other, each gun must display the same speed on its readout. Reciprocity implies that no one is *really* living in faster or slower time. This is an important point because in later chapters we *will* encounter asymmetric situations, and we will need to appreciate how that changes the dynamic.

Aggregation of Doppler ratios. Figure 9.2 repeats Figure 9.1 with one addition: wiggles in Bob's worldline representing Bob ducking his head to allow each flash from Alice to pass by. These wiggles make it clear that "Bob receives two flashes from Alice per second on his clock" must be a frame independent statement: no matter how we fold Bob's grid (i.e., view it in another frame) there will always be two head-ducks per grid line. Once we deduce a Doppler

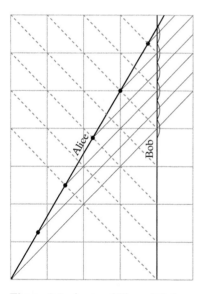

Figure 9.2 *A copy of Figure 9.2 showing Bob ducking his head (twice per Bob-second) to let Alice's flashes pass. This makes it clear that redrawing the diagram in another frame would not change the Doppler ratio.*

ratio such as $\frac{f_{BA}}{f_{emit}}$, we can be sure that all observers agree that this is indeed the Alice-Bob ratio.

Now, imagine Carol off the diagram to the right; she is purposely not shown because I want you to imagine her moving at any velocity you wish. Perhaps Carol is approaching Bob rapidly and observes Bob with a Doppler ratio of three. By the argument in the previous paragraph, Carol sees two Alice flashes passing Bob for each of Bob's clock ticks. If Carol sees Bob's clock ticking at three times its rest rate, and Alice's clock ticking twice as fast as *that*, then Carol sees Alice's clock ticking at *six* times its rest rate. If that sounds complicated, a money analogy may help: if Bob has three times as much money as Carol, and Alice has twice as much money as *that*, then Alice has $3 \times 2 = 6$ times as much money as Carol. Doppler ratios compound each other:

$$\frac{f_{AC}}{f_{emit}} = \frac{f_{AB}}{f_{emit}}\frac{f_{BC}}{f_{emit}}. \tag{9.1}$$

One of the first things we learned about special relativity is that Alice's velocity relative to Carol, v_{AC}, cannot be the simple sum of Alice's velocity relative to Bob, v_{AB}, and Bob's velocity relative to Alice, v_{BC} (Chapter 5). *But their Doppler ratios are simply multiplied.* We will exploit this in Section 9.4 to deduce the correct form of the velocity addition law.

Approaching versus receding. Now imagine a special case of this scenario in which Carol recedes from Bob at the same speed with which Alice approaches him (Figure 9.3). Carol sees Alice's clock at rest, so $\frac{f_{AC}}{f_{emit}} = 1$. Plugging this into the left side of Equation 9.1 we have $1 = \frac{f_{AB}}{f_{emit}}\frac{f_{BC}}{f_{emit}}$, which implies $\frac{f_{AB}}{f_{emit}} = \frac{f_{emit}}{f_{BC}}$. Thus, the observer receding from Bob has a Doppler ratio that is the reciprocal of the observer approaching him at the same speed. This thought experiment allows us to predict something about the formula for Doppler ratio as a function of relative velocity v (which we will derive in Section 9.3): it must turn into its own reciprocal when we change the direction of v from approaching to receding.

Check your understanding. (a) Extend Figure 9.2 for a bit more time to verify that when Alice recedes from Bob she measures a Doppler ratio of ½ rather than 2. (b) In this scenario, $v = 0.6c$ so time dilation is not a large effect ($\gamma = 1.25$). What then is the primary reason that Alice sees Bob's clock tick so slowly as she recedes?

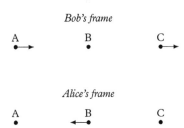

Figure 9.3 *Alice and Carol share a frame but one approaches Bob while the other recedes from him. The Doppler ratio between Bob and the approaching observer is the reciprocal of the Doppler ratio between Bob and the receding observer.*

9.2 Doppler effect and special relativity

Most aspects of the Doppler effect may be visualized as in Figure 9.4. A moving emitter E emits a flash of light in all directions. The light moves outward at speed *c* in all directions, so some time later the location of the flash may be drawn in our frame as a large circle centered on the point where the light was emitted in our frame—but the emitting object has moved on so it is no longer at the center of the circle. At a given instant in our frame, successively later flashes are represented by successively smaller circles centered successively closer to the current position of the emitter. Flashes are emitted at a regular frequency f_{emit} as measured in the rest frame of the emitter.

An observer who sees the emitter approaching (e.g., *A* in Figure 9.4) will encounter flashes more frequently than f_{emit}; an observer *R* who sees the emitter receding will encounter flashes less frequently than f_{emit}; and observers for whom the motion of the emitter is transverse to the line of sight (marked *T* in Figure 9.4) will encounter flashes about as frequently as f_{emit}. This basic picture does not rely on time dilation or any other special relativistic effect but it does rely on the invariance of *c*.

A similar picture arises when we consider sound traveling through air, but for different reasons. Sound waves travel at a fixed speed through air (determined by the air temperature) so an air-frame picture of sound waves emitted by a moving vehicle looks much like Figure 9.4. A pedestrian hears a rapidly approaching vehicle at a higher-than-typical frequency (also known as pitch), then hears it at a lower-than-typical frequency as the vehicle recedes. These effects in succession make the *nnnnyeowwww* sound so characteristic of passing vehicles, with the vehicle's rest-frame frequency audible only at the instant the vehicle passes (because motion is entirely transverse at that instant). The arrow in Figure 9.4 is about half the radius of the largest circle, so the emitter moves at half the sound speed if we think of the circles as sound emissions, or $\frac{c}{2}$ if we think of the circles as light emissions. Because everyday vehicles reach a noticeable fraction of sound speed but less than one millionth of *c*, we notice the effect on vehicle sound but not on vehicle appearance.

Figure 9.4 illustrates qualitatively how the Doppler ratio varies with observer position. Time dilation of the emitter's clock will always make the Doppler ratio somewhat smaller than one might guess based on this figure alone (we consider this quantitatively in Section 9.3). Time dilation applies equally to all observers in the figure—it has no directionality—so Figure 9.4 is still qualitatively correct, except for one detail. In the figure, observers *T* see no Doppler effect ($f_{receive} = f_{emit}$) because for them the emitter motion is purely transverse. This is close to true at everyday speeds. But at high speed, time dilation should result in a Doppler ratio measurably less than one *even for transverse observers*. This transverse Doppler effect is a hallmark of special relativity because it can only result from time dilation. Astronomers have indeed observed this effect in high-speed jets of plasma, providing additional proof that time dilation really exists.

Check your understanding. Redraw Figure 9.4 in the frame of the emitter, with arrows indicating the velocity of each observer. Satisfy yourself that this picture supports the conclusion that compared to f_{emit} observer *A* receives more flashes per unit time; observer *R*, fewer; and observers *T*, roughly the same (unless time dilation is a large factor).

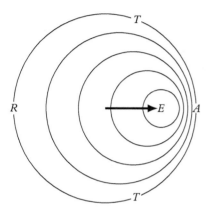

Figure 9.4 *Visualization of the Doppler effect. The motion of the emitter* E *is such that (at the instant depicted) it approaches* A, *recedes from* R, *and is transverse to either* T. *Circles depict flashes of light centered on their emission events. The approaching observer intercepts flashes more frequently than they are emitted, the receding observer intercepts them less frequently, and the transverse observers see little effect unless time dilation (not depicted) is substantial.*

Think about it

How would we draw the circles in Figure 9.4 if we had simply added the velocity of the light to the velocity of the emitter using the Galilean law?

Box 9.1 Deducing time dilation with the Doppler effect

We can *deduce* time dilation from the Doppler effect. Forget what you know about time dilation for a moment and return to the Alice-Bob-Carol setup in Figure 9.3. Absent time dilation, we focus on the distance effect: in the time between flashes emitted by Bob, the Bob-Carol distance increases by $v\Delta t_{\text{emit}}$ so the light travel time increases by $\frac{v}{c}\Delta t_{\text{emit}}$. In other words, $\Delta t_{BC} = \Delta t_{\text{emit}} + \frac{v}{c}\Delta t_{\text{emit}}$, which can be rearranged to read $\frac{\Delta t_{BC}}{\Delta t_{\text{emit}}} = 1 + \frac{v}{c}$. Meanwhile, the Alice-Bob light travel time continually *decreases* by the same amount so $\frac{\Delta t_{AB}}{\Delta t_{\text{emit}}} = 1 - \frac{v}{c}$.

Let us now admit that time dilation may be possible. We posit an unknown time dilation factor γ and solve for γ to find out how much, if any, time dilation there is. The key is that any dilation of Bob's clock lengthens Δt_{AB} and Δt_{BC} by the same factor:

$$\frac{\Delta t_{BC}}{\Delta t_{\text{emit}}} = \gamma(1 + \frac{v}{c}). \tag{9.2}$$

$$\frac{\Delta t_{AB}}{\Delta t_{\text{emit}}} = \gamma(1 - \frac{v}{c}) \tag{9.3}$$

Now, Section 9.1 showed us that the Doppler ratio Alice observes for Bob must be the reciprocal of the Doppler ratio Bob observes for Carol: $\frac{\Delta t_{AB}}{\Delta t_{\text{emit}}} = \frac{\Delta t_{\text{emit}}}{\Delta t_{BC}}$. Substituting in Equations 9.2 and 9.3 we find that

$$\frac{\Delta t_{AB}}{\Delta t_{\text{emit}}} = \frac{\Delta t_{\text{emit}}}{\Delta t_{BC}}$$

$$\gamma(1 + v/c) = \frac{1}{\gamma(1 - v/c)}$$

$$\gamma^2 = \frac{1}{(1 - v/c)(1 + v/c)} = \frac{1}{1 - v^2/c^2}$$

$$\gamma = \frac{1}{\sqrt{1 - v^2/c^2}} \tag{9.4}$$

Thus—even if we forget previous special relativistic arguments and start anew with no preconception about time dilation—we must conclude that moving clocks exhibit time dilation by the factor $\frac{1}{\sqrt{1-v^2/c^2}}$. Starting from the premises of reciprocity and invariance of c there seem to be many different routes to deduce the existence of time dilation.

9.3 Doppler law and applications

Now the pieces are in place to find the law relating velocity to Doppler ratio.
Starting with Equation 9.2 we simply plug in $\gamma = \frac{1}{\sqrt{1-v^2/c^2}}$ and find

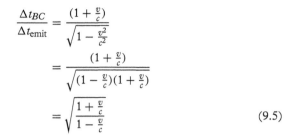

$$\frac{\Delta t_{BC}}{\Delta t_{\text{emit}}} = \frac{(1 + \frac{v}{c})}{\sqrt{1 - \frac{v^2}{c^2}}}$$

$$= \frac{(1 + \frac{v}{c})}{\sqrt{(1 - \frac{v}{c})(1 + \frac{v}{c})}}$$

$$= \sqrt{\frac{1 + \frac{v}{c}}{1 - \frac{v}{c}}} \tag{9.5}$$

Because frequency is the inverse of time,

$$\frac{f_{BC}}{f_{\text{emit}}} = \sqrt{\frac{1 - \frac{v}{c}}{1 + \frac{v}{c}}}. \tag{9.6}$$

Recall that Bob and Carol recede from each other. In contrast, Bob and Alice approach each other, so we follow a similar process starting from Equation 9.3 and find

$$\frac{f_{AB}}{f_{\text{emit}}} = \sqrt{\frac{1 + \frac{v}{c}}{1 - \frac{v}{c}}}. \tag{9.7}$$

In Section 9.1 we deduced that the Doppler ratio for a receding observer must be the reciprocal of the Doppler ratio for an observer approaching at the same speed. That prediction is confirmed here.

Equations 9.6 and 9.7 are plotted numerically in Figure 9.5. Focus on the ratio for approaching observers: it equals one at $v = 0$ and increases without bound as v approaches c. This may remind you of γ (Figure 7.6) but unlike γ the Doppler ratio ramps up linearly starting at the lowest speeds. This is because even low speeds have the direct effect of decreasing the travel distance of each successive flash. At the high-speed end, note that any Doppler ratio, no matter how large, always corresponds to a speed less than c. This will become important in Section 9.4.

To this point we have imagined flashes of light, but we may also think of light as a wave. The regular spacing between wave crests or troughs is conceptually identical to the regular spacing between the flashes we have been imagining. Wavecrests are intercepted more frequently if the source of light is approaching; although this effect is too small to be observed with the naked eye, this is called a **blueshift** because higher-frequency light is perceived as bluer. Conversely, wavecrests are intercepted less frequently if the source of light is receding, and this is called a **redshift** because lower-frequency light is perceived as redder. However, the effect is so small that we do not perceive the color equivalent of *nnnnnyeowwww*—blue to red—in passing vehicles; special devices must be constructed to measure the wavelength changes. Radar speed guns do this with the low-frequency light called radio waves; they make the target "emit" radio waves by bouncing radio waves off the target. Similarly, Doppler weather radar is designed to bounce off raindrops and has improved forecasting by revealing the velocity patterns in weather systems.

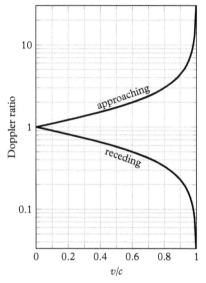

Figure 9.5 *Doppler ratios for approaching and receding observers, as a function of speed* v. *The former increases without bound as* v *nears* c.

Think about it

If a star appears red or blue to the naked eye, you are seeing its temperature—blue-hot is hotter than red-hot—rather than any Doppler effect, which is tiny at typical star speeds of 100 km/s, or ⅓₀₀₀ of c.

Astronomers use the Doppler effect as their primary source of information about the motions of all types of stars, galaxies, and gas clouds. We mount devices called spectrometers on telescopes to measure $f_{receive}$, but how do we know f_{emit}? It turns out that atoms of hydrogen emit only at certain specific frequencies, atoms of helium emit only at other specific frequencies, and so on. This gives each element a distinct frequency "bar code." This bar code remains recognizable regardless of Doppler ratio, because each frequency in the bar code is *equally* affected by the Doppler ratio (Figure 9.6). Comparing observed frequencies with laboratory samples thus reveals the elemental composition of the star or galaxy *and* its Doppler ratio, hence its velocity toward or away from us. Astronomers can use these velocities to learn much more about the cosmos, such as the masses of stars (Section 17.6) and of galaxies and galaxy clusters (Section 17.5). Spectrometers are now precise enough to detect even the tiny (roughly 1 m/s) changes in the velocity of a star caused by gravitational tugs from its planets, and as a result many planets have been discovered around other stars (Section 17.7).

Check your understanding. A police officer stops a driver who ran a red light. The driver explains that he observed it as green because of the Doppler effect. *(a)* Is this the direction of change in color we expected for a car approaching a light? *(b)* Why can we ignore this effect in real life?

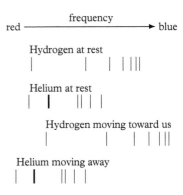

Figure 9.6 *Each element emits a unique set of frequencies, making it recognizable even after applying a substantial Doppler ratio. The effect is minute for everyday speeds (0.1c is shown here).*

Box 9.2 Low-speed approximation

If you have used the low-speed approximation to the Doppler effect in some other context, this box will help connect your prior experience to this chapter. Otherwise, you should skip this box.

The low-speed parts of the curves in Figure 9.5 can be reasonably well approximated by the linear functions $\frac{f_{receive}}{f_{emit}} \approx (1 + v/c)$ (approaching) and $(1 - v/c)$ (receding). Focusing on the former, we can write $f_{receive} \approx (1 + v/c)f_{emit} = f_{emit} + \frac{v}{c}f_{emit}$. Because v/c is very small, this indicates that $f_{receive}$ differs from f_{emit} by a small amount, hence the term *Doppler shift*. You should avoid this widely used term because it conveys the impression that Doppler effects are additive, when in fact they are multiplicative. You should always think in terms of a *Doppler ratio* rather than a shift. When you hear *redshift* or *blueshift* make the effort to think *redstretch* or *bluesqueeze*.

The low-speed approximation is misleading if applied indiscriminately: for oncoming speeds near c it implies $f_{receive} \approx 2f_{emit}$ but the full expression tells us that $f_{receive}$ increases without limit at speeds near c. The low-speed approximation also obscures the reciprocal nature of Doppler ratios for approaching and receding emitters. I encourage you to unlearn this approximation.

9.4 Einstein velocity addition law

We now have all the tools to work out the velocity addition law in special relativity, also known as the Einstein velocity addition law. To review the need for such a law, suppose Bob observes Carol and Alice approaching him from opposite directions; Carol with speed $v_{CB} = 0.6c$ and Alice with speed $v_{BA} = 0.8c$. In Bob's frame the distance between Alice and Carol decreases at a rate $v_{CB} + v_{BA} = 1.4c$, but what would Alice and Carol measure for their relative velocity v_{CA}? In Galilean relativity they would measure $1.4c$, but in Chapter 5 we saw that the invariance of c must cause some kind of diminishing return such that $v_{CA} < v_{CB} + v_{BA}$. The Doppler effect provides two useful tools for addressing this problem: a relationship between relative velocities and Doppler ratios, and the fact that adding velocities must be equivalent to *multiplying* Doppler ratios.

According to the Doppler law (Section 9.3) Carol observes Bob with a Doppler ratio $\frac{f_{CB}}{f_{emit}} = 2$, and Bob observes Alice with a Doppler ratio $\frac{f_{BA}}{f_{emit}} = 3$. And because Doppler ratios multiply (compound) each other (Section 9.1), Carol must observe Alice with a Doppler ratio $\frac{f_{CA}}{f_{emit}} = \frac{f_{CB}}{f_{emit}} \frac{f_{BA}}{f_{emit}} = 6$. All that is left is to convert this Doppler ratio between Carol and Alice back to a relative velocity using the Doppler law. Figure 9.5 shows that $v_{BC} \approx 0.95c$ would yield a Doppler ratio of 6, but let us try for a more exact solution.

Writing out the expressions for f_{CA}, f_{CB}, and f_{BA}, we see that

$$\frac{f_{CA}}{f_{emit}} = \frac{f_{CB}}{f_{emit}} \frac{f_{BA}}{f_{emit}}$$

$$\sqrt{\frac{1 + \frac{v_{CA}}{c}}{1 - \frac{v_{CA}}{c}}} = \sqrt{\frac{1 + \frac{v_{CB}}{c}}{1 - \frac{v_{CB}}{c}}} \sqrt{\frac{1 + \frac{v_{BA}}{c}}{1 - \frac{v_{BA}}{c}}}$$

$$\frac{c + v_{CA}}{c - v_{CA}} = (\frac{c + v_{CB}}{c - v_{CB}})(\frac{c + v_{BA}}{c - v_{BA}}) \tag{9.8}$$

This expresses the relation between v_{BA}, v_{CB}, and v_{CA} in a multiplicative way, and in a way that makes it clear that all these velocities share the same relationship with c. A bit more algebra to isolate v_{CA} on one side and everything else on the other side yields

$$v_{CA} = \frac{v_{CB} + v_{BA}}{1 + \frac{v_{CB}}{c} \frac{v_{BA}}{c}} = \frac{v_{CB} + v_{BA}}{1 + \frac{v_{CB} v_{BA}}{c^2}} \tag{9.9}$$

This is the Einstein velocity addition law. Equation 9.9 is the most practical form of the law, but Equation 9.8 more elegantly displays the logic behind the relationship: *adding two velocities is equivalent to compounding two Doppler ratios*. Equation 9.8 has the additional advantage that the cumulative effect of any number of "boosts" is found by simply chaining more terms onto the end, whereas the structure of Equation 9.9 prevents this.

We now see a complementary explanation for the "diminishing returns" effect first noted in Chapter 5 and explained in Chapter 6 by the skewed grids: adding velocities is equivalent to multiplying Doppler ratios, but the final Doppler ratio, no matter how large, will always map back to a relative velocity less than c. To deduce all this we needed only the invariance of c and reciprocity (which ultimately comes from the principle of relativity).

Think about it

Plugging the speeds in our story, $0.6c$ and $0.8c$, into Equation 9.9 yields $v_{CA} \approx 0.946c$, confirming the graphical result from Figure 9.5 with more precision.

Check your understanding. (a) The Galilean velocity addition law works well at everyday speeds, so for $v \ll c$ the Einstein law must closely mimic the Galilean law. Confirm this for one specific example such as $v_{CB} = v_{BA} = 0.1c$. *(b)* The Einstein law must predict that c added to any velocity still yields only c. Confirm this for one specific example, such as adding c to $0.5c$.

Box 9.3 Time runs more slowly in the basement

To highlight the importance of the Doppler effect as a thinking tool, here is a quick preview of Chapter 13, where we use the Doppler effect to show that gravity makes time run slowly. Imagine being on the floor (F) of an accelerating rocket, with a flashing light source in the nose cone (N). The acceleration causes you to gain speed on the emitting frame while each flash is in transit, so $f_{FN} > f_{\text{emit}}$. The flashes correspond to clock ticks, so floor observers see nose-time as running more quickly than their own. Observers in the nose accelerate *away* from the emission frame of a light on the floor, so $f_{NF} < f_{\text{emit}}$; nose observers see floor-time running slowly. All observers agree that time runs more slowly on the floor of the rocket—admittedly by an extremely small amount for today's rockets.

Back on Earth, gravity accelerates everything in your vicinity equally, so your vicinity can be considered an accelerated frame just like the rocket. Therefore, time runs slower in the basement—by an extremely small amount, but this effect *has* been detected in the laboratory.

This is far from a complete discussion, but it should whet your appetite for things to come.

CHAPTER SUMMARY

- The frequency of an approaching source of light is measured to be $\sqrt{\frac{1+v/c}{1-v/c}}$ times the frequency measured in the rest frame of the source. For a receding source, reverse the signs on v; this leads to the reciprocal ratio.

- This further implies that *all* time-dependent phenomena in the source frame appear to speed up or slow down by the same ratio.

- The Doppler effect is distinct from time dilation. Time dilation is measured by a network of synchronized clocks through which the test clock passes, whereas the Doppler ratio is what a single observer sees from a single position.

- The Einstein velocity addition law $v_{CA} = \frac{v_{CB}+v_{BA}}{1+v_{CB}\,v_{BA}/c^2}$ describes the relative velocity measured by two observers, A and C, given each of their velocities relative to some third observer B. At low speeds ($v_{CB} \ll c$ and $v_{BA} \ll c$) the second term in the denominator is negligible so this reduces to the familiar Galilean law.

☰ FURTHER READING

Astronomy abounds with applications of the Doppler effect. Almost any introductory astronomy text or online educational resource will explore those applications in more detail, along with pretty pictures.

It's About Time by N. David Mermin provides a completely different proof of the velocity addition law based on careful consideration of distance and time measurements of a race between a flash of light and a ball. In a separate chapter, Mermin takes the reader through a particularly elegant treatment of the Doppler effect.

Spacetime Physics by Edwin F. Taylor and John Archibald Wheeler is a classic comprehensive introduction to special relativity for mathematically proficient students. Taylor and Wheeler work out the velocity addition law in much more detail, including versions that specify how angles add on spacetime diagrams in Galilean relativity as well as special relativity.

◪ CHECK YOUR UNDERSTANDING: EXPLANATIONS

9.1   (a) The black dots in the diagram below mark Alice's clock ticks. After crossing Bob's worldline she receives light rays half as frequently as her clock ticks.

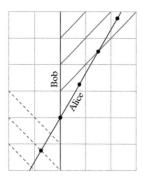

(b) Each flash must cover a successively longer distance from Bob to Alice.

9.2   In the emitter frame the circles are concentric but the labeled observers are moving to the left. Observer A intercepts more circles per unit time because she is moving toward the expanding circles, while the opposite is true for observer R. The motion of the T observers is neither toward nor away from the expanding circles, so the rate of flash reception is unaffected.

9.3   (a) Yes. Moving toward the light increases the observed frequency, and higher frequencies appear bluer to the eye. (b) The speed of a car is so much less than c that the effect is tiny. If cars moved literally one million times faster, this would be a substantial fraction of c and would cause a substantial change in apparent color.

9.4   (a) $\frac{0.1c+0.1c}{1+0.1\times0.1} = \frac{0.2c}{1.01} = 0.198c$. This is very close to the $0.2c$ predicted by the Galilean law. (b) $\frac{0.5c+c}{1+0.5\times1} = \frac{1.5c}{1.5} = c$.

? EXERCISES

9.1 Does the transverse Doppler effect always cause a redshift, always a blueshift, or does it depend on the situation?

9.2 Some roads have electronic signs that display your driving speed prominently. These are presumably connected to radar speed guns. *(a)* Should the speed guns be pointed parallel to the road, perpendicular to the road, or something in between? *(b)* A speed gun is placed in the orientation you chose, and a car drives with constant velocity. Describe how the displayed speed varies with the car's position. *(c)* Why do drivers see very little variation in the readout?

9.3 In deriving the Einstein addition law, we assumed the two velocities to be added were parallel or antiparallel (i.e., in opposite directions). Explain how the reasoning leading to Equation 9.9 is no longer valid if the two velocities are at some other angle.

9.4 Adding velocities is equivalent to multiplying Doppler ratios. Use this to explain why the Doppler ratios of approaching and receding observers must be multiplicative inverses.

9.5 Explain why an addition law of the form $v_{CA} = v_{CB} + \frac{v_{BA}}{1+v_{CB}\,v_{BA}\,/\,c^2}$ will not work even though it enforces the "diminishing returns" idea emphasized in Chapter 5. *Hint:* think about reciprocity.

9.6 In Figure 9.1 the light rays from Alice seem closely packed, as if she is emitting them quite frequently. Explain how to read the diagram to verify that (as expected from time dilation) more than one Bob-frame second passes between the emission of Alice flashes.

9.7 *(a)* On Figure 9.1 add a worldline for another character at rest relative to Bob. Show that this character has a Doppler ratio of 1 with Bob and 2 with Alice. *(b)* On Figure 9.1 add a worldline for another character at rest relative to Alice. Show that this character has a Doppler ratio of 1 with Alice and 2 with Bob.

9.8 Show that the Einstein velocity addition law yields a final velocity of c if c is added to *any* v. This requires algebra, but only a little.

+ PROBLEMS

9.1 Show how to obtain Equation 9.9 from Equation 9.8.

9.2 Reproduce the logic of Section 9.3 to fill in all the steps leading to Equation 9.7.

9.3 Show that at low v the approximation $1 - v/c$ produces results very close to those of Equation 9.3. You may use an empirical approach (computing a few examples) or a mathematical approach if you know how to do a Taylor expansion. At what v does the approximation begin to differ appreciably from the true value?

9.4 Using a graphical approach, argue that *(a)* the Doppler ratio must become arbitrarily small as an emitter's recession velocity approaches c; *(b)* the

Doppler ratio must become arbitrarily large as an emitter's approach velocity approaches c.

9.5 You are standing by the side of an interstellar highway when a fast spaceship ($\gamma = 2$) speeds by. You have a speed gun that receives a known frequency broadcast by the ship, and you keep the gun pointed at the ship at all times as it approaches, passes, and recedes. Graph the Doppler ratio you measure as a function of time, centering your graph on the moment the ship passes you (call this $t = 0$). Think carefully about the correct numerical values at $t = 0$ and at the earliest and latest times; in between those points, fill in the transition qualitatively. *Optional variation:* write a code to graph this exactly.

9.6   Draw a spacetime diagram of this situation:

A	B	C
•—→	•	←—•
0.6c		0.6c

including regular light rays ($\Delta t_{emit} = 1$) from Alice to the right. Verify that the Alice-Bob and Alice-Carol Doppler ratios match the values you would predict from the equations in this chapter.

9.7   Draw a spacetime diagram of this situation:

A	B	C
←—•	•	•—→
0.6c		0.6c

including regular light rays ($\Delta t_{emit} = 1$) from Alice to the right. Verify that the Alice-Bob and Alice-Carol Doppler ratios match the values you would predict from the equations in this chapter.

9.8   If you know Carol's γ factor relative to Bob, and Bob's γ factor relative to Alice, is there a way to predict Carol's γ factor relative to Alice? First, make a prediction as to whether the γ factors would add, multiply, or something else. Any prediction is acceptable as long as you state your assumptions about the directions of motion, make a specific prediction, and present an argument that it is plausible. Next, do some research on the Internet, find the correct law, and explain the extent to which you were right or wrong.

The Twin Paradox

10

Alice and Bob are the same age. Alice climbs aboard a rocket and travels across the galaxy at a speed close to c while Bob stays home. Alice measures Bob's clocks as ticking slowly, and vice versa. At a certain point, Alice returns to Earth at the same speed. As in the outbound journey, she measures Bob's clocks as ticking slowly, and vice versa. So when she returns to Earth, who is younger (i.e., whose clock *really* ticked slowly), or are they the same age? If they are the same age, how can it be that anyone's clock ticked slowly? In this chapter we will come to understand this most famous "paradox" of special relativity.

10.1 Alice and Bob communicate

This story feels like a paradox only if we mistakenly apply rules we deduced for inertial frames to Alice's distinctly *non*inertial motion. Because Alice changed velocity (at her turnaround) and Bob did not, basic thinking tools such as reciprocity will fail. (Reminder: Alice cannot claim "Bob is the one that accelerated relative to me" because we have objective tests for acceleration such as engines firing, coffee sloshing, etc.) So we can quickly answer "who is really younger" by sticking to an inertial frame such as Bob's, where the answer is clearly that Alice is younger due to time dilation. But this is far from a complete explanation; we would like to know what Alice observes along the way and whether she is surprised at the reunion to see that Bob is older rather than younger. Along the way, we will develop thinking tools for accelerated frames.

We start by looking at how Alice and Bob would see each other age throughout the story. Imagine that Alice and Bob send each other birthday party invitations (using radio waves traveling at c) each time one year elapses in their frame. Alice travels for four years in her frame at $v = 0.6c$ before turning around. Because $\gamma = 5/4$ at this speed, Alice's turnaround occurs at $t_B = 5$ as shown in Figure 10.1. Please study this figure to verify that Alice sends an invitation every $5/4$ years in Bob's frame.

Figure 10.1 makes it clear that well before the reunion (where the two worldlines meet at the top of the diagram) Alice has received nine invitations from Bob (red lines) but sent only seven (blue lines), and that Bob has sent nine but received only seven. The reunion event itself coincides with Bob's tenth birthday. At this event Bob hands Alice the tenth invitation; she is not surprised because she has received nine previous invitations and this one arrives right

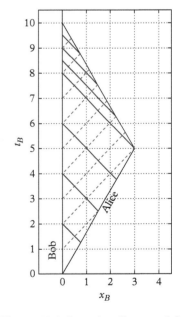

Figure 10.1 *Spacetime diagram of the twin paradox, with Alice and Bob each emitting birthday party invitations.*

The Elements of Relativity. David M. Wittman, Oxford University Press (2018).
© David M. Wittman 2018. DOI 10.1093/oso/9780199658633.001.0001

on schedule, judging by the frequency of invitations she received on the return journey. Similarly, the reunion coincides with Alice's eighth birthday, and Bob is not surprised because he was expecting the eighth invitation at this time. The ratio of birthdays is $10/8 = 5/4$, and *neither twin is surprised that Alice is younger when they reunite*. Alice did receive Bob's invitations at only half the normal rate (a Doppler ratio of $1/2$) on the outbound leg of her journey: only two birthday cards in four years. Up to her turnaround Alice could legitimately claim that Bob was aging more slowly. But on her return journey the Doppler ratio was two (the inverse of the outbound ratio), so she received eight invitations in four years for a grand total of ten (counting the one hand-delivered at the reunion event, with a worldline too short to be seen on the spacetime diagram). According to Alice, Bob's apparently high birthday rate during her return leg more than made up for his apparently low birthday rate on the outbound leg.

To recap, Alice received invitations once every two years for the first half of her journey, and two invitations per year for the second half; this adds up to more birthdays than a steady one per year. In contrast, Bob received invitations once every two years for *most* of the time (his first eight years) and then twice per year for *some* of the time (his final two years); this adds up to *fewer* birthdays than a steady one per year. There is no reciprocity because Alice's bent path through spacetime is qualitatively different than Bob's; it allows her to capture more of Bob's invitations than vice versa as illustrated by Figure 10.1. Furthermore, Alice's path will be bent no matter how we imagine "tilting" the spacetime diagram to represent any other inertial frame, so this is a frame-independent result.

Another useful thinking tool is having a triplet, Carol, ship out with Alice and continue forever at the same velocity. Carol sees Bob aging consistently slowly, with a steady Doppler ratio of $1/2$. Up to Alice's turnaround, Carol agrees that she and Alice are aging more rapidly than Bob. However, on Alice's return journey her invitations to Carol experience a *compounded* Doppler ratio of $1/4$. Carol sees Alice age *so* slowly on the return leg that Bob's birthdays soon outnumber Alice's. So *no* character is surprised that Alice is younger than Bob at the reunion.

Tracking the communications gives us greater confidence that we have the correct answer, but it may be difficult to shake the feeling that *something* is not right. The asymmetry feels disturbing, but this is only because we have not yet practiced analyzing asymmetric situations. In the next section, we will home in on the asymmetry by examining Alice's experience in detail.

Check your understanding. On a copy of Figure 10.1, draw Carol's worldline as well as the birthday invitations Carol receives from Alice on her return journey. Convince yourself that over the entire journey Carol receives two invitations from Bob for each one she receives from Alice.

10.2 What Alice observes

The key to understanding Alice's experience, as in so many "paradoxes" of relativity, is time skew. Before reading this section, you may wish to review Figure 6.5 and/or Section 8.5, which show how, at a given instant in Alice's frame, trailing clocks in another frame read progressively later times. In particular, Figure 8.11 is conceptually the same as each half of the journey. Once you have mastered that figure, this section will be straightforward.

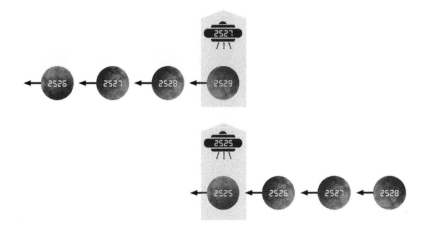

Figure 10.2 *Alice's ship clock and Galactic Standard clocks (planets) at one instant in Alice's frame at the outset (bottom) and at the end (top) of her outbound journey. Shaded areas highlight the planet clock most directly comparable to Alice. Between snapshots, two years pass on Alice's clock, versus one on each planet; Alice measures planet clocks as dilated. Nevertheless, the planet frame also measures Alice as time-dilated: the planet clock nearest Alice at the end reads four years more than the planet clock nearest Alice at the start.*

Imagine Alice passing a series of planets at rest relative to Earth; these planets synchronize their clocks with Earth to form Galactic Standard Time. In this section, each planet will display its time to the nearest year. Alice will move east on her outbound journey and west on her return, so Earth will always be the westernmost planet shown. Furthermore, each planet maintains unique surface markings throughout the figures in this section, so you can track any planet you wish. The story in this section will be that Alice travels each leg of her journey at constant velocity for four years on Galactic Standard clocks, while two years elapse on Alice's clock. The ratio of coordinate time to Alice's proper time (Section 7.2) is therefore $\gamma = 2$. To keep things simple, the figures will render the time skew as one year per planet spacing; we will not worry about what planet spacing is necessary to achieve this.

Alice boards her space yacht in the year 2525. As soon as she is moving at $0.866c$, the readings of Galactic Standard clocks at one instant in her frame are shown in the bottom snapshot of Figure 10.2. Alice cannot *see* all these clocks at that instant, but in principle she can construct this diagram after the fact by analyzing data collected *at the location of each clock* by hypothetical assistants in her frame. Two Alice-years later (top snapshot in Figure 10.2), Alice finds that the

Think about it

In practice, planets have some relative motion that prevents full synchronization, but their speeds are so slow compared to c that we can ignore this here.

Think about it

Details for aficionados: $\gamma = 2$ implies $v = 0.866c$, which further implies a hefty amount of time skew—in Alice's frame successive planet clocks will differ by one year if they are separated by little more than one light-year. This is a reasonable distance between solar systems, but the figures are not to scale; the planets should be roughly one billion planet diameters apart. You can find the exact (galaxy-frame) planet spacing from Alice's one-way travel distance of $0.866 \times 4 = 3.464$ light-years.

easternmost planet is now at her position, and all planet clocks have advanced by *one* year because in Alice's frame the *planet* time is dilated. Study this figure closely, and verify that *each* planet clock advanced one year even though Galactic Standard observers measure *four* years advancing on the clock nearest Alice. This measurement involves *different* planet clocks, but in the planet frame that is perfectly acceptable; the clocks are synchronized so it cannot matter which clock is used.

Figure 10.3 *Momentarily stopped at the turnaround, Alice finds all planet clocks synchronized again at the value of the local planet clock. They read four years since her journey started, despite only two years elapsing on her own clock.*

So far, we have followed the concepts discussed in Section 8.5 and illustrated in Figure 8.11, but when Alice turns around for the return journey we see something qualitatively new. In the midst of her turnaround Alice is briefly at rest relative to the planet frame. This eliminates time skew, so at this Alice-instant all planet clocks read 2529 just as the local clock does (Figure 10.3). Therefore, Alice *really is* in the year 2529 despite having traveled only two years of her time. Alice is already younger than Bob, and the return journey will do nothing more than double the size of the effect by repeating the process. The key action was changing frames, also known as accelerating. The acceleration did not actually affect any planet clock, but it changed the set of distant events Alice considers simultaneous to the "here and now" event.

Figure 10.4 reviews the clock readings on the return journey, in the same format as Figure 10.2 did for the outbound journey. When Earth arrives back at

Figure 10.4 *Alice and planet clocks at one instant in Alice's frame at the start (bottom snapshot) and end (top snapshot) of her return journey. As on the outbound leg, in this frame two years pass on Alice's clock and one year on each planet clock, but you can also see that the planet clock closest to Alice advances by four years.*

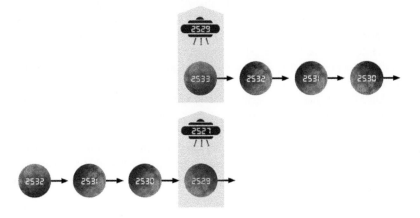

Alice, its clock reads 2533 (four years since the turnaround and eight years since the start) while Alice's has advanced only to 2529 (two years since the turnaround and four years since the start). Alice fires her engine to come to rest relative to Earth. After stopping, Earth's clock still reads 2533—and Alice now considers distant clock readings of 2533 to be simultaneous with "here and now" because she is at rest in the galaxy frame (Figure 10.5). She has traveled to the galaxy-frame future faster than those who remained continuously at rest in the galaxy frame.

As a reminder, Alice does not personally observe all the clock readings depicted in this section. The point is that she does observe a consistent picture: each successive planet passed displays a successively later time due to time skew. This makes no physical difference until she comes to rest in the planet frame, which makes time skew disappear and rejoins her to the planet frame in the advanced year displayed by the nearest planet clock. She is therefore not surprised at Earth's clock reading and Bob's advanced age.

Check your understanding. Explain why time skew on the return journey did not cancel out the effect of time skew on the outbound journey.

Figure 10.5 *Earth's clock reads 2533 as at the end of Figure 10.4, but with Alice at rest in the planet frame, other planet readings of 2533 are now simultaneous. In the round trip, planet clocks have advanced from 2525 to 2533 (eight years) while Alice has aged only four years.*

10.3 Changing frames

It seems like something magical happened during Alice's accelerations. How, for example, could the planets' clocks suddenly become synchronized when she decelerated at the turnaround? This happens because changing frames amounts to skewing the spacetime grid. By changing the skew of the grid *in the midst of the story*, Alice introduces an effect we have not seen in the other "paradoxes" of relativity.

Figure 10.6 illustrates the effect on a spacetime diagram. Focus first on the left panel, which shows only the outbound part of the story, with Bob's spacetime grid dashed and Alice's solid. (A few details of the story have been changed to make the red grid more readable. First, Alice's speed is now "only" $0.7c$ because higher speeds stretch the grid beyond readability. Second, Alice moves outbound for only one of her years so we can focus on just a few grid cells.) Confirm that the turnaround event occurs at $t_A = 1$ by finding event T and noting that it occurs at the first solid grid intersection after the origin. The line running through T and T' is the set of events simultaneous with T in Alice's outbound frame; in particular, T' is at Bob's location ($x_B = 0$) and simultaneous with T in Alice's outbound frame. So far, this is similar to other high-speed situations we have examined.

Now for the new part: the right panel shows the entirely new grid of Alice's returning frame (the outbound grid remains but is faded). In this frame the lines connecting simultaneous events skew in the opposite direction, and event T'' is the event at $x_B = 0$ that is simultaneous with T. When Alice changes frames at event T, she switches from finding that event T' (at which only 0.7 years have elapsed for Bob, $t_B = 0.7$) is happening "now" at Bob's location, to finding that

Figure 10.6 *Spacetime diagram of Alice moving away from Bob at 0.7c for one of her years (left) and then returning (right). The turnaround event* T *is simultaneous with* T′ *in Alice's outbound frame but in her returning frame the turnaround is simultaneous with* T″. *If Alice pivots instantly at* T, *then for her no time elapses between* T′ *and* T″ *while Bob ages a great deal. In practice, the pivot cannot be instantaneous but still results in a "shortcut through time."*

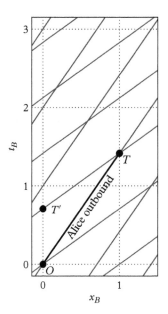
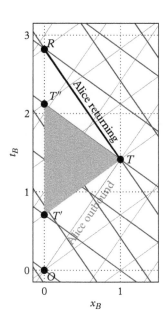

Think about it

Why did Alice age only 0.8 years less than Bob if 1.4 planet-years were skipped between T′ and T″? Alice would say that Bob aged less rapidly than she did (due to time dilation) throughout the *other* parts of the story. In Bob's frame the question never arises because Alice's time dilation explains everything.

event T'' (at which 2.1 years have elapsed for Bob, $t_B = 2.1$) is happening "now" at Bob's location. In Bob's frame, 1.4 years elapses between T′ and T″, which is why Bob is older at the end. He experienced that time and Alice did not. The shaded area illustrates how Alice's pivot allowed her to "fast-forward" through a segment of Bob's life.

Alice did not personally skip from T′ to T″; she was nowhere near those events. But changing frames did allow her to take a shortcut through time in the sense that she traveled 2.8 years into Bob's future by traveling 2.0 years into her own future. This particular shortcut through time required her to go very far out of her way in space! We will explore the connection between "speed through time" and speed through space in Section 10.4.

In practice, Alice does not *instantly* transition between outbound and returning frames. Changing frames means undergoing an acceleration, which means applying a force. A large change in velocity requires applying a substantial force for a substantial period of time, even though this is represented schematically in Figure 10.6 by the single point T. As Alice accelerates, the spacetime grid representing her frame folds like an accordion so the skewed grid in the left panel of that figure becomes the differently skewed grid in the right panel of that figure. This process requires effort, and we have a word for the difficulty of changing frames: inertia.

Did the acceleration *cause* Alice to be younger? Alice certainly needed the acceleration to become younger than Bob in any frame-independent sense, but the amount of acceleration is unimportant by itself. To see this, imagine that Carol initially goes with Alice but soon changes her mind; she returns to Earth promptly and remains with Bob until the Alice-Bob reunion event *R*, as shown in Figure 10.7. Carol undergoes the same accelerations as Alice, but ages nearly as much as Bob. For a fixed velocity change at the turnaround, the size of the "bite out of time" depicted by the shaded areas in Figures 10.6 and 10.7 is determined entirely by the length of the coasting legs of the journey. Conversely, Carol can match Alice's aging while accelerating much more if she changes her mind repeatedly and makes a sawtooth pattern between events *O* and *R* on Figure 10.7. Clearly, the connection between the amount (and/or duration) of acceleration and the size of the "shortcut through time" is a loose one; saying that acceleration *causes* less aging may imply too strong a connection. It is more accurate to say that acceleration *enabled* Alice to age less than Bob between *O* and *R*.

Check your understanding. (a) Verify on Figure 10.6 that Alice aged one year in 1.4 planet-years on the outbound journey. *(b)* Verify on Figure 10.6 that Alice aged one year in 1.4 planet-years on the return journey.

10.4 Principle of longest proper time

Section 8.5 showed how time dilation reciprocity works at constant velocity: Bob uses two of his clocks to determine Alice's clock tick rate while Alice uses two of her clocks to determine Bob's clock rate. In this chapter, Alice's change of velocity finally allows the same two clocks to be compared at the departure and reunion events, but it also destroys the symmetry of the comparison. This highlights an important distinction—first mentioned in Section 7.2—between elapsed proper time $\Delta\tau$ and elapsed coordinate time Δt. My wristwatch moves around with me even if I accelerate, and thus does not necessarily measure the time in any consistent coordinate system: this is my proper time. In contrast, coordinate time by definition follows some consistent coordinate system such as the square or skewed grids in our spacetime diagrams.

Until this chapter, the moving clocks we studied were always at rest in *some* inertial coordinate system, so the distinction between proper and coordinate time was blurred. Now it becomes crucial: if a clock accelerates it fails to measure the time in *any* inertial coordinate system, so the time it displays can *only* be interpreted as its proper time. The flip side of this is that its proper time is defined *only* at events along its worldline.

Figure 10.8 illustrates how proper time accumulates for Alice if she accelerates in various ways. The key is to break any journey into segments with (at least approximately) constant velocity, because we know how to compute ticks on a constant-velocity clock: Equation 7.4 tells us that $\Delta\tau=\sqrt{1-v^2/c^2}(\Delta t)$. To simplify our computation we note that $v^2\equiv\left(\frac{\Delta x}{\Delta t}\right)^2$ so we can write $\Delta\tau=\sqrt{(\Delta t)^2-(\Delta x)^2/c^2}$. Furthermore, we will use units of years for t and

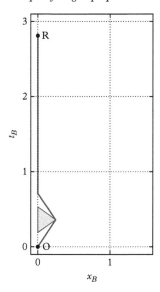

Figure 10.7 *Carol launches with Alice but turns around sooner to demonstrate that the size of the shortcut through time is not determined by the amount of acceleration. Carol undergoes the same acceleration as Alice (compare to Figure 10.6) but ages nearly as much as Bob.*

Confusion alert

In addition to distinguishing between proper and coordinate time, a problem may require us to think about coordinate times in different coordinate systems, and/or proper times measured by different characters.

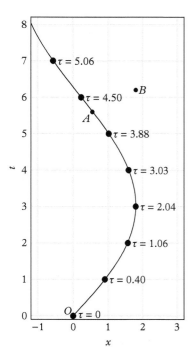

Figure 10.8 *A noninertial path through an inertial coordinate system. Numbers along the path indicate accumulated proper time τ at Δt = 1 year intervals of coordinate time; this is computed by applying the rules for inertial clocks along approximately straight (inertial) worldline segments. Events* A *and* B *are discussed in* Check Your Understanding *at the end of the section.*

Think about it

The concepts of speed through time and through space become even more parallel if we define the latter as $\frac{\Delta x}{\Delta \tau}$ rather than $\frac{\Delta x}{\Delta t}$. This redefinition makes no difference at everyday speeds where t and τ are indistinguishable, but physicists find it useful at high speed.

light-years for x so $c = 1$ and we can write simply $\Delta \tau = \sqrt{(\Delta t)^2 - (\Delta x)^2}$. Setting Alice's clock to read $\tau = 0$ at the origin O for convenience, we split her journey into segments of $\Delta t = 1$ year. In the first square of time in Figure 10.8 she crosses $\Delta x = 0.92$, which yields $\Delta \tau = 0.40$ (you may also interpret this a speed of $0.92c$ causing substantial time dilation). In the next unit of coordinate time she crosses $\Delta x = 0.75$, which yields $\Delta \tau = 0.66$, for an accumulated total of 1.06 years of proper time. In the third year of coordinate time, Δx is quite small so Alice ages nearly a full additional year of proper time. This computation continues all along the worldline in Figure 10.8. Modeling a curved worldline as a set of straight segments is admittedly an approximation, but we can calculate the proper time to any precision we desire by further subdividing the path into more segments that more closely track the curve. Make sure you understand this divide-and-conquer strategy, called **integration**, because it will appear in future chapters as well.

Two aspects of this process are remarkable. First, the relation $\Delta \tau = \sqrt{1 - v^2/c^2}(\Delta t)$ ensures that elapsed proper time is always less than (or at most equal to) the elapsed coordinate time. Because this is true for each segment, it is also true for the total journey. This extends our twin-paradox conclusion (the accelerating twin ages less) to arbitrarily complicated motions. The accelerating twin can even travel in circles because the direction of motion does not affect the v^2 term. In fact, particle accelerators force subatomic particles to travel in circles rapidly, and those particles do age slowly as measured by their radioactive decay rates.

Second, Alice moves through the time coordinate in a way that is linked to her motion through the space coordinate. In a given bit of proper time, a greater speed through space results in a greater advancement through the time coordinate as well. For example, Alice crosses a substantial amount of space in her first year of proper time in Figure 10.8, and as a result she crosses nearly *two* full years of the time coordinate. In her second year of proper time, she crosses very little space, and as a result her "speed through time" falls to approximately one year of coordinate time per year of proper time. Just as speed through space involves displacement in the space coordinate divided by some measure of time, we define speed through time as $\frac{\Delta t}{\Delta \tau}$; you may recognize this as γ. Your γ describes your progress through the time coordinate per unit proper time, so increasing your γ allows you to fast-forward into the future. This is a limited form of time travel; proper time moves inexorably forward so you cannot travel to the past.

Your speed through the t coordinate depends on the coordinate system chosen, so a large γ does not by itself mean that you can quickly see your twin as an old man. You have to *change* your velocity at some point to do that. And that leads to another important property of inertial frames. The Earth-frame explanation of the twin story is that Bob ages more because by keeping $v = 0$ he keeps $\Delta \tau = \sqrt{1 - v^2/c^2}(\Delta t)$ as large as possible. All other paths between the departure and reunion events involve *some* nonzero v for some part of the time,

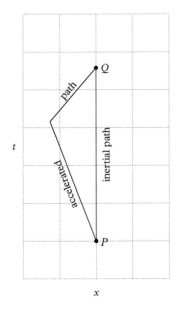
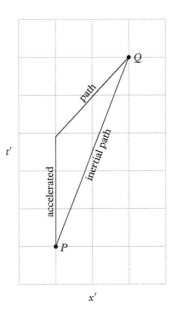

Figure 10.9 *Comparison of inertial and accelerated journeys in two frames. In the frame on the left, we know the stationary observer records more proper time from event* P *to* Q; *reframing it on the right makes it clear that maximizing proper time is actually a property of inertial paths, not just stationary paths.*

and therefore a smaller $\Delta\tau$. Of all possible travelers between those events, Bob has the maximum proper time. Now, it is tempting to say that Bob maximized proper time by remaining stationary, but "stationary" is a frame-dependent statement. He was stationary in one inertial frame, so more generally we can say he followed an inertial path. Paths that deviated from his and then veered back to the reunion event are inevitably noninertial. So *inertial paths are paths of longest proper time.*

Figure 10.9 recaps this reasoning graphically. First, convince yourself that the two panels illustrate the same story—two routes from event P to Q—in two different frames. Now, imagine you were given the panel on the right and asked to determine which route has more proper time: the answer may not be obvious without doing a multistep calculation. But the answer *is* obvious in the left panel, where it looks like the twin story. The story may look different in frame on the right, but it has exactly the same physical structure. This thinking tool helps us see that maximizing proper time has nothing to do with remaining stationary— rather, it is the hallmark of *inertial* paths. This principle will guide us when we think about motion in the presence of gravity.

Check your understanding. (a) For event A in Figure 10.8, what is the proper time on the moving clock and what is the coordinate time? *(b)* In the same figure, estimate the proper time elapsed on a hypothetical clock traveling *inertially* from O to A. *(c)* What happens if we try to estimate the proper time on an inertial clock traveling from event A to event B?

Confusion alert

"Maximizing" proper time refers to selecting from all possible worldlines from event A to event B; it does *not* imply any comparison with coordinate time.

Box 10.1 Interstellar travel

The Milky Way galaxy is 100,000 light-years across, so you might think you could cross it in no fewer than 100,000 years, even traveling near the speed of light. Indeed, such a trip would take 100,000 years in the frame of the galaxy, but because of time dilation it could take far less time in your frame. To you, the trip would be short, not because you were traveling faster than the speed of light, but because the galaxy was contracted.

This makes interstellar travel seem more practical than it might be without relativity. The nearest star is a bit more than four light-years away so the trip would take roughly four years in the galaxy frame for speeds near c. But a time dilation factor of $\gamma = 12$ (requiring $v = 0.9965c$) would cut this to four *months* in the crew frame. The same factor could put more distant stars in reach without the social complexity of a mission spanning multiple generations.

But nothing is free. We will see in Chapter 12 that even reaching $\gamma = 2$ requires an enormous amount of energy (a 5-ton spacecraft would require the current annual world energy consumption, with no energy left for slowing down at the end of the journey), and larger γ requires proportionally more energy. At the very least, it would be impractical to store the fuel *on* the ship. The acceleration technology would have to rely on lasers pushing from Earth, scooped-up interstellar gas, or even more fanciful schemes. And in the ship's frame, bits of space debris will have enormous incoming energy, so the ship will require a great deal of shielding.

Box 10.2 Relativity in popular culture

The song *'39* by Queen is about a group of brave explorers who leave in the year '39 (2039? 2139?) to find new planets for mankind. After a journey of less than one of their years, they return to a much older Earth and reunite with loved ones:

> *But my love this cannot be*
> *Oh so many years have gone*
> *Though I'm older but a year*
> *Your mother's eyes from your eyes cry to me*

10.5 Faster-than-light speeds and time travel*

Think about it

We are all constantly traveling into the future, but Alice is more recognizable as a time traveler because she controls the *rate* of travel.

In the twin paradox Alice engages in a form of time travel, to the future. Can anyone travel to the past, or at least send a message back in time? Messages could hypothetically be sent if faster-than-light particles existed. In Chapter 5, we built an argument that proved that no particle with initial sublight speed could be boosted all the way to or beyond c, but this argument did not address the possibility that a particle is "born" with superluminal speed. Here we show that if such particles exist, they would allow communication with the past.

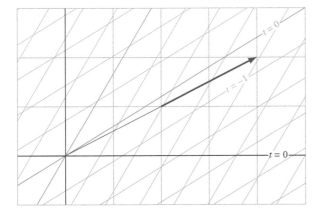

Figure 10.10 *A hypothetical particle moving at* v $= 2c$ *moves forward through time in the black frame, but backward through time in the red frame. The particle passes through a wet paintbrush (not shown) at* x$_{black} = 2$ *and leaves a paint trail thereafter; this is difficult to explain in the red frame.*

First, let's debunk the myth that any faster-than-light particle is *automatically* traveling backward in time. Figure 10.10 shows the worldline of a hypothetical faster-than-light particle (called a *tachyon*) launched at the origin and traveling at $2c$. In the black frame, this particle moves two units of space for each unit of time (the definition of $v = 2c$) but there is nothing backward-in-time about it. To see this, imagine the particle passing through a wet paintbrush located at $x = 2$, becoming coated with paint and leaving a thicker worldline thereafter. The painted particle passes through later events, so it moves forward in time.

However, this particle *is* moving backward in time in *some other frames*. Relative to the red grid in Figure 10.10, the tachyon launched at the origin actually proceeds toward earlier and earlier values of t_{red}; verify this by noting that the worldline moves away from $t_{red} = 0$ and toward $t_{red} = -1$. Red-frame observers focusing on positions and times alone may claim that the particle was first observed on the right and then moved forward in time to the left. But the paint trail provides a counterargument: the particle can paint only "after" the collision with the brush, not before. So red-frame observers could reasonably conclude that this is a faster-than-light particle traveling to the right and backward in time. Regardless of what they think, as black time moves forward the tachyon does pass through events with earlier and earlier t_{red} coordinates.

And therein lies a problem. Imagine that you and an accomplice in the red frame have mastered tachyon technology and want to use it to for profit. You wait for an extremely improbable event—such as the Cubs winning the World Series—and then use tachyons to place a sure bet as follows. You code a message describing the result of the game into a stream of $v = 7c$ tachyons aimed at your accomplice. In Figure 10.11 the transmission begins at the origin of both coordinate systems, and shows the first message as the upper blue line; verify that this tachyon moves at $7c$ in the black frame by counting black squares from the origin. Your accomplice receives the message and instantly emits another coded tachyon back in your direction. This one travels at speed $7c$ in the *red* frame, as

Figure 10.11 *A hypothetical tachyon is sent from the origin at* v = 7c *in the black frame. A red-frame accomplice receives the tachyon and replies with a tachyon traveling at 7c in the red frame. The second tachyon arrives and reaches the original emitter before the first message is sent. Tachyons can thus violate causality and would stir great scientific interest if they were discovered, but there is no evidence for their existence.*

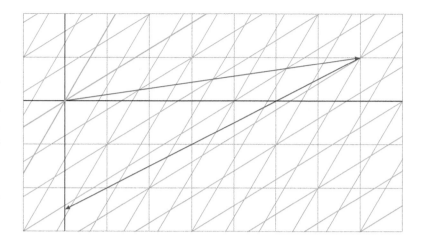

you can verify by counting red grid cells. But the skew of the red grid allows the tachyon to cross your location at $x_{\text{black}} = 0$ (the heavy black vertical line) *before* you sent it, at $t_{\text{black}} = -2.5$ or so. You receive the message before the Cubs win, and in time to place a winning bet. If tachyons exist, this would be a great way to profit. Here's the problem: you can now collect the money without bothering to send the message. Never sending the message means never receiving it, which either changes the past after you already lived it, or yields an effect without a cause.

To remove the human element, we can imagine instead a device programmed to send a message at $t = 0$ if and only if it receives no message before that, and an automatic message-relay device in place of the accomplice. Your device, having received no message, at $t = 0$ launches a message to the relay device, which ensures that your device receives a message at $t = -2.5$. This prevents your device from sending its message, but that implies that it does not receive a message, which implies that it *does* send a message, and so on. Cause and effect no longer make sense.

Relativity does not mean that anything goes. We saw in Chapter 6 that the order of events can be different in different frames, but we also saw that causality is preserved because the ordering is still unambiguous for events that could be causally related. The existence of tachyons would imply something much more astounding about causality. Ultimately, the existence of tachyons should be decided by experiment rather than by our preconceptions, but extraordinary claims require extraordinary evidence. Given the lack of experimental evidence for tachyons there is little reason to start re-examining causality.

Recall from Chapter 5 that nothing can be accelerated to or beyond c; if tachyons exist they must be *born* moving faster than c rather than accelerated to that state. So even if tachyons exist, *people* cannot visit the past using the mechanism described here.

CHAPTER SUMMARY

- The twin paradox is perfectly explicable in terms of time dilation and time skew. Most "paradoxes" in relativity result from neglecting to consider time skew.

- Inertial paths are paths of maximum proper time. Chapter 4 identified inertial frames as those in which Newton's first law is respected; this is still true, but the maximum-proper-time criterion allows us to determine inertial paths without thinking about forces or accelerations.

- The twin paradox highlights how inertial frames differ from accelerated frames. Accelerated frames allow far more varied behaviors and consequences.

- We are all traveling into the future. Acceleration enables you to travel to the future more quickly, but travel to the past is not possible.

CHECK YOUR UNDERSTANDING: EXPLANATIONS

10.1 To draw Carol's worldline, extend Alice's outbound worldline indefinitely. For each blue birthday message Alice emits toward Bob on the return journey, draw a second line emitted at the same event but tilted 45° to the right. These will catch up with Carol *very* infrequently. Alice's birthdays add up to half of Bob's over the whole journey only because we must also count the invitations that Alice hands personally to Carol during their time together.

10.2 If you think in terms of east *vs.* west, time *does* skew in the opposite direction on the return journey. But on either leg it is true that trailing clocks read a later time; in that sense, time skew always does the same thing. Because a traveler always passes ever more trailing clocks, time skew always facilitates travel into the future regardless of the direction through space.

10.3 (a) The Bob-frame time coordinate of the turnaround event T is 1.4; read this as 1.4 years in the planet frame. The fact that one year passed in Alice's frame can be read off the red grid (T occurs at the first intersection after the origin) or can be calculated starting from $v = 0.7c$: γ at this speed is 1.4. (b) The reasoning exactly parallels part (a), but now uses event T as the starting point.

10.4 (a) The coordinate time is about 5.6 and the proper time on the moving clock is about 4.25. (b) An inertial clock must move from O to A at very low speed, making time dilation nearly negligible. Therefore, the time elapsed on the clock is barely under the elapsed coordinate time of 5.6. (c) Moving from A to B would require a speed greater than c, so a clock cannot do it and the proper time between these events cannot be defined.

10.1   Building on the *Check Your Understanding* exercise at the end of Section 10.1, how many Carol-years pass between receiving birthday invitations sent by Alice on her return journey? Read the interval in Bob-years off the Bob-frame diagram and account for $\gamma = 5/4$.

10.2   *(a)* Use the velocity addition law to find Alice's velocity relative to Carol after the turnaround in the story in Section 10.1. *(b)* Compute γ for this velocity to find how slowly Alice's clock ticks on the return journey according to Carol. *(c)* Compute γ for Bob's Carol-frame velocity to find how slowly Bob's clock ticks according to Carol. (d) Using the results of parts (b) and (c), explain why Carol is not surprised to hear that Alice is younger than Bob at the reunion, even though Carol considered Bob to be the one aging more slowly until Alice's turnaround.

10.3   *(a)* Can Alice "time travel" to the past? *(b)* Can she make herself younger than she was when she started?

10.4   Analyze the statement "if you start running at the speed of light and then you stop, you are younger." In what way is this true and in what way is this misleading?

10.5   In Section 10.2, if we were to add a planet to the left of Earth it would read 2525 in panel 1 but 2524 in panel 2. Can Alice use her acceleration to see the year 2524 unfold on that planet more than once?

10.6   A crew performs interstellar travel at $\gamma = 12$ and arrives at Alpha Centauri (about four light-years from Earth) after four months of proper time, compared to four years of planet-frame time. While en route, do they find themselves traveling four-light-years in four months and thus moving at 12 times the speed of light? Explain your reasoning.

10.7   In one episode of *Futurama*, Professor Farnsworth builds a time machine, but it can only go forward in time. Explain why this makes perfect sense. (Writer David X. Cohen earned a degree in physics and crafts science gags with large elements of truth. In this story, Farnsworth intended to go forward one minute as a test, but mistakenly went way too far. To get back, he cleverly continued foward to search for the more advanced technology of a backward time machine. However, if a backward time machine is *ever* invented, users of such machines should *already* be here.)

10.8   Research the "grandfather paradox" and explain how it relates to Section 10.5.

10.1   Draw a complete spacetime diagram of the story in Section 10.1 in Carol's frame. Instead of using the Einstein velocity addition law to compute Alice's return velocity in Carol's frame, accurately locate the turnaround and reunion events in Carol's frame, and then draw a straight (inertial) worldline between them. Measure the velocity of this worldline and compare to the prediction of the Einstein velocity addition law.

10.2   Draw a complete spacetime diagram of the twin "paradox" in Bob's frame assuming Alice travels at $v = 0.8c$ for five Bob-years before turning around and heading back at $v = -0.8c$. Mark each of Alice's and Bob's birthday parties. How many are there of each?

10.3   You and two friends, Alice and Bob, are running late for a meeting; the meeting starts in five minutes and you are ten light-minutes from the meeting location, and you have not yet started moving toward the meeting. Bob has a car that can accelerate to near c and then stop at the meeting place so that only one minute passes on your watch and you will not be late for the meeting. Alice says this is impossible; if you are ten light-minutes away you cannot possibly get there in less than ten minutes, but you try it anyway because Bob's car is the fastest around. Are you late for the meeting? Identify the faulty reasoning in one of the arguments.

10.4 Redraw Figure 10.6 for a version of the story in which Alices changes her mind and returns to Bob soon after departing. By marking the new locations of events T, T′, and T″ show that Alice takes a very small "shortcut through time" even though she performed the same acceleration as in the original story.

10.5 Redraw Figure 10.6 for a version of the story in which Alice makes two half-size trips between events O and R (at the same $v = 0.7c$). By marking the locations of events analogous to events T, T′, and T″ on each trip, show that Alice's total "shortcut through time" with these additional accelerations is the same as in the original story.

10.6 Draw a complete spacetime diagram representing this story: Bob stays on Earth while Alice and Carol travel to the right at $\frac{3}{4}c$ ($\gamma = 1.5$) for one of their years. Then Carol continues while Alice turns around, instantly changing to a velocity that is $\frac{3}{4}c$ to the *left* relative to Bob. She eventually reunites with Bob back on Earth. Draw this in Carol's frame, according to the steps below. *(a)* Draw Bob's and Carol's worldlines and mark each of Bob's and Carol's birthdays with big dots. *(b)* Draw Alice's worldline leaving Carol on their first birthday of the trip. (Alice, Bob, and Carol are triplets, and they celebrated a birthday together just before starting the trip.) Think carefully about where Alice's worldline must intersect Bob's, using your knowledge of Bob's age at the reunion. Mark all of Alice's birthdays. *(c)* Measure Alice's velocity relative to Carol (after they separate) by counting squares on the graph paper. How fast is it, and is that consistent with the velocity addition law? (Rather than compute a velocity with the addition law, just explain roughly what the result must be and whether that is consistent with your graph.) *(d)* In Carol's frame, when Bob and Alice reunited, how many years had Bob aged? How many years had Alice aged? How many years had Carol aged? *(e)* Mark "R" at the event at which Alice and Bob reunite. Also mark an event called R' that, *in Bob's frame*, is simultaneous with R and on Carol's worldline. How old is Carol at that event? Explain conceptually why she must be that age.

11

Spacetime Geometry

Why are some quantities frame-dependent and others frame-dependent? This chapter will reveal a deep connection between this question and the geometry of spacetime.

11.1 Geometry of space

Before tackling spacetime, we review and discuss some properties of space that are so fundamental we rarely think about them. *Space* here refers to the familiar three dimensions physicists call x, y, and z, and need not involve interstellar space. In fact, if the room you are in has a tiled floor, the floor is an excellent example of a grid laid out in two-dimensional space, and two dimensions are all we need to make the necessary points. Some of the points in this section may seem obvious, but we must define some terms on familiar territory before we make the leap to the more abstract territory of spacetime.

Comparing two coordinate systems in two-dimensional space is a useful way to illustrate invariance and frame-dependence. Imagine two locations on a map as in Figure 11.1; we put location A at the origin for convenience. If the black grid defines the east-west and north-south coordinates, traveling from A to B requires a displacement of three miles in the east-west direction and four miles in the north-south direction. Collectively, the components $(3, 4)$ form a **displacement vector** that tells you how to get from one location to the other.

Now imagine that your compass is off so that your "east" actually points a bit north of east as shown by the gray grid in Figure 11.1. You find that to get from A to B using this compass, you actually need to go 4 miles "east" and then 3 miles "north." In this new coordinate system the displacement vector is $(4, 3)$. Nothing about the relationship of locations A and B has changed, only the language used to describe it. We can easily imagine other coordinate systems too. If someone's compass were so far off that "north" pointed straight from A to B, the displacement vector would be $(0, 5)$. Other coordinate system rotations would produce displacements like $(2.397, 4.388)$ or $(2.992, -4.006)$. The specific numbers for east and north displacements are mere artifacts of the way we set up the coordinate system. The only quantity relating locations A and B that does not vary with coordinate system is the distance: five miles. (*Distance* here will always refer to straight-line distance.)

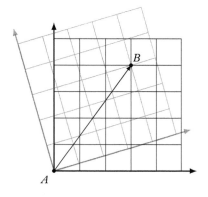

Figure 11.1 *Two coordinate systems laid over the same map.*

The Elements of Relativity. David M. Wittman, Oxford University Press (2018).
© David M. Wittman 2018. DOI 10.1093/oso/9780199658633.001.0001

How does distance relate to the displacement vector? The displacement components form two legs of a right triangle and the distance is the length of the hypotenuse, so we can use the Pythagorean theorem. This is usually written $a^2 + b^2 = c^2$ but we will write it as $(\Delta x)^2 + (\Delta y)^2 = (\Delta s)^2$ where (Δs) is the distance between initial and final points. This works in *any* of these coordinate systems: $3^2 + 4^2 = 5^2$, $4^2 + 3^2 = 5^2$, $0^2 + 5^2 = 5^2$, $2.397^2 + 4.388^2 = 5^2$, and $2.992^2 + (-4.006)^2 = 5^2$. The numbers representing the components of the displacement vector, Δx and Δy, are artifacts of the coordinate system, but the quantity $(\Delta x)^2 + (\Delta y)^2$ does *not* depend on the coordinate system. Such **invariant** quantities are the most physically meaningful quantities. In this example, we can use the distance to predict the time required to walk directly from A to B, how tired we will be at the end of the walk, and so on.

You may also think of distance as the size or length of the displacement vector. Physicists use the term **magnitude** to denote the invariant "size" of a vector (any vector, not only the displacement vector). The components of a vector are frame-dependent, but the magnitude is invariant. The equation $(\Delta x)^2 + (\Delta y)^2 = (\Delta s)^2$ provides what is called a **metric** for tabulating the contributions of the components Δx and Δy to the invariant magnitude of the displacement. This particular metric is known as the **Euclidean** or **flat-space metric** because the plane geometry developed by the ancient Greek mathematician Euclid follows this metric. However, this metric is not restricted to planes. If you extend the grid of your tile floor upward with a z coordinate as in Figure 11.2 you will find that the Euclidean metric $(\Delta x)^2 + (\Delta y)^2 + (\Delta z)^2 = (\Delta s)^2$ describes the space in your room. A surface or space is called "flat" if the Euclidean metric applies, regardless of the number of dimensions.

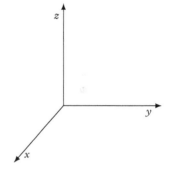

Figure 11.2 *Axes of an* xyz *coordinate system (grid not shown).*

For simplicity this chapter deals with only two dimensions in any particular situation: x and y when analyzing space, or x and t when analyzing spacetime. Spacetime maps show the relationships of *events*, and the time dimension is necessary to show these relationships. We have seen that time and space are tightly related, but clearly there is something very different about time and space dimensions. This chapter aims to capture the essence of that difference.

You may recall that $(\Delta x)^2 + (\Delta y)^2 = (\Delta s)^2$ is also the equation for a circle of radius Δs. In other words, a circle is the set of points *equidistant* from some reference point, with the distance given by Δs. This definition of a circle is so obvious that we never think about it, but we are about to encounter a situation where a set of equidistant points is *not* a circle, so it is worth thinking about now. Equidistance defines a circle exactly because we define distance with the metric equation $(\Delta x)^2 + (\Delta y)^2 = (\Delta s)^2$. If we use a different metric equation such as $(\Delta x)^2 + 2(\Delta y)^2 = (\Delta s)^2$ then displacements in y would have a bigger effect than displacements in x, and the set of points with equal (Δs) would be an ellipse rather than a circle.

The circle therefore serves as a useful icon for the Euclidean metric. Figure 11.3 shows a circle of words plotted in one coordinate system, with the axes of another coordinate system displayed for reference; rotate your view to see what the

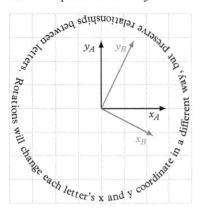

Figure 11.3 *Rotating your head to view this word circle in the "B" coordinate system changes the coordinates of each letter, but in a way that preserves the order of letters, distances between letters, and the distance between letters and the center. This is the familiar geometry of space; space*time *will differ.*

circle looks like in the second coordinate system. This is analogous to a change of frames in relativity, so note carefully what changed and what remained invariant. The x and y coordinates of each letter change, as do Δx and Δy between any two letters; while the radius (Δs) and the distance between any two letters is invariant. Furthermore, the order of the letters does not change so the message is still readable. Changing coordinate systems carries each point on the circle to a different point on the circle in an orderly way; the mathematical description of this "orderly way" is called a **coordinate transformation**.

Pick any two letters in Figure 11.3 and consider the displacement vector from one to the other in the "A" coordinate system. How does this compare to the displacement between the same two letters in the "B" coordinate system? If switching coordinate systems increased the size of Δx, then it *decreased* the size of Δy, and vice versa. Figure 11.1 offers a simple example of this phenomenon: the displacement $(3, 4)$ became $(4, 3)$ in the second coordinate system. This is a basic property of coordinate systems describing *space* and is reflected in the metric: given that the combination $(\Delta x)^2 + (\Delta y)^2$ is invariant, an increase in the size (absolute value) of Δx implies a decrease in the size of Δy and vice versa. Coordinate system rotations never increase the size of both components simultaneously because the *distance* between the points must remain invariant.

The remainder of this chapter generalizes these ideas to spacetime. By focusing on invariance, we will tease out exactly how spacetime differs from space.

Check your understanding. Find the displacement from the period to the comma in each frame in Figure 11.3. *(a)* Did the size of Δx increase or decrease when changing from the "A" to the "B" coordinate system? *(b)* Did the size of Δy increase or decrease? *(c)* Did the distance increase, decrease, or stay the same?

11.2 The spacetime metric

The examples of coordinate transformations in Section 11.1 were of a limited variety: the "north" and "east" axes always rotated in the same direction by the same amount. The axes therefore remain perpendicular; the fact that these axes form a right angle is invariant. This in turn causes the pattern we noticed regarding the displacement components: if in one of these transformations the size of Δx increases then the size of Δy decreases and vice versa.

Spacetime is different because the space and time axes cannot remain perpendicular when we change frames. The key invariant feature of spacetime is that c is the same speed in all frames. As first shown in Figure 6.6, the space and time axes fold in a way that keeps light rays midway between them. If Bob's time axis is rotated clockwise from Alice's, his space axis rotates *counterclockwise* as in the left panel of Figure 11.4. A clockwise rotation of the t_B axis is much less mysterious than it sounds: it simply means that Bob moves to the east (the t_B axis is simply

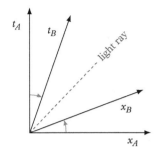
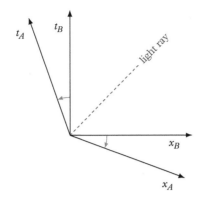

Figure 11.4 Left: *If Bob travels east in Alice's frame, the* t_B *axis rotates clockwise from the* t_A *axis and the* x_B *axis rotates the opposite way so that a light ray remains midway between his space and time axes.* Right: *the same situation drawn in Bob's frame, to illustrate west-moving axes. This key result of Chapter 6 illustrates how spacetime coordinate systems differ from coordinate systems describing space alone.*

Bob's worldline). This much is true even in Galilean relativity; the counterintuitive part is that (due to invariance of c) the x_B axis rotates in the *opposite* direction.

Because space and time axes rotate differently, the Euclidean metric fails to describe the situation. To find the metric that describes spacetime, let us first identify the quantities analogous to those we used in Section 11.1. The displacement between two events involves Δt and Δx, so let us call $(\Delta t, \Delta x)$ the **spacetime displacement vector**. Can we identify anything involving these quantities that is invariant? We know (Chapter 10) that a clock traveling inertially between the events will display a time elapsed between the events (the proper time $\Delta \tau$) that can be read by all observers and can thus be used as an invariant. Equation 7.4 tells us that $(\Delta \tau) = \sqrt{1 - v^2/c^2}(\Delta t)$. This equation does serve the purpose of a metric equation, which is to identify a *frame-independent combination of frame-dependent quantities*. So let us see if we can make this equation look a bit more like the Euclidean metric equation, to better highlight the similarities and differences.

One thing we can do based on our experience with the Euclidean metric equation is to square both sides. It makes sense that the displacement components would be squared, so the direction of the displacement (positive or negative in some coordinate) will not matter when calculating the amount of spacetime separation between events. Squaring both sides yields

$$(\Delta \tau)^2 = \left(1 - \frac{v^2}{c^2}\right)(\Delta t)^2$$

$$= \left(1 - \frac{(\Delta x)^2}{(\Delta t)^2 c^2}\right)(\Delta t)^2$$

where the second step simply plugged in the definition $v = \frac{\Delta x}{\Delta t}$. If we multiply out the right side, we get

$$(\Delta \tau)^2 = (\Delta t)^2 - \frac{(\Delta x)^2}{c^2}. \tag{11.1}$$

Think about it

Equation 11.1 carefully isolates the frame-dependent quantities on one side; multiplying through by c^2 preserves this isolation because c^2 is invariant.

We can multiply both sides by c^2 to make a nicer-looking equation:

$$c^2(\Delta\tau)^2 = c^2(\Delta t)^2 - (\Delta x)^2. \qquad (11.2)$$

Equation 11.2 is the **spacetime metric** we have been seeking. This version applies in the absence of complicating factors, such as gravity, and is known as the **Minkowski metric** after Hermann Minkowski (1864–1909) who pioneered the geometric approach to special relativity, including the use of spacetime diagrams. You may also see this metric written with the cs omitted, because it is taken for granted that you will use units (such as light-years and years) in which $c = 1$. Beginners are often uncomfortable with dropping the c entirely, though, so we will generally write it in even if we are assuming $c = 1$.

Equation 11.2 forms the central conclusion of this chapter: the frame-independent measure of spacetime separation between two events is the combination $c^2(\Delta t)^2 - (\Delta x)^2$. The remainder of the chapter is devoted to interpreting this statement. But before moving on, we should note that Equation 11.2 oversimplifies one thing: while the right-hand side is indeed an invariant combination, the resulting number is not *always* interpretable as a proper time increment. Take, for example, two events separated by three years of time and five light-years of space: the right-hand side works out to -16, which as a negative number cannot possibly be the square of the elapsed proper time. This apparent conflict arises because we developed Equation 11.2 by analyzing a clock traveling inertially from one event to another, but no clock can travel five light-years in three years! Box 11.1 explains how to interpret the invariant quantity when it works out to be negative, while the main text assumes it will be positive. This is because we tend to analyze events along worldlines, which move more through time than space so $c^2(\Delta t)^2$ is greater than $(\Delta x)^2$. A borderline case to keep in mind is that light moves equally through time and space, reducing the proper time to zero.

The spacetime metric can be the fastest way to solve problems when you are given event coordinates. Suppose Carol leaves Earth and travels at constant velocity to the planet Zork three light-years away, arriving after five years of Galactic Standard Time; what is the elapsed time on her watch? Using our previous thinking tools we would compute her velocity, then compute γ, then compute $\frac{5}{\gamma}$. With the metric, we note that departure and arrival events are separated by $\Delta x = 3$ and $\Delta t = 5$, so we compute $(\Delta\tau)^2 = 5^2 - 3^2 = 16 = 4^2$. The elapsed proper time is four years.

Check your understanding. (a) Continuing this last story, Alice leaves Earth simultaneously with Carol but has a faster ship and arrives at Zork after four years GST. What is the time elapsed on her watch? Does it match what you know about time dilation? *(b)* Bob wants to outrace Carol, so he leaves simultaneously with the others but plans to arrive at Zork after three years GST. Is this realistic?

Box 11.1 The spacetime interval

The thinking behind Equation 11.2 was ultimately based on a clock moving from one event to another, and clocks move at sublight speeds. We can express this mathematically as $\frac{(\Delta x)^2}{(\Delta t)^2} < c^2$ or $c^2(\Delta t)^2 > (\Delta x)^2$, and informally as "more time than space between the events." So we must ask whether Equation 11.2 works for other types of event pairs.

Let us first try the case $c^2(\Delta t)^2 = (\Delta x)^2$. This describes two events along the path of a light ray because $\frac{\Delta x}{\Delta t} = c$ for a light ray. Because c is the same in any frame, the combination $c^2(\Delta t)^2 - (\Delta x)^2$ can be written as $(\Delta x)^2 - (\Delta x)^2$ or zero in any frame. This confirms that Equation 11.2 yields an invariant quantity for these types of event pairs. Calling this invariant number (zero) the proper time between the events may be considered an abuse of the term "proper time" because no clock can travel between the relevant events. So physicists adopted the more neutral term "spacetime interval" for the number that results from the combination $c^2(\Delta t)^2 - (\Delta x)^2$. Note that the interval between two events can be zero even if they are separated by a large Δx *and* a large Δt! This is quite unlike distance in space, which can be zero only if each and every component of the displacement is zero.

The spacetime interval between two events can even be negative. Consider simultaneous events at the ends of a ruler of rest length L: $\Delta t = 0$ and $\Delta x = L$ so the right-hand side of Equation 11.2 evaluates to $-L^2$. In this case we interpret the result as related to proper length (also called rest length or proper distance) rather than proper time.

Physicists condense these three cases into one equation as follows:

$$c^2(\Delta t)^2 - (\Delta x)^2 = \begin{cases} c^2(\Delta \tau)^2 & \text{if } c^2(\Delta t)^2 > (\Delta x)^2 \\ -L^2 & \text{if } c^2(\Delta t)^2 < (\Delta x)^2 \\ 0 & \text{otherwise} \end{cases} \qquad (11.3)$$

The right-hand side here becomes unwieldy if you have to write it often, so physicists further define a new symbol, Δs^2, to serve as a compact way of representing everything on the right side of this equation. If you study relativity further you will often see the spacetime interval written as Δs^2, but the main text avoids all this notation by focusing on the first case on the right side, which is nothing more than Equation 11.2.

The appearance of three kinds of "distance" distinguishes spacetime coordinate systems from coordinate systems describing space alone, and is entirely due to the minus sign in the metric. The three cases listed in Equation 11.3 are also known as timelike, spacelike, and lightlike intervals, respectively. A timelike interval simply means that two events are separated by more time than space (allowing enough time for an object to move from one event to the other), whereas a spacelike interval describes two events separated by more space than time. This is just a way of restating Section 6.4, which found that the spacetime around a given event A can be divided into a light cone (with timelike-separated events forming the interior of the cone and lightlike-separated events forming the boundary) and "elsewhere" (events that are spacelike-separated from A).

11.3 Understanding the metric

The properties of the spacetime metric

$$c^2(\Delta\tau)^2 = c^2(\Delta t)^2 - (\Delta x)^2$$

are best exposed by comparison with the Euclidean metric

$$(\Delta s)^2 = (\Delta x)^2 + (\Delta y)^2$$

as follows:

- Just as the Euclidean metric involves the *squares* of Δx and Δy, the spacetime metric involves the *squares* of Δt and Δx. The reason these displacements must be squared when we compute the distance stems from the distinction between a *displacement* and a *distance*. If Δx represents the displacement from A to B, then the displacement from B to A is $-\Delta x$. But surely the *distance* is the same in either case! The squaring of a displacement makes its sign irrelevant.

- Squaring the displacements erases their signs, so some information is lost. Given only $(\Delta s)^2$ or $c^2(\Delta\tau)^2$, we cannot infer which of the points is more east or which event occurred first. These details are not the job of the metric, which tells us only what is invariant.

- $(\Delta s)^2$ is always positive for distinct locations in space, but the minus sign in the spacetime metric enables an expanded repertoire of positive, negative, or zero intervals for distinct spacetime events (see Box 11.1).

- Crucially, the minus sign in the spacetime metric makes $(\Delta t)^2$ and $(\Delta x)^2$ cancel each other (at least partially) rather than add. Thus a frame that measures a larger $(\Delta t)^2$ between two given events must *also* measure a larger $(\Delta x)^2$ between the same two events, to make the difference between these quantities invariant. This is in stark contrast with spatial coordinate systems: a larger $(\Delta y)^2$ between given locations implies a *smaller* $(\Delta x)^2$ to make the sum invariant.

The last point highlights how time and space contribute in fundamentally opposite ways to the spacetime interval. This is graphically illustrated by the stretching triangles introduced in Section 7.2. The next few paragraphs recap that concept with a new story.

Consider looking down on a train car with eastbound velocity v (relative to the tracks), and let x increase to the east. A flash of light is emitted from the south wall of the train car. In the train frame (upper panel of Figure 11.5) the light travels directly north and hits the wall directly opposite. The fact that light hits that particular spot on the north wall must be frame-independent—the response

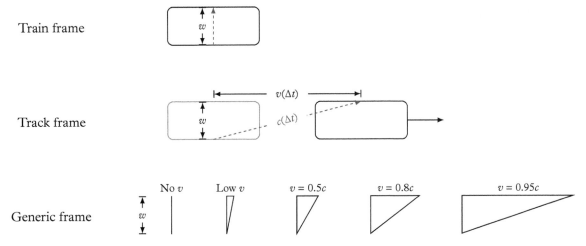

Figure 11.5 Top: *light crossing a train car in the train frame.* Middle: *the same story in the track frame.* Bottom: *abstract representations of the same story in a variety of frames. In each case the vertical leg is invariant, the hypotenuse (representing speed through time) is γ times longer than this, and the horizontal leg (representing speed through space) is $\frac{v}{c}$ times the hypotenuse. As speeds approach* c *the horizontal leg becomes nearly as long as the hypotenuse length, stretching the triangle greatly and making γ very large.*

of a light sensor on that spot cannot depend on the frame. But in the track frame (middle panel of Figure 11.5), that sensor has moved to the east by an amount $\Delta x = v(\Delta t)$ while the light was en route.

The distance light must travel to reach that spot (and, correspondingly, the time between emission and reception events) is *strongly* frame-dependent. But the width of the train (call it w) is the same in both frames, by the argument in Section 4.4. We can therefore represent the situation with a right triangle with a vertical left leg of invariant length w, a horizontal top leg of length $\Delta x = v(\Delta t)$, and a hypotenuse (the path of the light) of length $c(\Delta t)$ as marked in the middle panel of Figure 11.5. Now, imagine the same physical situation measured in a variety of inertial frames: the triangle will morph from a thin sliver (in frames measuring the train to be moving slowly) to enormously wide (in frames measuring the train speed to be near c) as illustrated by the series of triangles in the bottom panel.

Frames in which the train moves nearly at c can stretch the triangle nearly infinitely. To see this, recall that the horizontal leg represents the distance traveled by the train and the hypotenuse represents the distance traveled by the light in the same time, so their ratio represents v/c. Frames in which v is nearly as large as c are those in which the triangle is stretched so far that the horizontal leg is nearly as long as the hypotenuse. And for that to happen, the ratio of hypotenuse to invariant leg (γ, the time dilation factor) must be extremely large. In the rest frame, in contrast, γ is exactly one because the horizontal leg disappears. The geometry is such that even if we were able to launch a person at, say, $v = 0.2c$, γ is barely greater than one so there would not be much time dilation.

Think about it

The direction of the triangle is irrelevant to the thought process here. There are frames in which the train is measured as moving west, and those would yield a left-pointing triangle with the same geometric properties.

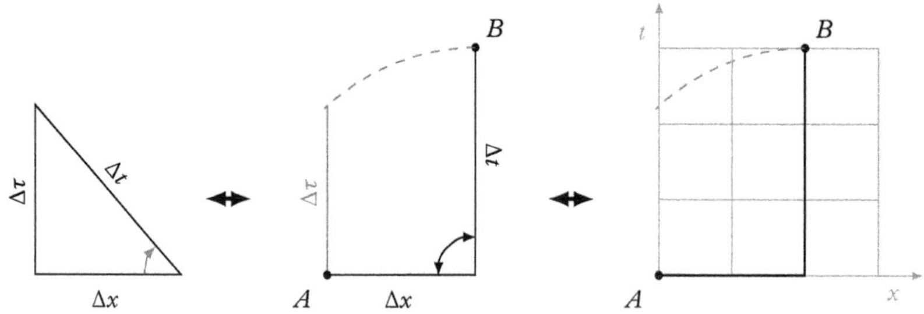

Figure 11.6 *To convert from the triangle representation of a spacetime displacement to a spacetime diagram, unfold the triangle as shown from left to right. To graphically find the proper time between two events on a spacetime diagram, follow the steps from right to left; once you get the idea, simply follow the dashed arc as a shortcut. This shows how frames that stretch the triangle place* A *and* B *far apart on a spacetime diagram without affecting the proper time between them.*

Think about it

The train car acts as a light clock with the invariant width w equal to $c(\Delta\tau)$, so the equation $w^2 = c^2(\Delta t)^2 - (\Delta x)^2$ in the text is simply the metric equation, Equation 11.2. The stretching triangle thus graphically illustrates the properties of the metric.

That last paragraph can be captured mathematically as follows. The Pythagorean theorem tells us that $w^2 + (\Delta x)^2 = c^2(\Delta t)^2$. If we rearrange this to separate the invariant quantity w from the other quantities, we find $w^2 = c^2(\Delta t)^2 - (\Delta x)^2$. This is where the minus sign in the metric came from: separating the invariant and frame-dependent quantities required rearranging the Pythagorean theorem with its familiar plus sign. The graphical expression of this is that the invariant feature of the triangle is a particular leg rather than the hypotenuse.

Furthermore, the metric shows us how to turn the triangle representation into a spacetime diagram. The horizontal leg of the triangle represents the Δx component of the displacement between two events, and the hypotenuse represents the (coordinate) time component Δt. Therefore, to view these components perpendicularly as on a spacetime diagram, we simply need to break the triangle and stand the hypotenuse straight up as in Figure 11.6, reading left to right. To convert the displacement between events A and B on a spacetime diagram to the triangle representation, read the figure right to left: first, erase the axes and grid, then fold the Δt leg over until its free end is directly above event A. By the Pythagorean theorem, the length of the newly created leg is $\sqrt{(\Delta t)^2 - (\Delta x)^2}$, or $\Delta\tau$ (neglecting factors of c because spacetime diagrams always use units in which $c = 1$). Spacetime vectors thus have the strange property that their magnitude is smaller than one—and often both—of their components! The stretching triangle picture captures this property perfectly because the invariant leg is always shorter than the hypotenuse representing the Δt component—and often also shorter than the leg representing the Δx component.

The triangle, the skewed grid, and the spacetime metric are inextricably related: they are just different ways of describing the same kind of geometry. Each representation has pros and cons. The triangle is compact and—unlike the spacetime diagram—clearly displays both the invariant and frame-dependent aspects of a displacement. The metric is the algebraic representation of this triangle—more abstract, but also more useful for computation. Only the spacetime diagram can illustrate relationships between three or more events, or between coordinate systems, but it can be tedious to fully show the relationships by drawing skewed grids.

So far we have discussed only *displacements* in spacetime, but other vectors use the displacement vector as a foundation. For example, velocity is displacement divided by elapsed time. The *same* transformation and invariance properties that apply to displacement vectors therefore apply also to velocity and other vectors. Chapter 12 shows how this leads to new ideas about mass and energy.

Check your understanding. Convince yourself of the following properties of the stretching triangle: *(a)* the horizontal length divided by that of the hypotenuse equals $\frac{v}{c}$; *(b)* the hypotenuse is γ times longer than the invariant vertical leg; *(c)* γ is barely more than one if the speed is low; *(d)* γ grows without bound as v approaches c.

Box 11.2 The full Minkowski metric

We have been assuming that x is the only relevant spatial dimension as we examine how it interacts with the time dimension t. But we know of three spatial dimensions, and the other two must have the same properties as the one we call x. Therefore, a more complete version of the Minkowski metric is

$$c^2(\Delta t)^2 - (\Delta x)^2 - (\Delta y)^2 - (\Delta z)^2 = (\Delta s)^2$$

where y and z refer to the additional spatial dimensions, and $(\Delta s)^2$ serves as shorthand for everything written on the right side of Equation 11.3. Some authors write the metric with the opposite sign:

$$-c^2(\Delta t)^2 + (\Delta x)^2 + (\Delta y)^2 + (\Delta z)^2 = (\Delta s)^2$$

with corresponding changes to the first two cases listed on the right side of Equation 11.3. The important thing is not the overall sign convention, but the fact that the time displacement and space displacements contribute to the spacetime interval in *opposite* ways (i.e., they have different signs in the equation). This is what distinguishes spacetime from plain old space and what makes the stretching triangle an accurate model for the behavior of spacetime displacement vectors.

Spacetime vectors are also called **4-vectors** because (with the inclusion of all spatial components) they have four components. But the crucial property of these vectors is how they transform rather than their number of components.

11.4 Spacetime geometry is hyperbolic*

In Section 11.1 we argued that a circle is an apt symbol for the Euclidean metric because a circle is a set of points that are—according to the Euclidean metric $(\Delta s)^2 = (\Delta x)^2 + (\Delta y)^2$—equidistant from a given point. What is the corresponding symbol for the spacetime metric? This section provides an answer,

Figure 11.7 *A series of triangles with identical proper times is unfolded onto a spacetime diagram. The events marked on the diagram thus share the same proper time from the origin. These events fall on a curve called a hyperbola (dashed).*

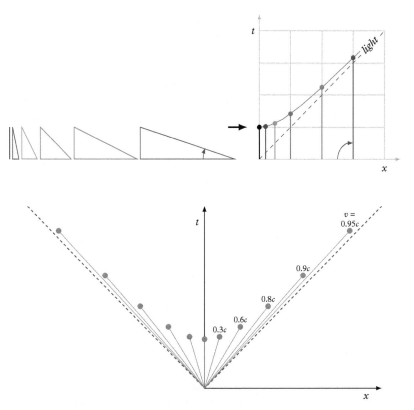

Figure 11.8 *Inertial clocks leaving the origin at a variety of velocities read $\Delta\tau = 1$ at the dotted events. Connecting the dots (i.e., imagining an infinitude of clocks at all possible velocities) yields a hyperbola that traces the set of events "equidistant" (in terms of proper time) from the origin.*

Think about it

Figure 11.7 implicitly contains the skewed grid studied in previous chapters, as follows. The black dot marks a particular intersection in a square grid. When measured in other frames that grid folds in a way that carries the black dot up and to the right, to the position of a different dot.

---

but is marked as optional because students should strive foremost to understand the stretching-triangle representation, which forms the basis for understanding Chapter 12.

Figure 11.7 unfolds a series of triangles with identical proper times onto a spacetime diagram; the events marked on the diagram thus share the same proper time from the origin. Highly stretched triangles place events only slightly above the path of a light ray (dashed line) emitted at the origin. Thus, events equidistant (in terms of proper time) from the origin lie along a curve with a flattish bottom and sides that rise to nearly meet (but never quite touch) the straight lines that represent light rays on either side. This kind of curve is known as a **hyperbola**.

Check this conclusion with a thought experiment: launch a suite of inertial clocks (each initially reading $\tau = 0$) from one event at a variety of speeds. Place the launching event at the origin of a spacetime diagram (Figure 11.8), and mark each event where a clock ticks $\tau = 1$ s. Clocks with low speed in either direction suffer a small amount of time dilation and tick just after the stationary clock ticks ($t = 1$). Clocks with larger speed suffer more time dilation, placing the $\tau = 1$ events substantially higher (and further left or right because of the motion of the clock). The clocks cannot quite reach speed c so the $\tau = 1$ events can approach but never quite fall on a light-ray worldline. Figure 11.8 shows the resulting worldlines and events, again tracing out a hyperbola. Figure 11.8 is thus a group portrait of spacetime displacement vectors with identical magnitudes. Note that the same set of points could also represent the ticking of *one* clock as measured in a variety of frames, and thus represent what is invariant about measurements of that clock.

The general equation for a hyperbola in the x, y plane is usually written as $a^2 = y^2 - x^2$, where a is a constant indicating where the curve crosses the y axis. In a spacetime diagram, the vertical ("y") axis is actually ct, so we can see that $c^2(\Delta\tau)^2 = c^2(\Delta t)^2 - (\Delta x)^2$ indeed follows the form of the equation for a hyperbola, only with different symbols. The minus sign makes spacetime geometry hyperbolic rather than Euclidean: any change of frame that increases $(\Delta x)^2$ between a pair of events must *also* increase $(\Delta t)^2$.

Nothing in this chapter touches on the geometry of *space*; the set of locations in space equidistant from a given location is still a circle. What is hyperbolic here is the relationship between space and time. A consequence of this relationship is that indirect spacetime paths between events—such as the path taken by the traveler in the twin "paradox"—involve *less* proper time than straight-line spacetime paths. (Contrast this with paths through space: indirect paths involve longer travel distances.) Physically, this means that inertial paths are paths of *maximum* rather than minimum proper time. The geometry of spacetime differs radically from the geometry of space!

Check your understanding. If we redraw the events of Figure 11.8 in another inertial frame, what pattern will be produced? How will the relationships of the dots change or remain the same?

Think about it

The group portrait in Figure 11.8 is incomplete. Events occurring 1 s of proper time *before* the origin also qualify as 1 s from the origin, and form an additional downward-facing hyperbola.

CHAPTER SUMMARY

- Between any two events the components of the spacetime displacement vector $(\Delta t, \Delta x)$ are frame-dependent, and the spacetime metric or interval $c^2(\Delta t)^2 - (\Delta x)^2$ is the unique combination of these quantities that remains invariant. If positive, this quantity is interpreted as $c^2(\Delta\tau)^2$ where $\Delta\tau$ is the proper time elapsed on a clock traveling inertially from one event to the other.

- This quantity also serves as the magnitude squared of the spacetime displacement vector, because the magnitude of a vector is what is invariant across coordinate systems.

- The properties of a spacetime displacement vector are best captured by the stretching triangle picture. The hypotenuse represents the Δt component of the displacement (γ times the invariant leg), one leg represents the Δx component of the displacement ($\frac{v}{c}$ times the hypotenuse), and these stretch or shrink together, depending on frame, while keeping the other leg invariant.

- The counterintuitive properties of spacetime stem from the fact that the invariant part the triangle (see Figure 11.6) is one leg rather than the hypotenuse. This leads to a minus sign in the metric, which gives spacetime a hyperbolic geometry. A consequence of this geometry is that an inertial path between two events is the *longest* possible path (as measured by proper time) between the events.

FURTHER READING

For those who wish to pursue the transformation properties of spacetime vectors, there are many books offering much more mathematical detail without requiring proficiency beyond algebra. *Spacetime Physics* by Edwin F. Taylor and John Archibald Wheeler is a widely used text in this category.

It's About Time by N. David Mermin emphasizes the special role of time in spacetime using more detailed arguments than are presented here, but with much less mathematics than *Spacetime Physics*.

CHECK YOUR UNDERSTANDING: EXPLANATIONS

11.1   (a,b) In the "A" coordinate system, the displacement is mostly in the x_A direction; Δy_A is approximately zero and Δx_A is nearly as large as the diameter of the circle. In the "B" coordinate system the two displacement components are more nearly equal. Thus, Δx decreased in size and Δy increased in size when switching from the "A" to the "B" system. (c) The distance between the period and comma is the same in either coordinate system.

11.2   (a) 2.65 years; as expected, Carol's clock exhibits more time dilation. (b) No, he would have to travel at speed c to do this. The metric tells us this is impossible by yielding a result of zero proper time; only light can do this.

11.3   (a) Consider the extremes: at $\frac{v}{c} = 0$ the horizontal length is zero, and at $\frac{v}{c} \approx 1$ the train moves nearly

as much as the light, thus making the horizontal leg nearly as long as the hypotenuse. At these extremes $\frac{v}{c}$ certainly matches the ratio of horizontal leg to hypotenuse. (b) Again consider the extremes: at $\frac{v}{c} = 0$ the hypotenuse equals the vertical leg, and at at $\frac{v}{c} \approx 1$ the hypotenuse is dramatically longer. Thus the hypotenuse to vertical leg ratio matches the behavior of γ. (c) At low speeds the horizontal leg is so short that the hypotenuse is barely longer than the vertical leg. (d) Building on the answer to (a), the only way to make the hypotenuse no longer than the horizontal leg is to stretch the triangle infinitely.

11.4   The dots will still form a hyperbola. They will all shift along the hyperbola in order, so neighboring dots in this frame will remain neighboring in other frames.

11.1 Why can the rotation depicted in Figure 11.1 *not* describe the "rotation" of spacetime coordinates that occurs when we change frames?

11.2 Using graph paper, redraw each panel of Figure 11.4 to depict Bob moving at $0.9c$ to the right relative to Alice.

11.3 This exercise guides you through the steps of redrawing the middle panel of Figure 11.5 to depict the train moving at $0.5c$ to the right. First, faintly draw a bird's-eye view of a train car. *(a)* Calculate γ for $0.5c$ and draw a hypotenuse γ times longer than the invariant leg. Use the upper right end of the hypotenuse as the starting point for drawing a faint outline of the car at its new position. *(b)* Fill in the missing horizontal leg of the triangle. *(c)* How long is the horizontal leg, as a fraction of the length of the hypotenuse? Give a number. If this number is not 0.5 (the given velocity as a fraction of c), find where you went wrong and fix it.

11.4 *(a–c)* Repeat Exercise 11.3 for a frame in which the train moves at $0.95c$ to the right. *(d)* Estimate this frame's speed and direction relative to the frame used in Exercise 11.3, keeping in mind that the velocity addition law is not entirely linear.

11.5 *(a)* Referring to Figure 11.9, use graph paper to accurately draw the triangle representation of the spacetime displacement between events A and B. *(b)* Do the same for the displacement between events A and C. *(c)* What happens if you try to draw the triangle representation of the displacement between events A and D? How do you interpret this?

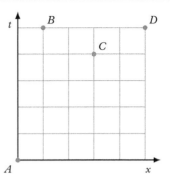

Figure 11.9 *Spacetime diagram for Exercise 11.5.*

11.6 *(a)* Referring to the triangle representation of a spacetime displacement vector in Figure 11.10, use graph paper to accurately draw the same displacement on a spacetime diagram. *(b)* What is the speed of a an inertial particle that performs such a displacement?

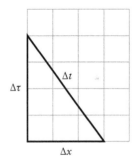

Figure 11.10 *Triangle representation of a spacetime displacement vector, for Exercise 11.6.*

11.1 On Valentine's Day, you get a message sent from a secret admirer. You want to find out where it came from. You are able to determine that the message was written one day ago, in the message's frame, and that it was sent at one of five possible constant velocities: $-0.9c$, $-0.5c$, 0, $+0.5c$, or $+0.9c$. *(a)* On graph paper, draw a spacetime diagram showing the possible worldlines starting from the event it which the message could have been written and ending at the origin (take the origin to be the event at which you received the message). Label the worldlines 1–5. Show your calculations for $+0.5c$ and $+0.9c$. *(b)* Connect the possible message-writing events with a curve. What is the technical name for what you just drew? *(c)* Redraw everything in the frame of the $-0.5c$ message. Make sure the labels 1–5 identify the same messages as in the other frame. You may need to use the velocity addition law to determine some of the velocities in the new frame.

11.2   Consider this version of the twin paradox. Alice leaves Earth at the start of the year 2525 GST and arrives at the planet Zork four light-years away, at the start of the year 2530 GST. She immediately turns around and arrives back at Earth at the start of the year 2535 GST. *(a)* Use the metric to determine how much Alice aged on the outbound leg. *(b)* Use the metric to determine how much Alice aged on the return leg and therefore how much she aged over the entire journey. *(c)* Bob remained on Earth throughout the story. Use the metric to determine how much he aged between the events of Alice's departure and return. *(d)* Alice was also present at the departure and return events, so why is it incorrect to use the metric with those events alone to determine how much she aged between the events?

11.3   *(a)* Building on Problem 11.2, use time dilation to compute Alice's aging; you will need to infer her velocity from the positions and times. *(b)* Which way of computing Alice's aging is easier? Does it depend on what information you are given?

11.4   Figure 11.8 shows some of the events that are separated from the origin by a spacetime interval of $+1$. What does the set of events separated from the origin by a spacetime interval of -1? Explain why it is not a downward-facing hyperbola. *Hint:* review Box 11.1.

11.5   On a piece of graph paper, draw a spacetime diagram showing a one-meter stick at rest, at $t = 0$ in your frame, starting at the origin. *(a)* Draw four other sticks of rest length 1 m, at rest in four other frames: two moving at moderate velocity in either direction, and two moving at high velocity in either direction. In each case, the meter stick should be drawn as being at rest in its frame, at time 0 in its frame. For the first two you draw, write an explanation of *why* you draw them that way, and show your calculations. *(b)* Connect the far ends of the sticks. What have you just drawn?

Energy and Momentum

12

In this chapter we will come to understand the famous equation $E = mc^2$. It is actually part of a wider relationship between energy, mass, and momentum so we will start by defining energy and momentum in the everyday sense. We will then build on the properties of spacetime vectors explored in Chapter 11 to see how energy, mass, and momentum have a deep relationship that is not obvious at everyday low speeds.

12.1 Energy and momentum (Galilean)

Figure 12.1 reproduces a spacetime diagram we discussed in Section 4.5, with two equal-mass billiards bouncing off each other. This particular collision swaps the two velocities of the two billiards, so the average velocity cannot be affected: the same two numbers would be averaged pre-collision as post-collision. This statement is frame-independent, because we saw in Section 4.5 that the velocity-swapping property is preserved in all frames. Figure 12.2 shows that the average velocity is preserved even if the incoming particles stick together at the collision: the combined particle follows a worldline midway between the two outgoing worldlines in Figure 12.1. This statement is frame-independent as well, because tilting the diagram affects the pre- and post-collision worldlines equally, at least at everyday speeds.

Do collisions always conserve average velocity? If you collide billiards of *unequal* mass, you will quickly find the answer is no. For example, if a massive stationary billiard is hit by a fast low-mass billiard, the latter loses most of its velocity but the former gains little velocity. Physicists have found, through exhaustive experimentation, that what is conserved in every collision is *mass times velocity*. This quantity is so important that it has a specific name: **momentum**. The total momentum of all particles entering a collision always equals the total momentum of all particles exiting. In fact momentum is conserved not only in collisions, but in every physical process. *Conserved* means not changing over time; do not confuse this with invariant, which means frame-independent. Because momentum is defined as mass times velocity, it is most definitely frame-dependent.

Important as it is, momentum cannot be the only word we need to describe the net motion of a set of particles. In Figure 12.2 *something* about the motion is changed by the collision even if the momentum is not. Another example: a set of two stationary particles have the same total momentum (zero) as a set of two particles with equal and opposite momenta, yet something differs between these

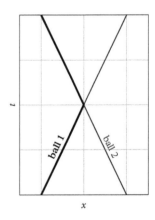

Figure 12.1 *Two equal-mass billiards with equal and opposite velocities have zero total momentum before they collide—and* after *they collide. In some other frame, the momentum is not zero, but it is nevertheless unchanged by the collision.*

The Elements of Relativity. David M. Wittman, Oxford University Press (2018).
© David M. Wittman 2018. DOI 10.1093/oso/9780199658633.001.0001

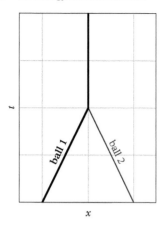

Figure 12.2 *Momentum is conserved even in a sticky collision; it is always conserved. What is not conserved here is kinetic energy—the pre-collision kinetic energy turns into some other form of energy such as heat.*

Think about it

The definition of energy here is not rigorous because I have not defined work. Work does have a specific definition in physics, but defining it rigorously here would take us too far afield.

Think about it

The scalar (nondirectional) nature of energy becomes even more obvious if we think about non-kinetic forms of energy, such as the chemical energy stored in a chocolate chip cookie.

two situations. We need a term to capture the "total amount" of motion, regardless of direction; physicists call this **kinetic energy**, meaning energy of motion.

Physicists define **energy** as the ability to do work. Energy comes in many apparently different forms that can be exchanged back and forth without ever destroying the energy or creating new energy. For example, when you throw a ball you increase its kinetic energy by a certain amount, and you expend an equal amount of the chemical energy stored in your body. Similarly, if the ball hits a wall and stops, that amount of kinetic energy is turned into an equal amount of sound and/or heat energy (also known as thermal energy). Some forms of energy are rather subtle, which leads to an analogy with money. Most children first enounter money in the form of coins, and tend to assume that money *means* coins. But over time we began to understand that certain pieces of paper could be exchanged for coins. This made money a concept that superseded mere coins. As we grew older, we began to appreciate more and more forms of money: checks, stocks, bonds, and other forms that would not even have been recognized as money a century ago. Similarly, kinetic energy is the most visible form of energy, but some aspect of kinetic energy can be "stored"; for example, in a compressed spring or by storing a projectile at the top of a hill. This makes energy a concept that supersedes motion, which in turn helps us recognize additional forms of energy interchange and storage. Just as a single check efficiently stores a lot of money compared to coins, we can recognize that a chocolate chip cookie efficiently stores a lot of energy compared to a rolling ball—with the energy in a single cookie we can roll a ball many times.

Energy is conserved, but this is not obvious in a spacetime diagram. What happened to the pre-collision kinetic energy in Figure 12.2? It went into heat—the final product in Figure 12.2 is likely to be warm to the touch. Heat is a peculiar form of energy, because it tends to spread out and is not always easy to collect for exchange back into other forms of energy. As a result, all forms of energy tend to eventually end up as heat, but careful experiments show that when heat is accounted for, energy is conserved.

To recap, momentum and energy are each conserved, but otherwise have very different properties. Momentum is a vector—it points in the same direction as the velocity. Energy has no direction—the classical formula for kinetic energy is $\frac{1}{2}mv^2$, and squaring v removes information about its direction. Momentum can only arise through motion, whereas energy ties together many processes that may or may not involve motion. Momentum and energy do have some properties in common. First, they are both frame-dependent because velocity is frame-dependent. Every object has a frame—its rest frame—in which its momentum and kinetic (but not necessarily total) energy are both zero. Second, the total momentum (or energy) of a set of particles is simply the sum of the individual particle momenta (or energies). Finally, the reason both concepts are so important is that they are conserved. We will use this property to predict the outcomes of physical processes such as collisions.

Check your understanding. (a) Which of the following are frame-dependent: momentum; kinetic energy; total energy. *(b)* A bullet with positive momentum in Frame *A* has greater positive momentum in Frame *B*. How does the kinetic energy of the bullet in Frame *A* relate to its kinetic energy in Frame *B*?

12.2 Energy and momentum (including speeds near *c*)

Figure 12.3 shows that if we insist on defining momentum as mass times velocity, or *mv*, it is no longer conserved for particle speeds near *c*. The left panel portrays the equal-mass collision shown in Figure 12.2, again in a frame where the combined particle is at rest. The right panel makes a guess as to how this would look in another frame; the dashed lines echo the worldlines in the original frame for reference. Arcs are marked to remind you that from one frame to the other, the worldlines of particle 2 and the combined particle must rotate by the same amount. (This follows from reciprocity: if particle 2 has a speed of *v* through the combined particle's rest frame, the combined particle must have speed *v* through particle 2's rest frame.) Now, in the second frame all the post-collision momentum is carried by the combined particle and all the pre-collision momentum is carried by the particle 1. The combined particle has twice the mass of particle 1, so if *mv* is conserved particle 1 must have twice the velocity of the combined particle. But particle 1 cannot have *v = c* as shown! Even if we collide slower particles, the diminishing returns of velocity addition ensure that particle 1 has less than twice the speed of the combined particle in the right panel of Figure 12.3. Therefore, *mv* is not the same before and after the collision.

This suggests that *mv* is merely a low-speed approximation to the true definition of momentum. We can expose the approximation by starting at rest. We

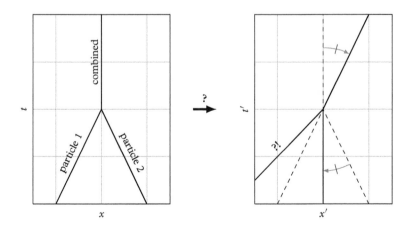

Figure 12.3 *By reframing the collision in Figure 12.2, we see that* mv *cannot be conserved because that would force particle 1 to have* v = c *in the second frame. We infer that* mv *as a definition of momentum works well only at low speeds. The more complete definition* mvγ *works at high speeds as well.*

Confusion alert

We can increase the momentum of a particle by pushing on it, but this does not violate the conservation of momentum. One particle can gain if others lose; conservation applies only to the complete *set* of interacting particles, often called a "closed system."

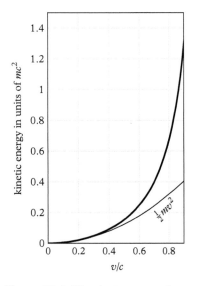

Figure 12.4 *Kinetic energy as a function of speed* v *(expressed as a fraction of c). Only* v *up to 0.9c is plotted here, because kinetic energy grows without bound thereafter. The lower curve illustrates how the classic formula for kinetic energy works well up to about 0.4c.*

give an initially stationary object momentum by changing its velocity: $v = a\Delta t$ if we apply a constant acceleration over some time Δt. This makes its momentum $ma\Delta t = F\Delta t$ where F is the force we apply in order to make the acceleration happen. The Einstein velocity addition law, however, tells us that at high speeds we can keep pushing forever (i.e., increase $F\Delta t$ without limit) and still not quite reach c. This argument suggests that momentum should increase without limit as v approaches c; a detailed calculation using the $F\Delta t$ argument suggests that momentum should be defined as $mv\gamma$. Test-drive this claim in light of the behavior of γ: it predicts that momentum increases without limit as v approaches c, and *also* that momentum is very close to mv at everyday speeds ($\gamma = 1$). High-speed experiments show that indeed $mv\gamma$ is conserved in collisions and other physical processes, so this is universally adopted as the modern definition of momentum.

The modern definition of kinetic energy follows a similar path: the low-speed definition $\frac{1}{2}mv^2$ is clearly wrong at speeds approaching c, where you can push with arbitrarily large amounts of energy and still not reach c. If we upgrade the reasoning process that led to $\frac{1}{2}mv^2$ (which involves $F\Delta x$ rather than $F\Delta t$) to include the diminishing returns of velocity addition, we find that kinetic energy is equal to $mc^2(\gamma - 1)$. Again, test-drive this claim by taking it to its limits: at rest $\gamma = 1$ so this expression is zero (as we expect for kinetic energy at rest), and as v approaches c this expression increases without limit, as we expect for the energy required to push an object arbitrarily close to c. Figure 12.4 shows this quantitatively, up to $v = 0.9c$. The figure also shows that $\frac{1}{2}mv^2$ remains a good approximation for kinetic energy up to $v \approx 0.4c$, far faster than anything most of us have observed. But subatomic particles routinely achieve speeds near c; their energies far exceed the lower curve in Figure 12.4 and confirm the upper curve.

Let us test this line of reasoning on one more old friend: mass. According to Section 2.2, if I apply a force F and observe an acceleration a, mass is the ratio of force to acceleration: $m = \frac{F}{a}$. Imagine I perform this test on an initially stationary object, then repeat it when the same object is moving at v near c. The diminishing returns of velocity addition tell us that same amount of force must be less effective at increasing the speed when the speed is already near c. So, the same F yields a smaller a, and the ratio $m = \frac{F}{a}$ must be larger at high speed—even though it is the *same* particle! Does the mass of the object really increase at high speed?

Nature does not tell us how to define mass; it is up to us to define terms that help us think clearly. If we define mass as $m = \frac{F}{a}$ we must accept that it is frame-dependent and remain mindful of this frame-dependence whenever we refer to mass. Or, we can choose to define mass as $m = \frac{F}{a}$ *in the object's rest frame* so that mass is a property of the object alone, and not the frame. Choosing one option or the other does *not* affect the laws of physics, in the important sense that our final prediction for the behavior of matter and energy must be the same in either case. But the first of our two options makes the laws of physics sound more complicated than they really are. Acceleration is small at high speed because of time skew, length contraction, and time dilation, which have nothing to do with properties of the object itself. So let us define mass as a property of the object alone: $\frac{E}{a}$ *in the object's rest frame.*

Check your understanding. *(a)* According to Figure 12.4, the kinetic energy of a rocket traveling at $0.9c$ to the east in your frame is approximately how many times as large as predicted by the low-v approximation? What does this imply for the difficulty of high-speed space travel? *(b)* Imagine that Planet X also travels at $0.9c$ to the east in your frame. What is the kinetic energy of the rocket in Planet X's frame?

12.3 Energy-momentum relation

The relationships between mass, kinetic energy, and momentum nearly match the triangle picture we developed for spacetime displacement vectors (Sections 7.2 and 11.3), but not quite:

- Mass as we have defined it is a property of the object, not its velocity. Thus it is frame-independent, like the invariant vertical leg of the triangle.

- The horizontal leg in our triangle picture (Figure 7.7) is $\frac{v}{c}\gamma$ times the invariant leg. Apart from a factor of c, this matches the fact that momentum is $v\gamma$ times the invariant mass: $p = mv\gamma$. (Physicists always use p for momentum, presumably because m would get confused with mass.)

- The hypotenuse in our triangle picture (Figure 7.7) is γ times the invariant leg. This nearly matches the fact that, apart from a factor of c^2, kinetic energy is $\gamma - 1$ times the invariant mass.

The fact that the triangle picture does not quite fit is actually surprising, because all vectors are ultimately built on a displacement vector. A velocity vector is a displacement vector divided by a time interval, a momentum vector is that velocity vector multiplied by mass, and so on. If the relationships between mass, kinetic energy, and momentum do not quite fit the triangle picture, our understanding of those quantities must not be complete. This section will complete our understanding.

One thing we know for sure about spacetime displacement vectors is:

$$c^2(\Delta\tau)^2 = c^2(\Delta t)^2 - (\Delta x)^2.$$

We can multiply this equation through by any invariant quantity and the resulting relationship will still be true. We think mass may form the invariant leg of the triangle so we would like to see mass, and no other variables, on the left side of the new equation. Let us therefore multiply through by the invariant quantity $\frac{m^2}{(\Delta\tau)^2}$:

Think about it

If an object is moving east at a speed near c, an additional eastward acceleration is much more difficult than a north-south acceleration. So *if* we defined a frame-dependent $m = \frac{E}{a}$ we would need to think about *two* kinds of mass: one for accelerations along the direction of motion and another for transverse accelerations. This is an example of making the laws of physics sound more complicated than they really are. Beware that some older texts do adopt this approach!

$$c^2 m^2 = m^2 c^2 \left(\frac{\Delta t}{\Delta \tau} \right)^2 - m^2 \left(\frac{\Delta x}{\Delta \tau} \right)^2 . \tag{12.1}$$

Now $\frac{\Delta x}{\Delta \tau} = \frac{\Delta x}{\Delta t} \frac{\Delta t}{\Delta \tau} = v \gamma$ so we can rewrite the second term on the right:

$$c^2 m^2 = m^2 c^2 \left(\frac{\Delta t}{\Delta \tau} \right)^2 - m^2 (v \gamma)^2 .$$

You may recognize $m^2 (v \gamma)^2 = (m v \gamma)^2$ as the square of the momentum. Furthermore, the remaining $\frac{\Delta t}{\Delta \tau}$ can be written as γ, so

$$c^2 m^2 = m^2 c^2 \gamma^2 - p^2 . \tag{12.2}$$

To clarify the meaning of the first term on the right, multiply through by another c^2 and simplify:

$$c^4 m^2 = m^2 c^4 \gamma^2 - p^2 c^2 \tag{12.3}$$

$$(m c^2)^2 = (m c^2 \gamma)^2 - (p c)^2 \tag{12.4}$$

Compare the term $m c^2 \gamma$ with the kinetic energy, $m c^2 (\gamma - 1)$. These two expressions are quite similar, but $m c^2 \gamma$ is always larger. This suggests that $m c^2 \gamma$ represents some kind of energy that is always larger than kinetic energy. But kinetic energy becomes arbitrarily large as v approaches c, so surely it cannot be true that there is some specific type of energy, such as chemical energy, that is *always* larger than kinetic energy. Thus, $m c^2 \gamma$ can only represent the *sum* of all kinds of energy, kinetic and otherwise; this total is always larger than kinetic energy because any increase in kinetic energy increases the total as well. Furthermore, experiments show that $m c^2 \gamma$ is conserved in collisions and other physical processes—meaning that the sum of all the interacting particles' $m c^2 \gamma$ is unchanged by the interaction. Thus, $m c^2 \gamma$ has all the properties we expect of the total energy E of the particle. We can therefore write

$$(m c^2)^2 = E^2 - (p c)^2 . \tag{12.5}$$

This is the **energy-momentum relation**.

We can therefore represent the mass, energy, and momentum with the same stretching triangle we used to represent spacetime displacements in Section 7.2. Figure 12.5 shows a series of triangles as in Figure 11.5, but now when you look at the triangles think of the invariant vertical leg as mass, the horizontal leg as momentum, and the hypotenuse as energy. This figure ignores factors of c so you can focus on the conceptual relationships. In practice, measurements of mass, momentum, and energy often use units where c is not one, but $3 \times 10^8 \frac{\text{m}}{\text{s}}$. If you are making quantitative calculations, the factors of c in Equation 12.5 cannot be ignored. Energy, for example, is most often measured in units of $\text{kg} \frac{\text{m}^2}{\text{s}^2}$ or joules (abbreviated J). To compare this to a mass, Equation 12.5 requires you to multiply the mass in kg by c^2 or $9 \times 10^{16} \frac{\text{m}^2}{\text{s}^2}$, an enormous factor with ramifications to be discussed in Section 12.4.

Confusion alert

"Energy" with no modifier includes all kinds of energy, not just kinetic. I will occasionally add the modifier "total" for emphasis.

Think about it

Rewriting the energy-momentum relation as $E^2 = (m c^2)^2 + (p c)^2$ may help you see how mass and momentum each contribute to the total energy of a particle.

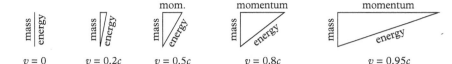

Figure 12.5 *The energy and momentum of a particle as seen in a series of frames ranging from one in which the particle is stationary (left) to one in which it moves at 0.95c (right). The relations between hypotenuse (energy), horizontal leg (momentum), and vertical leg (mass) are the same as for the triangles representing spacetime displacements in Chapters 7 and 11.*

Study the two extremes in Figure 11.5. A triangle with hypotenuse nearly equal to its invariant leg represents a slowly moving particle with a small momentum leg and total energy only slightly larger than its mass. At the other extreme, a triangle with hypotenuse nearly equal to its momentum leg represents a particle moving near c: $\gamma \gg 1$, and the particle energy and momentum are enormous. Of course, these extremes bracket a continuous range of possibilities.

Check your understanding. (a) Visually estimate the value of γ for the $v = 0.2c$ triangle in Figure 12.5, then calculate γ to check your estimate. (b) Draw the triangle for a 500 m/s (0.000002c) bullet. Compare the total energy of this bullet to the energy it has at rest. The result may surprise you.

12.4 E $= mc^2$

For a particle at rest, $p = 0$ so the energy-momentum relation becomes $(mc^2)^2 = E^2$ or simply $E = mc^2$. This is the *minimum* energy a particle of mass m can have, so we call it *rest energy*. This has two surprising implications. First, because c^2 is such large number, everyday objects must somehow store enormous energy even at rest. Second, any process that reduces the energy of an object at rest—such as cooling it down—must reduce its mass! This is an important conclusion so we will double-check the reasoning before addressing the energy-storage aspect.

A simple thought experiment demonstrates that mass cannnot be conserved in many physical processes. Figure 12.6 shows the collision of two clay balls, each with mass m, and with equal and opposite velocities so their net momentum is zero and they have the same pre-collision value of γ. The clay balls stick together, producing a single stationary object of mass M (we write M rather than $2m$ to allow for the possibility that $M \neq 2m$). The substantial kinetic energy that existed before the collision has vanished; where did it go? The collision heats the clay, so the standard answer is that the kinetic energy was converted to thermal energy. But relativity makes a further claim: the post-collision mass M must be greater than $2m$. The proof is simple. Energy is conserved—take this as an experimental fact for the moment—so we can equate the pre- and post-collision energies:

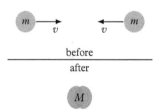

Figure 12.6 *Collision of equal-mass clay balls in the frame in which their net momentum is zero. The claim of relativity is that the final mass* M *is greater than* 2m.

$$2mc^2\gamma_{\text{pre}} = Mc^2\gamma_{\text{post}} \tag{12.6}$$

Because γ_{post} is smaller than γ_{pre}, the only way energy can be conserved is if M is *larger* than $2m$. The extra thermal energy in the stationary clay ball somehow corresponds to an increase in mass!

We verify this with a triangle representation of the collision:

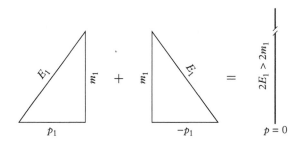

The incoming momenta are equal and opposite so the combined particle has $p = 0$ and its "triangle" is just a vertical line with energy equal to its mass (in units where $c = 1$ as always with the triangles). If we take conservation of energy as an empirical fact, the post-collision energy is $2E_1$ so the post-collision mass is $2E_1$. As we can see by looking at the incoming triangles, this is greater than $2m_1$.

Now that you can use the triangle representation as a thinking tool in collisions, Box 12.2 asks you to push it further by considering the same collision in *two* frames. The extra information we get by making sure the collision inputs and outputs are consistent across frames allows us to drop the assumption that energy is conserved, and instead prove that energy *must* be conserved if momentum is conserved.

We now turn to the implications of the fact that mass is equal to rest energy divided by c^2 and thus can be changed by any physical process that changes the rest energy. First, mass *is* a form of energy: the energy a particle retains when at rest. This does not "look" like energy at a casual glance, but it is a full-fledged form of energy because it can be converted to kinetic energy and vice versa. The conversion has a very favorable exchange rate: according to $E = mc^2$ one unit of mass can be converted into c^2 units of other forms of energy. One kilogram of mass, if entirely converted, would yield 9×10^{16} joules of other forms of energy—roughly the entire annual output of a large electric power plant.

Second, the mass (inertia) of an object is not the "amount of stuff" in it, but the sum of all the nonkinetic forms of energy in it. Take two otherwise identical objects at rest and add energy (but not material) to one of the objects, say by heating it. Because the hot object has more energy, $E = mc^2$ implies that it has more mass, and because $m = \frac{F}{a}$ (Chapter 2) that object will require more force to accelerate by a given amount; it has more inertia. This is the reasoning behind the suggestion in Chapter 2 that the mass of an object depends not just on the mass of its parts but also on how those parts are arranged or interact with each other. For a second

Think about it

Energy is proportional to γ, which represents "speed through time." Everyday objects thus have rest energies much larger than their kinetic energies simply because their speed through time—one second per second—is much larger than their speed through space, which is perhaps one millionth of a light-second per second.

example, consider identical uncompressed metal springs at rest. Compressing one increases its stored energy even while it is at rest, and this increase in rest energy must correspond to an increase in mass. The increase is extraordinarily small because $m = \frac{E}{c^2}$; each one-joule increase in energy is divided by the enormous number c^2 (3×10^8 m/s squared equals 9×10^{16} m²/s²) to obtain the increase in kilograms. Such tiny changes in mass are hardly of practical importance, but the concept that *energy has inertia* is a deep insight that will reappear later in the book.

Third, there are important practical implications of 1 kg of mass storing so much energy. Contrast this with our society's most common energy source, chemical energy from fossil fuels. These fuels release energy when their chemical bonds are rearranged, but the corresponding drop in mass is less than a part per billion. Converting *all* the mass of a fuel would therefore release billions of times more energy. Nuclear reactions do convert close to 1% of the mass of their fuel; this is why nuclear bombs are so powerful, why nuclear reactors produce so much energy with small amounts of fuel, and why research continues into cleaner types of nuclear reactors. This is also how the Sun can release so much energy continuously for billions of years without running out of fuel (Box 12.1).

Check your understanding. (a) In Box 12.2, add the pre-collision triangles in Frame 1 to fill in the details of the correct post-collision triangle. Do the same for Frame 2. *(b)* Explain in more detail what I mean by "add" here.

Think about it

The favorable exchange rate between mass and energy explains why antimatter appears so often in science fiction. When a particle meets its corresponding antiparticle they both *annihilate*, or release all their rest energy as other forms of energy. So an antimatter tank could store vast amounts of energy compactly—but any contact between the antimatter and the tank walls would be catastrophic. Furthermore, nature contains relatively little antimatter so one cannot just mine a planet for antimatter as can be done for other fuels.

Box 12.1 Nuclear fusion in the Sun

The Sun turns four hydrogen nuclei (protons) into one helium nucleus through a series of collisions. The mass of a helium nucleus is 6.645×10^{-27} kg, or 0.7% less than the total mass of four protons (6.690×10^{-27} kg); these numbers are measured precisely in the laboratory by measuring the force required to accelerate these particles. Did 0.7% of the mass really disappear? Although other particles are released, their masses add up to little more than a rounding error in these numbers. The missing rest energy is mostly converted to light (radiative energy) and kinetic energy of the particles flying out of the collisions. This energy heats the Sun and makes it shine; the luminosity of the Sun matches the energy production rate we predict from nuclear reactions in its core. Each second 636 million metric tons of hydrogen are converted to 632 million metric tons of helium. Four million metric tons of rest energy per second yields a power output of 3.8×10^{26} watts. That is one powerful light!

Box 12.2 Graphical conservation of energy and momentum

In this box, we graphically collide the triangle representations of particles to show two things: energy is conserved, and mass cannot be. We need only assume conservation of momentum (which is well tested experimentally) and frame-based thinking tools such as adding velocities and computing γ. To keep it simple, this box analyzes two equal-mass particles colliding to form one combined particle; uses two frames that are particularly convenient; and omits c on the assumption that we are using units where $c = 1$. But the resulting truth is not limited to this particular setup.

The figure below shows a symmetric frame with incoming particles moving at $\frac{v}{c} = \pm\frac{3}{5}$, which makes $\gamma = \frac{5}{4}$. Expressing γ as a fraction helps us verify each pre-collision triangle; the hypotenuse is $\frac{5}{4}$ the length of the invariant mass leg, making classic 3-4-5 triangles:

Frame 1

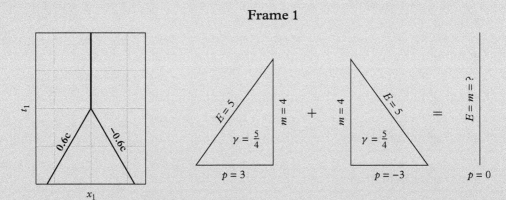

The incoming momenta cancel, leaving $p = 0$ for the combined particle; its "triangle" must be a vertical line with $E = m$. But how much mass and energy does this particle have? If mass is conserved, $E = m = 8$ (the sum of the incoming masses); but if energy is conserved, $E = m = 10$ (the sum of the incoming energies). How can we deduce which (if either) is conserved? We can use a second frame to ensure consistency. We will redraw *each* of the three triangles in a second frame using known transformation rules; this yields a model of the collision process in Frame 2. That model may help us deduce what is and what is not conserved in Frame 2. We will then double-check that the conclusion makes sense back in Frame 1 as well.

We could work out the result for a generic Frame 2 with algebra, but the task is greatly simplified for a frame attached to one of the incoming particles as in the diagram below. In this frame, one incoming particle is quite simple to describe ($v = 0$ and $\gamma = 1$), and so is the *combined* particle: $v = \frac{3}{5}c$ corresponds to a triangle with 3-4-5 proportions as we saw above. As we learned from Figure 12.3, though, we must apply the Einstein velocity addition law to find the speed of the other incoming particle in this frame. That yields $v = \frac{15}{17}c \approx 0.88c$. The spacetime diagram below summarizes Frame 2, echoing the old Frame 1 worldlines with dashed lines for reference:

Box 12.2 *continued*

Frame 2

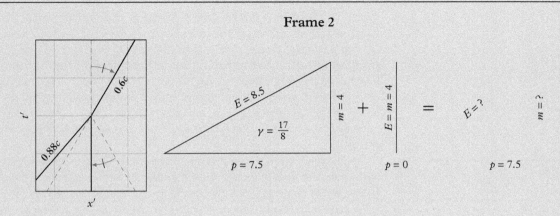

Now we draw the triangle representations for each particle in the spacetime diagram. Each incoming particle has $m = 4$ (mass is invariant), and the stationary particle is just a line with $E = m$ and $p = 0$. For the other incoming particle, its speed implies $\gamma = \frac{17}{8} = 2.25$; this means the energy leg of its triangle is $E = \frac{17}{8} \times 4 = 8.5$. The momentum leg is $\frac{v}{c}$ times this, or $p = 7.5$. Conservation of momentum then implies that the combined particle has $p = 7.5 + 0 = 7.5$.

We still need to deduce E and m for the combined particle in Frame 2. What makes this frame so convenient is that we already know the triangle representation must have 3-4-5 p-m-E proportions. Knowing $p = 7.5$, these proportions immediately give us $m = 10$ and $E = 12.5$. This energy is exactly the sum of the incoming energies ($8.5 + 4 = 12.5$), so we conclude that energy is conserved. Mass is *not* conserved: $m = 10$ is more than the sum of the incoming masses. In fact, because the invariance of mass was a key step in setting up the Frame 2 triangles we could even say that part of the reason mass is not conserved is *because* it is invariant. (Recall the difference between these concepts: conservation applies over time, while invariance applies across frames at each instant.)

Deducing conservation of energy required knowing not only the triangle proportions but also $p = 7.5$, which stemmed from conservation of momentum. Therefore, it is no exaggeration to say that energy is conserved *because* momentum is. In fact, energy and momentum are so tightly coupled that physicists view them as components of a single spacetime energy-momentum vector, just as Δt and Δx are components of a single spacetime displacement vector (Chapter 11). Energy is mass times speed through time ($\frac{\Delta t}{\Delta \tau} = \gamma$), and momentum is mass times speed through space ($\frac{\Delta x}{\Delta \tau} = \frac{v}{c}\gamma$).

Check your understanding. (a) Check that $m = 10$ post-collision also satisfies conservation of energy (and nonconservation of mass) back in Frame 1. (b) What contradiction would you encounter if you tried to conserve mass in Frame 2 and thus tried $m = 8$ for the resultant triangle there?

12.5 Energy budget for particles with mass*

Figure 12.7 provides a breakdown for the energy of an object or particle as a function of its speed, for particles with nonzero mass m. (Particles with $m = 0$ fall into a different category and will be discussed in Section 12.6.) Of course, speed is frame-dependent so we may also interpret this figure as illustrating the frame-dependence of particle energy. As an example of parsing this figure, locate $v/c = 0.6$ on the horizontal axis then go straight up until you hit the curve. This occurs at 1.25 on the vertical axis, indicating that the (total) energy of a particle traveling at $v = 0.6c$ is $1.25mc^2$. Of this, mc^2 is rest energy and $0.25mc^2$ is kinetic energy.

The heavily shaded region indicates that the rest energy mc^2 is always a part of the energy budget. In frames where the object moves, we add to this its kinetic energy, which makes the total energy add up to $mc^2\gamma$. Kinetic energy increases with speed, rather slowly until $v \approx 0.5c$ or so, but more sharply at higher speed. Most notably, the energy increases without bound as v approaches c, and the graph in Figure 12.7 is not tall enough to show the energy for speeds greater than $0.98c$.

The amount of rest energy may look small on Figure 12.7 but (as explained in Section 12.4) is enormous by everyday standards. To put this in context, it equals the enormous amount of *kinetic* energy an object has if $\gamma = 2$, which happens at $v = 0.87c$. (Take a moment to verify this on Figure 12.7: the total energy curve crosses two units of energy at $v = 0.87c$, and these two units are split evenly between rest and kinetic energy.) Thus, accelerating even a miniature 1 kg space probe from rest to $v = 0.87c$ requires as much energy as is locked up in 1 kg of rest energy, which is roughly the total annual output of a large power plant (Section 12.4). This is one reason why the time-dilated interstellar travel discussed in Box 10.1 is much more difficult than it seems; halving the proper time of a trip by accelerating a loaded ship to $\gamma = 2$ would require an enormous amount of energy input. Even this enormous energy would cut the journey time only in half, so even the nearest stars would still require journeys of a few years. If you want your trip to Alpha Centauri to take only months of your time you need $\gamma = 10$ or so, which requires nine times the rest energy for each kilogram in your ship—plus an equal amount of energy to decelerate at the end of the trip.

Is there any way we can reduce the proper time of a journey *without* expending this enormous amount of energy? No. Your speed through time *is* γ (Section 10.4), and large γ means large kinetic energies. The best we can do is find some way to recycle the energy. If we could decelerate returning ships by having them compress and latch a giant spring, for example, the same spring could be used to launch outgoing ships. Unfortunately, springs cannot store that much energy without breaking.

Nevertheless, this thought experiment leads us to the idea that energy can be stored in the *interactions* between parts of an object, such as the forces that atoms in a compressed spring exert on each other. This stored energy is present

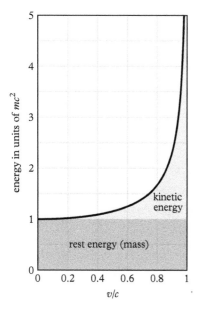

Figure 12.7 *Energy budget for particles with nonzero mass, as a function of speed. The energy is literally off the chart before v reaches 0.98c.*

Think about it

If a spring *could* store this much energy without breaking, its mass when compressed would be substantially larger than its initial mass.

when the object is at rest, and therefore contributes to its mass. In fact, internal interactions comprise *most* of the mass of a typical object, but we must drill down to submicroscopic scales to see this. Everyday objects have thermal energy (the kinetic energy of atoms and molecules moving in different directions with zero net momemtum) and chemical energy (stored in the bonds between atoms). These both contribute to the rest energy, but very little. The vast majority of the rest energy (mass) of an object resides in the rest energy of its atomic nuclei. Even within a nucleus, only about 1% of the rest energy resides in the bonds between protons and neutrons—this is the energy released by nuclear reactions. So far, then, the mass of an object is mostly explained by the masses of its constituent protons and neutrons. Individual protons and neutrons finally provide us with great examples of storing mass in interactions: *most* of their rest energy comes from interactions between their constituent quarks and gluons.

Particles like quarks, gluons, and the electrons that surround atomic nuclei are called *elementary* particles because they are not composites of smaller particles. For elementary particles, we cannot invoke internal interactions as the source of mass. Why then do these particles have mass? Physicists have recently shown that elementary particles acquire mass through something called the Higgs mechanism. You may have read in the popular press that "the Higgs mechanism explains where mass comes from." But this is an oversimplification: the Higgs mechanism explains mass only for *elementary* particles. Most of the mass of everyday objects comes from interactions within the composite particles called protons and neutrons.

Check your understanding. Use Figure 12.7 to verify that a jumbo jet at cruising speed has very little kinetic energy compared to its rest energy. *(a)* Would you say this is because its kinetic energy is small, or because its kinetic energy is large but its rest energy is even larger? *(b)* Rest energy is not easily visible, so how can you determine the rest energy of an object?

> **Confusion alert**
> _____
>
> I will use *mass*, *rest mass*, and *rest energy* interchangeably because they are the same thing. Some texts ues the term *mass-energy*. Variations in terminology should not obscure the central concept that mass measures how much energy is stored in an object.

12.6 Massless particles*

Figure 12.8 shows why particles with measurable energy but very low mass must travel at speeds near *c*. An example of such a particle is the neutrino, which has a mass so low scientists have not been able to measure it exactly. Neutrinos cannot be kept in a laboratory; with nearly zero rest energy, even a very small amount of total energy is sufficient to send a neutrino off at very nearly the speed of light. Even more extreme is a particle of zero mass, for which the triangle representation becomes a mere horizontal line. Lacking a vertical (mass) leg to the triangle, the horizontal (momentum) leg is the same length as the hypotenuse (energy), so the ratio $\frac{v}{c}$ is one. In other words, massless particles *must* travel at speed *c*! The momentum leg representing such a particle has different lengths in different

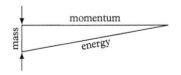

Figure 12.8 *At a given energy, particles with lower mass have momenta more nearly equal to their energy, and therefore travel at speeds closer to c. In the limit of zero mass, the particle* must *travel at c.*

frames, but can never quite be shortened to zero (corresponding to zero energy) because observers can never quite achieve speed c.

This argument can be turned around to look for tiny amounts of rest energy (mass). Imagine staging a race between neutrinos of identical mass but different energies. In the triangle representation the higher-energy neutrinos have a longer hypotenuse and the same vertical (mass) leg, so the triangle is more highly stretched and $\frac{v}{c}$ is closer to one. Therefore, higher-energy neutrinos will finish the race in slightly less time. If the race is too close to call, we must conclude that the neutrino mass is too small to measure. We can use this procedure to test whether particles believed to have zero mass really do. The most important such particle is the **photon**, which is the particle of light. (Readers may initially feel uncomfortable referring to light as a particle, but experiments reveal that on a microscopic scale light is indeed composed of indivisible units.) Tests show that high- and low-energy photons do have the same speed, so the rest energy (mass) of the photon is zero.

For massless particles such as photons, the energy-momentum relation $(mc^2)^2 = E^2 - (pc)^2$ becomes $0 = E^2 - (pc)^2$ or $pc = E$. So, special relativity predicts that massless particles carry momentum, something we would never have expected from earlier forms of reasoning about momentum. This prediction has been confirmed in a wide variety of contexts (see the last paragraph of Box 12.3). In fact, momentum transfer from photons provides the propulsion in some plans for future interplanetary and interstellar travel (Box 12.4).

Check your understanding. Explain why a massless particle must have momentum and move at speed c if it has any energy at all (i.e., if it exists).

Think about it

If photons *did* turn out to have nonzero rest energy, their speed would be slightly less than the invariant speed we call c. This would not invalidate relativity because there are so many lines of evidence that one speed in nature is indeed invariant. This is a reminder that c is a constant of nature that is logically distinct from the behavior of light.

Think about it

We obtained $mc^2 = E^2 - (pc)^2$ by multiplying the metric equation through by m, so you may question its validity for $m = 0$. A more rigorous analysis confirms that indeed $pc = E$ when $m = 0$; Problem 12.13 helps you construct an argument using the triangle representation.

Box 12.3 Experimental proof

Experimental particle physicists use the energy-momentum relation every day while analyzing billions of particles and collisions. That makes the energy-momentum relation, along with time dilation and length contraction, one of the most well-tested equations in all of science.

Without momentum, the energy-momentum relation reduces to $E = mc^2$, and the enormous amount of energy released by nuclear reactions (in reactors, bombs, and stars) is often cited as supporting this equation—and conversely, $E = mc^2$ is often cited as explaining those phenomena. The truth is more nuanced. $E = mc^2$ does predict that a small amount of mass represents a great deal of rest energy, but says nothing about how nature may or may not facilitate the conversion of rest energy into other forms of energy. And bombs, reactors, and stars are hardly environments in which the inputs and outputs are so well measured that they would prove the energy-momentum relation more precisely than a particle-physics experiment.

Still, the prediction that each kilogram of mass stores so much energy is a powerful symbol of the modern era. It captures in a nutshell how a small device can level a city, how a nuclear reactor can power a city while using very little fuel, and how stars can shine with tremendous power for billions of years. In the nineteenth century, astronomers had no good model to explain how the Sun could emit so much energy each second without running out of fuel; they

Box 12.3 *continued*

tentatively concluded that the Sun (and other stars) had very short lifetimes. $E = mc^2$ indicates that tapping the rest energy of a star *could* keep it shining for billions of years, but nuclear reactions explain *how* that rest energy is tapped. Astronomers now know in great detail the different mixes of nuclear reactions that make some stars extremely luminous and others less so. Those reactions have produced all the elements in the universe beyond hydrogen and helium, including the iron and carbon in your body. This is a rich scientific tapestry, and $E = mc^2$ is just one thread in it.

The low-mass limit of the energy-momentum relation says that photons must carry momentum. This has not only been verified in the lab but is also a routine part of exploring the solar system, as spacecraft trajectories must account for pushes by photons from the Sun—*and* by photons emitted from the spacecraft. Momentum transfer from light (also known as *radiation pressure*) also affects natural bodies in the solar system, particularly small ones such as dust grains and small asteroids. The pattern of light emitted from such a body depends on its size and rotation, so the cumulative effect on an orbit differs substantially from one body to another in a predictable manner. Radiation pressure also explains how massive, luminous stars are able to support themselves against gravity, as well as why there is an upper limit to the luminosity of a star—overluminous stars would burst from radiation pressure.

Box 12.4 Sailing on light

The fact that spacecraft trajectories are affected by momentum transfer from sunlight may seem like a nuisance at first, but with the right equipment this can be transformed into a source of propulsion. The Japanese "solar sail" named IKAROS successfully tested this idea in 2010, and NASA planned to launch a larger sail called Sunjammer in 2015 but cancelled the mission a year before launch. More speculative is the idea of using powerful lasers on Earth to provide more push than the Sun can. Spacecraft with sails need not carry any fuel—a huge advantage over rockets, which must start with an enormous supply of fuel that is used mostly to push the fuel that will be used later in the mission.

The Breakthrough Starshot project takes this idea to the next level. Their proposal is to focus an array of Earth-based lasers on a solar sail carrying a very low mass camera chip (a "starchip") so the craft can accelerate quickly to $0.2c$ and reach the nearest stars within twenty years of launch. The spacecraft mass will be minimized not only to allow faster acceleration, but also to minimize the required energy input of $mc^2(\gamma - 1)$. Any human travel using this method would require vastly more powerful lasers, and the starchip is already on the edge of what can be done this century. Another downside for travel this way is that from Earth we can push a spacecraft to higher speed, but not slow it down: the starchips will fly by stars and send back data but never return. For humans to travel and disembark, we would need counteracting lasers at the destination as well. Breakthrough Starshot is extremely ambitious in aiming for other stars, which are hundreds of thousands of times more distant than other planets in our solar system. If that goal proves too ambitious, propulsion from Earth-based lasers may still prove to be a useful tool for moving spacecraft around our solar system.

CHAPTER SUMMARY

- Just as we found an invariant quantity to represent spacetime displacements, there is an invariant quantity to represent the energy of an object with mass: its rest energy mc^2.

- The factor of c^2 in mc^2 implies that a small amount of mass stores a huge amount of energy. Mass is a very compact way to store energy, but it is not necessarily easy to convert it to other forms of energy.

- All nonkinetic forms of energy are stored in rest energy (mass) and thus contribute to the inertia of an object.

- Energy and momentum each depend on frame, but the combination $E^2 - (pc)^2 = (mc^2)^2$ is invariant. This is the energy-momentum relation.

- The energy-momentum relation can be represented by the stretching triangle picture because it is ultimately based on spacetime displacements. For a particle of unit mass, energy is the time component, and momentum the space component, of a vector describing spacetime displacement per unit proper time.

- Massless objects such as photons still carry energy and momentum: with $m = 0$ the energy-momentum relation reduces to $E = pc$.

≡ FURTHER READING

The *Minute Physics* video $E = mc^2$ *is Incomplete* provides an excellent visualization of stretching triangles illustrating the energy-momentum relation. Watch this video until the triangle relation is perfectly clear, then realize that *all* spacetime vectors, including the displacement vector, share these properties. This may give you a new perspective on Chapter 7 with its stretching triangles representing spacetime displacement.

Another video in the *Minute Physics* series, *Einstein's Proof of* $E = mc^2$, analyzes the emission of light from a mass in two different frames to prove the energy-momentum relation—

all in just two minutes. The steps are far too quick to follow on the first viewing, but will make increasing sense with subsequent views and after a review of this chapter.

The Making of the Atomic Bomb is an excellent (and Pulitzer Prize winning) book by Richard Rhodes, starting with the earliest days of particle physics as the enormous amount of energy locked up in atomic nuclei slowly dawned on physicists.

The March 2017 issue of *Scientific American* features a highly readable article on the Breakthrough Starshot project.

◪ CHECK YOUR UNDERSTANDING: EXPLANATIONS

12.1 (a) All these quantities are frame-dependent. Total energy may involve some forms of energy that are not, but it also includes kinetic energy, which *is* frame-dependent. (b) The momentum tells us that the bullet's speed is higher in Frame *B* so its kinetic energy is also higher in Frame *B*.

12.2 (a) Comparing the two curves in Figure 12.4, at 0.9*c* the kinetic energy (heavy curve) is about 1.5 while the low-*v* approximation for kinetic energy taught in introductory physics courses is about 0.4. Thus, the true kinetic energy is about four times the estimate from the low-*v* approximation, which quadruples the expense of pushing a rocket to that speed. (In fact the expense is even greater because storing extra fuel on the rocket increases its mass, which further increases the energy required to accelerate it from rest to any given speed.) (b) The rocket has zero kinetic energy in this frame because its velocity is zero.

12.3 (a) The hypotenuse is barely longer than the mass leg, maybe 5% longer, so I estimate $\gamma = 1.05$. The calculated value is $1/\sqrt{1 - 0.2^2} = 1.02$. (b) The horizontal leg is too small to draw, so a vertical line is a fairly accurate drawing. The energy (the hypotenuse) is then undetectably larger than the energy the bullet would have at rest (where the hypotenuse *exactly* coincides with the vertical leg). This is surprising if we think of the bullet as having substantial kinetic energy. This puzzle is resolved if we realize that everyday objects must quietly store enormous amounts of rest energy.

12.4 *Box 12.2:* (a) The combined particle has $m = E$ in Frame 1 (recall that for this box we let $c = 1$ and drop factors of c to keep the notation as simple as possible), so $m = 10$ implies that $E = 10$, which

is indeed the sum of the incoming energies. (b) In Frame 2, the combined particle is represented by a triangle with 3-4-5 $p - m - E$ proportions, so $m = 8$ implies $p = 6$, which contradicts conservation of momentum.

Section end: (a) In Frame 1, the "triangle" is a vertical line with length $m = E = 10$ (in units where $c = 1$), the point being that it is longer than the sum of the two incoming vertical legs. In Frame 2 the resultant triangle is:

(b) "Adding" here means adding the momenta to determine the horizontal leg, adding the energies to determine the hypotenuse, and then completing the triangle to determine the mass.

12.5 (a) There is no right answer; it is definitely true that the rest energy is much larger than the kinetic energy, but whether you consider the kinetic energy of a jumbo jet at cruising speed to be "large" depends on whether you are comparing it to other everyday phenomena or to the incredible amount of energy stored in mass. (b) Apply a force to the object, observe the acceleration, and take the ratio $\frac{F}{a}$; this is the mass. Multiply by c^2 to convert this number to units of energy.

12.6 Shrinking the vertical leg of the triangle representation to zero, we see that the horizontal leg (momentum) must equal the hypotenuse (energy). The ratio of these lengths is v/c, so equal lengths imply $v = c$.

? EXERCISES

12.1 (a) Referring to Section 12.1, explain the difference between kinetic energy and momentum. Why do we need two distinct ways to quantify the motion of a mass m? (b) Referring to later sections of this chapter, does momentum necessarily involve motion of a mass?

12.2 Redraw the collision diagrammed in the left panel of Figure 12.3, in the rest frame of the *eastbound* incoming particle, and show that this frame also demonstrates the impossibility of conserving the total mv.

12.3 Recall that γ represents a particle's "speed through time." Use this to explain why, even at rest, a particle has a large (by everyday standards) energy of mc^2.

12.4 This chapter states that energy and momentum grow without bound as v approaches c. Explain what this means using examples. Contrast this with how energy and momentum would behave as v approaches c if we were to extrapolate from the low-speed behavior discussed in Section 12.1.

12.5 (a) Use graph paper to accurately draw the triangle representation of a massive particle moving at $0.2c$. (b) Use graph paper to accurately draw the triangle representation of the same particle in a frame in which it moves at $0.9c$.

12.6 (a) Use $E = mc^2$ to determine how much energy would be released if you converted one gallon (3 kg) of gasoline entirely to other forms of energy, leaving no mass behind. (b) Compare this to the 120 million joules released by burning the same gallon in a chemical reaction. Why do automotive engineers not build an engine that releases all the rest energy?

12.7 Explain the resemblance of Figure 12.7 to Figure 7.6.

12.8 If a subatomic particle has $\gamma = 100$, what percentage of its energy is rest energy and what percentage is kinetic energy?

12.9 Consider two sets of magnets: set A consists of magnets glued together with opposite poles touching and set B consists of identical magnets glued with like poles touching. Which, if either, is easier to accelerate as a unit? Explain your reasoning.

12.10 If, hypothetically, high-energy photons were found to travel slightly more slowly than low-energy photons, would you be able to explain this by hypothesizing that photons have some small rest energy? Justify your reasoning.

+ PROBLEMS

12.1 When a sliding object slows down and stops due to friction, where did its momentum go? Explain why momentum transfer is difficult to observe in this and similar cases. *Hint:* imagine the object sliding on a table that itself can slide freely on the Earth.

12.2 Section 12.1 claims that mv is numerically equal to the "force times duration" required to push an object from stationary to velocity v. Show this mathematically, assuming the force is always in the same direction.

12.3 Section 12.1 claims that $\frac{1}{2}mv^2$ is numerically equal to the "force times displacement" required to push an object from stationary to velocity v. Show this mathematically, assuming the force and displacement are always in the same fixed direction.

12.4 The prevailing explanation for the extinction of the dinosaurs is an asteroid roughly 10 km across (implying a mass roughly 2×10^{12} kg) impacting Earth at a relative speed of roughly 40 km/s. *(a)* Compute the kinetic energy of this asteroid using the low-v approximation as well as the fully correct expression. *(b)* How good is the low-v approximation at solar system speeds? *(c)* Describe the triangle representation of this asteroid in Earth's frame. *(d)* Your answer to (c) should imply that most of the asteroid's energy is rest energy rather than kinetic energy. How do you reconcile this with the fact that the conversion of this kinetic energy to heat was enough to cause worldwide destruction?

12.5 According to Figure 12.4, the kinetic energy of a rocket traveling at $0.9c$ to the east in your frame is approximately $1.5mc^2$. What is the kinetic energy of the same rocket as measured in a frame that travels at $0.9c$ to the *west* relative to you?

12.6 Use graph paper to model a sticky collision with the triangle representation as in Box 12.2. In this collision, the particles have equal masses of 3 units and equal and opposite momenta of 4 units. *(a)* Draw the triangle representation of the inputs and outputs of this collision in this frame. *(b)* Do the same in a frame in which one of the original particles is stationary. *(c)* Comment on the similarities and differences with the collision modeled in Box 12.2.

12.7 Use graph paper to model a sticky collision with the triangle representation as in Box 12.2. In this collision, the particles have equal and opposite momenta of 4 units, but one has a mass of 3 units and the other has a mass of only 1 unit. *(a)* Draw the triangle representation of the inputs and outputs of this collision in this frame. *(b)* Do the same in a frame in which the $m = 1$ particle was originally stationary.

12.8 You want to travel to a nearby star (distance 8 light-years) in only one month of proper time. How much energy does this take? Assume for simplicity that you spend the voyage at constant velocity, so

you only have to compute one γ factor, but you do need energy to accelerate at the start and decelerate at the end. Also assume that through some miracle of technology your entire spacecraft has a fully loaded mass of only 10,000 kg and need not carry its own fuel. Compare the energy required with the current annual energy consumption of all of humanity.

12.9 Building on Problem 12.8, imagine that lasers from Earth are used to accelerate your ship away from Earth so that it need not carry its own fuel. *(a)* Realizing that the momentum of your ship at cruising speed is equal to the change in Earth's momentum, divide by the mass of the Earth (6×10^{24} kg) to find the change in Earth's velocity as a result of pushing your ship. Is there substantial environmental impact? *(b)* Can you use the Earth lasers to decelerate your ship upon arrival at the other star? If not, suggest an alternate plan.

12.10 One idea for changing the path of a hypothetical asteroid on a collision course with Earth is to paint it. *(a)* Explain why this strategy could work if done early enough. *(b)* Would you use an absorbing color such as black or a reflective paint such as silver?

12.11 Research the current state of solar sail technology. List the missions that have actually used the Sun for propulsion (as opposed to simply testing sail-unfurling technology). How far have these missions sailed?

12.12 Research the current state of the Breakthrough Starshot project. What hurdles have they overcome and what are the remaining hurdles?

12.13 A stationary object with mass m emits a massless particle to the right. Draw before-and-after triangle representations and explain *(a)* why the massless particle must carry momentum and *(b)* why the mass of the object must decrease in this interaction.

13

The Equivalence Principle

In our everyday lives, objects spontaneously accelerate downward unless they are supported by other objects. We infer that there is a force on them despite the lack of a visible mechanism, and we commonly call this force **gravity**. Over the next few chapters we will see that gravity is responsible for many other phenomena as well. The primary task of this chapter is to convince you that gravity makes time run more slowly in the basement than in the attic.

13.1 Gravity is special

Gravity is remarkable because—assuming we minimize air resistance—all objects in the vicinity of the Earth's surface fall with the *same* acceleration, 9.8 m/s^2, at all times. Contrast this behavior with that of other forces:

- contact forces: when a compressed spring pushes on a pinball, the force F exerted by the spring is determined only by properties of the spring, and has nothing to do with the mass m of the pinball. Because the pinball accelerates according to $a = \frac{F}{m}$ and F is determined entirely by the spring, more massive pinballs will accelerate more slowly. The same reasoning applies to all other examples of contact between two bodies: a truck accelerates more slowly when it is fully loaded, and so on.

- electric force: rub two balloons to charge them with static electricity, and they repel each other. F depends only on the amount of charge rubbed on, so the repulsive acceleration $\frac{F}{m}$ would be smaller for more massive balloons.

- magnetic force: the acceleration of two magnets toward or away from each other depends on their magnetic strength and their mass.

Gravity is different. It apparently arranges for its force to increase proportionally with mass, leaving the combination $a = \frac{F}{m}$ independent of mass (or any other property of the object). So mass is not *just* inertia but somehow also a generator of gravitational force; this is either a remarkable coincidence or a hint of a deeper connection.

Check your understanding. Compare a bowling ball and a marble in terms of *(a)* inertia, *(b)* the force of gravity on each when near the surface of the Earth, and *(c)* their acceleration due to gravity when near the surface of the Earth.

The Elements of Relativity. David M. Wittman, Oxford University Press (2018).
© David M. Wittman 2018. DOI 10.1093/oso/9780199658633.001.0001

Box 13.1 Air resistance

Air resistance contributed to unclear thinking about gravity for thousands of years. A dropped piece of paper accelerates at roughly 9.8 m/s^2 if crumpled tightly, but much less if flat. When the paper is flat, many more air molecules are able to push back on it, and that adds up to a substantial upward push that cancels much of the downward pull of gravity. Conversely, in an airless test chamber even a feather accelerates downward at 9.8 m/s^2, which makes for an impressive physics demonstration (many such videos are available on the internet).

But air resistance is not entirely to blame for the common misconception that heavier objects fall faster: when we hold a hammer we simply feel that it "wants" to move downward more than a pencil. This statement is consistent with objective measurements if we identify "wants to move downward" with the *force, not the acceleration* of gravity on the hammer. The force *is* more—but just enough more to preserve the ratio $a = \frac{F}{m}$. Consider an analogy with horizontal acceleration: a big truck engine generates much more force than a car engine, but it would be wrong to conclude that the truck accelerates faster—we must also account for the mass (inertia) of each vehicle. When we hold a heavy object we are impressed by the force (the "engine") and we forget about the inertia. Air resistance does reinforce this misconception by decreasing the *net* force F and therefore the acceleration $\frac{F}{m}$—and more so for light, fluffy objects.

Air resistance is simply force from collisions with air molecules. As a skydiver accelerates downward, these collisions become more frequent and collectively exert more upward force, so the *net* force (gravity minus air resistance) decreases over time. When the resistance grows so large that the net force is zero, there is no cause for further downward acceleration so the skydiver has reached *terminal velocity*. Most of a skydive is spent in this state, where divers feel like they are "swimming in air." Only the start of a dive is in true free fall—meaning no forces other than gravity—with its distinctive sinking sensation.

13.2 Equivalence principle

If gravity really provides the same acceleration to all objects in a given laboratory or region of space, then thinking of gravity as a *force* is a needless complication. Thinking of gravity as an *acceleration* is more straightforward. This idea had been around for hundreds of years, but Einstein developed it into a powerful thinking tool in the early twentieth century. With this tool we will determine the effect of gravity on light and on clocks, for which *force* is not a very useful concept.

Before using this thinking tool, let us declare exactly what we are assuming. For now let us focus on a given place such as the room in which I am sitting—later chapters will address how the acceleration due to gravity varies from place to place in the universe. The **equivalence principle** is the conjecture that at a given place gravity provides the same acceleration on all particles regardless of their mass, composition, velocity, or any other property (we will refer to this as *uniform acceleration* for brevity). The equivalence principle is essentially a guess about the way nature works, but if true it leads to some startling insights—so physicists have been motivated to extraordinary efforts to confirm it experimentally over the past century and a half. Modern tests confirm that the equivalence principle is true to

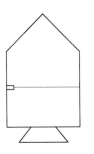

Figure 13.1 *A rocket traveling at constant velocity, with a laser beam crossing the cabin.*

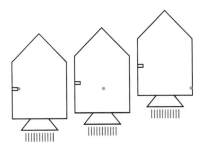

Figure 13.2 *A rocket accelerating upward—note the engine firing—with a laser beam crossing the cabin. Successive snapshots in time are displaced to the right for clarity.*

Figure 13.3 *As for Figure 13.2 but in the rocket frame.*

at least 1 part in 10 trillion. This section applies the equivalence principle to the question of whether light falls in gravity, as a warmup for the more challenging and astonishing application in Section 13.3.

Imagine you are in a rocket, and there is a laser beam crossing the cabin horizontally (shown in red in Figure 13.1). If the rocket is coasting at constant velocity, is there any way to determine your velocity by measuring what happens to the laser beam? By now, you should be comfortable enough with relativity to instantly answer *no*. You can go as fast as you like, but you will never measure a change in the light as long as the source is moving with you at constant velocity. (If you could, that would violate the principle of relativity by providing you with a speedometer that makes no reference to anything outside the rocket.) In fact, the phrase "the rocket's velocity" is meaningless in this context. The rocket may have one velocity relative to a star outside the left porthole, a different velocity relative to the planet outside the right porthole, and a third velocity relative to the oncoming Death Star, but no frame has a monopoly on determining the rocket's velocity in any absolute sense.

Now imagine that the rocket begins accelerating. To help think about accelerations, we introduce a new thinking tool called the **momentarily comoving reference frame** (MCRF). The MCRF will be an *inertial* frame so it will only momentarily be in the same frame as the rocket, which is accelerating. But it is useful nonetheless, because we can determine the result of the experiment in the MCRF using familiar thinking tools, and then translate that result to the accelerating frame. In this case, we imagine an inertial frame at rest relative to the rocket at the start of the experiment.

The MCRF must measure the light as moving straight across, because that is what happens at constant velocity. Focus on one particular photon or flash of light in the stream provided by the laser pointer: it moves straight across while the rocket accelerates upward. Successive images from a movie would appear as in Figure 13.2, where snapshots are displaced to the right to avoid overlap in the drawings. The rocket moves up only a small amount between the first two snapshots, and a larger amount between the last two snapshots. This is what acceleration looks like: velocity increasing with time. Clearly, the light hits the opposite wall lower than it would have in the absence of the rocket accelerating. We can assure ourselves that this conclusion is frame-independent, because we could cover the wall with light sensors and the triggering of any specific light sensor cannot be a frame-dependent question.

An observer in the rocket must agree that the light hits the wall lower down, but she may not agree with the inertial-frame explanation that it happened because the rocket is accelerating. The astronaut does *not* measure the rocket accelerating relative to her, so she has coined a special word to explain why anything that slips out of her hand accelerates downward relative to her and the rocket: gravity. In her frame, the laser hit the lower spot because its path curved downward (Figure 13.3), as does the path of *any* projectile. In other words, *light falls*. If the equivalence principle is true, this must also be how light behaves in the presence of gravity.

Let us be clear that the inertial frame is used only to determine what the outcome must be in the accelerating frame; we are not saying that inertial and accelerated frames are equivalent. What *is* equivalent is that accelerating frames provide free particles with uniform acceleration (independent of particle mass, velocity, and so on) just as gravity does. The astronaut must accept that the rocket is either in space and accelerating, or sitting on the launch pad back on Earth—the only two contexts in which free particles would uniformly accelerate downward. Onboard experiments cannot distinguish between these scenarios, but they *do* clearly reveal that the rocket is not an inertial frame.

Does light fall by a measurable amount? The distance the light "falls" is the distance the rocket moves relative to the constant-velocity case: $\frac{1}{2}a(\Delta t)^2$, where a is the acceleration and Δt is the time light needs to cross the rocket. If we call the width of the rocket w, then $\Delta t = \frac{w}{c}$ so the distance fallen is $\frac{1}{2}a\frac{w^2}{c^2}$. This turns out to be a very small number in most cases because c is so large. If $w = 5$ m and $a = 9.8$ m/s^2, the distance fallen is only about 10^{-16} m! Light falls with the same downward *acceleration* as any other particle, but c is so fast that light leaves the laboratory before it falls a measurable downward distance. In Chapter 18 we will scale this experiment up to a much more powerful source of gravity—the Sun— and a much wider laboratory—the solar system—and see that light indeed falls.

Check your understanding. Alongside the laser in this section, imagine firing a marble horizontally at 5 m/s and a bullet at 500 m/s. Sketch their trajectories as well as that of the laser. Do you see the pattern?

13.3 Slow time

What about a downward- or upward-pointing laser? Now there is no reason for the light to bend, but there may be other observable consequences. Imagine a downward-pointing laser in the nose cone of an *inertial* rocket that sends out a pulse of light every nanosecond. Light travels at about one foot per nanosecond, so at any given moment, pulses are lined up one foot apart between the laser and the floor as shown in Figure 13.4.

Does the distance between the pulses (or, equivalently, the time between pulses as measured by a detector on the floor) depend on the rocket's velocity as long as it is constant? According to the principle of relativity, a constant velocity can have no effect; we may as well say that the rocket is stationary and that outside observers are moving. But if the rocket accelerates, the floor comes upward (as measured in an MCRF) to meet each pulse before it otherwise would. Because *each* pulse hits the floor before it otherwise would, the time between pulse receptions on the floor is *always* less than one nanosecond. Equivalently, we can say that the floor receives more than one pulse per nanosecond. Figure 13.5 illustrates the situation.

Think about it

By the argument in this section, no room on Earth is an inertial frame: Newton's first law is constantly violated because free particles spontaneously accelerate downward. This is a radical shift from the everyday point of view that downward accelerations of "free" objects in your room are due to the force of gravity rather than violations of Newton's first law. This shift may seem uncomfortable at first, but it is worth pursuing because it leads to new insights.

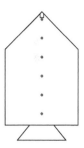

Figure 13.4 *Flashes of light emitted at regular intervals by a downward-pointing light source in an inertial rocket.*

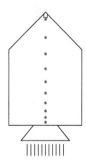

Figure 13.5 *Flashes of light emitted at regular intervals by a downward-pointing light source in an* accelerating *rocket.*

We reach a similar conclusion if we send the laser up from the floor instead. In this case the acceleration of the rocket carries the nose cone *away* from the pulses and makes them arrive spaced further apart. So, while an observer on the floor will claim that the observer in the nose cone is sending pulses more frequently than once per nanosecond, the observer in the nose cone will claim that the observer on the floor is sending pulses *less* frequently than once per nanosecond. In other words, both observers *agree* that the higher clock is ticking faster, even if they do not agree on which one is really correct. Because clocks measure time, we must come to the astonishing conclusion that *time itself runs more slowly on the floor than on the ceiling.*

This is difficult to believe at first, so it may help to think of the pulses of light as being coded to carry messages, such as *the time is now ten nanoseconds after the hour . . . the time is now eleven nanoseconds after the hour*, and so on. The floor would receive *more than one* of these messages per nanosecond and conclude that the nose cone clocks runs too fast, while the nose cone would receive less than one per nanosecond and claim that the floor clock runs too slowly. While they might disagree on which clock is correct, they do agree on which one is faster, and by what ratio. In other words, they agree that time runs more slowly further down the rocket. This is quite distinct from time dilation in special relativity, where each frame concluded that the *other's* clocks ran slowly. The Doppler effect in special relativity can make clocks appear to run either slowly or quickly, but there too each frame observes the same thing regarding the other. The acceleration here has an effect on the vertical direction unlike anything we have seen before—it breaks the symmetry.

You may think that this is just some weird thing about clocks, and that time cannot *really* run more slowly on the floor. To see that it affects humans as well, just imagine that Alice lives in the nose cone and Bob lives on the floor, and each of them set off fireworks on their birthdays. By replacing "nanosecond" with "year" in the preceding paragraphs, we must come to the conclusion that Alice has birthdays more often than Bob! Time is simply what clocks measure, so *all* time-dependent processes will be faster in the nose cone.

If the equivalence principle is true, then this must also happen in the presence of the uniform acceleration provided by gravity. Clocks in the basement of a building on Earth will run more slowly than clocks on the fifth floor. We do not notice this in everyday life because the effect is so small (Section 13.4) and remains small even for astronauts very high up. Even if the effect were large you would never *feel* time running slowly in the basement because your biological clock runs at the same rate as any other clock in your vicinity. To notice the effect, you would have to communicate with someone or observe some process that is higher up. Those processes will appear to be in fast-forward. Section 13.4 describes how physicists have verified that this really happens, and how it is now a part of everyday technology.

Check your understanding. (a) Would the effect on clock rates be more noticeable, less noticeable, or the same if we increase the acceleration of the rocket? *(b)* If a clock were placed not far below Alice would she see it ticking as slowly as Bob's?

13.4 Gravitational redshift

We can repeat and confirm the reasoning in Section 13.3 using the Doppler effect. We saw in Chapter 9 that the emission and reception of regularly spaced flashes of light occurs at different frequencies if the receiver and emitter are in relative motion. But any acceleration performed by the emitter *after* emitting the light cannot matter, so we must think about the emitter frame here as being the frame of the light source at the instant of emission. Our hypothetical MCRF observer shares this frame, so sees no Doppler effect. But while the light is in transit from nose to floor (which takes a time $\Delta t = \Delta h/c$ where Δh is the height difference), the floor-based observer accelerates upward relative to the MCRF, attaining a velocity $v = a(\Delta t) = \frac{a(\Delta h)}{c}$ toward the emitter. The floor-based observer must therefore receive the pulses at a Doppler ratio greater than one. The effect is small because v is small—for a 10-m laboratory t is only about 30 nanoseconds, and accelerating at $a = 9.8$ m s^{-2} for such a short time yields a velocity of only about 300 *billionths* of a meter per second—but determined experimenters can measure it (Box 13.2).

The vertical acceleration breaks the Doppler symmetry emphasized in Chapter 9: the floor-based observer accelerates *toward* the nose-based light while the nose-based observer accelerates *away from* the floor-based light. So observers in the nose receive flashes—think of them as clock ticks—from the floor at a Doppler ratio *less* than one even as floor-based observers receive flashes (ticks) from the nose at a Doppler ratio *greater* than one. This is functionally equivalent to floor-based clocks *really ticking less frequently* than nose-based clocks; time passes more slowly for clocks further down in the direction that free particles accelerate. This—along with violation of Newton's first law (Chapter 4)—is a distinctive feature of accelerated frames.

"Frequency" here need not refer to separate flashes of light. We may also think of light as a wave, and in any of the previous reasoning we can substitute "wavecrest" for "flash" or "pulse" and reach the same conclusion. Higher-frequency waves of light are perceived by the human eye as blue, and lower-frequency waves as red. So the effect of gravity on light is to shift its frequency ever so slightly blueward (if the receiver is lower than the emitter) or redward (if the receiver is higher than the emitter). This effect is called **gravitational redshift** ("blueshift" is also acceptable when you wish to call attention to the direction of the effect). The "shift" part is misleading—what happens is really a stretching or compressing—but the term is so entrenched that we will use it as well.

Think about it

The room you are sitting in is an accelerated frame, because all free particles in it experience the same downward acceleration. This is a difficult conceptual shift from the idea that a falling object is not "free" but is responding to the "force" of gravity. Try making the shift in stages: first, uniform acceleration is a simple model that matches observations and therefore deserves a hearing; second, the force model leaves unexplained how gravity arranges for larger forces on objects with more inertia; third, the accelerated-frame model predicts an effect on clock rates so we should at least test those predictions; fourth, those predictions turn out to be correct so we should accept the accelerated-frame model.

Gravitational redshift is small on Earth. Plugging the 300 billionths of a meter per second we calculated for the 10-m high laboratory into the Doppler formula, we find a ceiling-to-floor Doppler factor of about 1.000000000000001— a number so close to one that it is more conveniently written as $1 + 10^{-15}$. More generally, the clock tick rate for a frame with acceleration a increases with height h as $1 + \frac{ah}{c^2}$; our result is so close to one because c^2 is so large compared to a and h. We will use this formula again in Section 14.2, so the remainder of this paragraph justifies it further for those interested. We showed earlier in this section that a floor-based observer gains a velocity $v = \frac{ah}{c}$ while the light is in transit from a source of height h above the observer, so we need only plug this velocity into the Doppler formula to find the ratio of ceiling time to floor time. Although I warned you away from the low-speed approximation to the Doppler formula in Box 9.2, that approximation is actually well justified for the tiny velocities here. Using that approximation, $\frac{f_{\text{receive}}}{f_{\text{emit}}} = 1 + \frac{v}{c}$ with $v = \frac{ah}{c}$ yields a ratio of $1 + \frac{ah}{c^2}$.

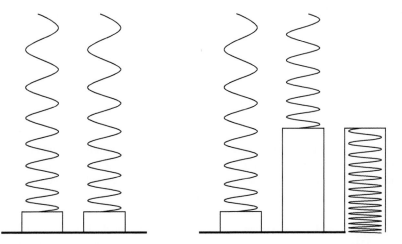

Figure 13.6 *Gravitational redshift as light climbs from the surface of the Earth.* Left, *two sources of light (serving as clocks) are compared side-by-side to ensure they are identical.* Right, *one is placed atop a tower. At that height, it emits more than one wavecrest (tick) for each passing wavecrest (tick) emitted by the lower clock. The second tower on the right shows the upper clock flipped over; all observers* still *agree that the lower clock ticks less frequently.*

Think about it

Note that each climbing wave in Figure 13.6 matches Figure 13.5 if you draw a wave there such that its crests match the dots.

Energy provides another way to think about gravitational redshift. A bullet fired upward loses kinetic energy as it climbs by slowing, but light cannot slow down. Instead, a wave loses energy by cresting less frequently. This makes the waves lengthen as they climb, as illustrated in Figure 13.6. The left panel shows two light sources side-by-side on the surface of the Earth to illustrate that they are identical—their clocks tick at the same rate. In the right panel, one source has been moved atop a tower. For each tick of the lower clock one long wave passes the upper clock, which emits *more than one* tick in that time. (The size of the effect is greatly exaggerated.) The upper clock therefore accumulates more ticks during any given experiment, and this can be verified by looking at the time displayed by each clock. The rightmost tower shows what happens if we flip the tower clock over; its light gains frequency as it falls, so everyone between and below the clocks also agrees that the lower clock ticks less frequently.

The clocks are physically identical, so this effect has nothing to do with the clocks themselves: time itself must run more slowly at the bottom of the tower. To illustrate this, we could bring the lower clock up to the top of the tower for a second side-by-side comparison. Under these conditions the formerly-lower clock will once again match the *rate* of the second clock, but will have recorded fewer total ticks over its lifetime. It recorded fewer ticks *only* when it was at a lower altitude than the second clock. This justifies again the statement that time runs more slowly at the base of the tower. The same argument could be repeated for *any* two heights, so the general conclusion is that time runs more slowly further down when "down" is defined by acceleration or gravity.

Gravitational redshift is a practical way to compare clock rates because the frequency of a light wave is easy to visualize and can be measured precisely. For this reason, the first experiment confirming slow time in gravity (the Pound–Rebka experiment in 1959) was a measurement of gravitational redshift using a several-story building. Since then more sophisticated experiments (such as launching a laser into space to achieve a vertical height difference of 10,000 km) have verified the gravitational redshift effect to very high precision (0.007%). And astronomers routinely detect much larger gravitational redshifts around celestial objects with stronger gravity than Earth. Gravitational redshift is thus an important tool for astronomers to study how well a theory of gravity describes nature. We will return to this theme when we study the celestial objects with the largest gravitational redshifts.

The term *gravitational redshift* is used because it describes the observation process, but when you read this term think *slow time*: gravitational redshift corresponds directly to time running slowly where the light was emitted. Well-known frequencies of light emitted by atoms serve as well-calibrated clocks (Section 9.3), so astronomers and physicists can directly measure how slowly time passes in different environments.

Check your understanding. A tall tower has a red light at the top. *(a)* When you stand on the ground, is the light you receive a tiny bit redder, bluer, or the same as the light that was emitted? *(b)* How would you answer the same question if the light were on the ground and you were atop the tower?

Think about it

In principle, light from celestial objects undergoes a small *blue*shift as it falls into our Earth-based telescopes, but Earth's gravity is so weak that this is a mere rounding error compared to the redshift light suffers as it climbs away from a massive celestial object.

Think about it

When light from a celestial object is redshifted, how do astronomers know if the cause is gravity or Doppler effect from overall motion? The latter would redshift all the light the same way, but gravity causes a *spread* of redshifts because the emitting gas occupies a range of heights in and above the object. This spread further allows astronomers to infer how the gas is distributed in relation to the central object.

Box 13.2 Experimental proof

Experimental tests of the equivalence principle either test directly whether all objects fall with the same acceleration, or test the logical consequences such as altitude-dependent clock rates and gravitational redshift. In the first category, Galileo is said to have dropped balls of different masses from the Leaning Tower of Pisa to prove that they accelerate equally. Galileo experts think this is an embellishment, but similar experiments were performed by others in sixteenth-century Italy. Later, Newton used pendula of differing masses and materials to demonstrate this principle to higher precision. Even more precise tests were conducted by Loránd (often anglicized to Roland) Eötvös in the late 1800s

continued

Box 13.2 *continued*

and early 1900s, ultimately finding that Earth's gravity accelerates different masses equally, to a precision of one part in 100 million. The cleverly named Eöt–Wash group at the University of Washington has pushed this to a precision of one part in ten trillion. Studies of the Earth-Moon orbit show to a similar precision that these two very different bodies accelerate equally in the Sun's gravity. Earth and Moon have slightly different mass budgets in terms of "stuff" (rest mass of atoms) vs. other forms of energy, so this last test can probe whether other forms of energy respond to gravity differently. No distinction has been found.

Clock rates provide another test of the equivalence principle. In the first and most famous such experiment, the Hafele–Keating experiment, atomic clocks flew around the world on commercial airliners (an inexpensive method of keeping them at high altitude for a long time) in October 1971 and were compared to stationary clocks. The results agreed with the time gain predicted by the equivalence principle, minus the loss due to time dilation, to a precision of about 10%. Subsequent iterations of this idea under more controlled conditions confirmed the prediction to a precision of 1.6%. Since then, atomic clocks have become so precise that slow time can be detected in the lab at height differences of *less than one meter* (*Science*, volume 329, pp. 1630–33). Clock rates can also be probed with gravitational redshift; this has been confirmed in the vicinity of Earth to a precision of 0.007% (Section 13.4). Astronomers looking at gravitational redshift throughout the universe have confirmed that time runs slowly in many places, and always by the ratio we would expect based on what we know about the gravitational acceleration there.

Altitude-dependent clock rates are now built into everyday technology. Your phone locates itself by listening to Global Positioning System (GPS) satellites broadcasting their current time and position. A GPS chip subtracts the time encoded in the broadcast from the current time to determine the time of flight of the message (traveling at *c*) and therefore your distance from that GPS satellite (Section 7.4). The clock aboard the satellite runs faster than Earthbound clocks, by several parts per trillion. This sounds negligible, but adds up to microseconds per day. If this still sounds negligible, recall that in a few microseconds light travels the better part of a kilometer. Thus, if the GPS clocks were not corrected for their faster ticking rate due to their altitude, your position would be off by about a kilometer *after the first day of operation of the satellite*. The system would be completely useless if this correction were not made.

13.5 Gravity disappears in freely falling frames

The implication of the equivalence principle, that *gravity makes time run more slowly further down*, is so astonishing that you will need some time to absorb it. The aim of this section is to increase your comfort level with the equivalence principle by applying it to some everyday situations.

First, we introduce the concept of a *freely falling* observer. This observer has no forces other than gravity applied to him. You know what this feels like: that "roller coaster drop" feeling. But your experiences with free fall have probably been too short in duration to actually perform the following experiments.

Confusion alert

As explained in Box 13.1, even skydivers experience little true free fall due to air resistance. But throughout this section, imagine that there is no air resistance and all objects continue to accelerate uniformly.

Given the uniform acceleration provided by gravity, two objects freely falling next to each other will remain next to each other. Imagine that Alice and Bob jump out of a plane together. While both are in free fall, each measures the other as stationary, with no downward acceleration. If Alice drops her phone, *it will not fall relative to her* (again, assuming no air resistance). Normally to prevent a phone from falling relative to you, you must apply a supporting force equal to the weight of the phone. In Alice's free fall the force she needs to apply is zero, so the phone is weightless. In fact Alice and Bob themselves are weightless; any scale they step on will read zero. If we put freely falling walls around Alice and Bob, they would say that their room simply has no gravity. Therefore, we can restate the equivalence principle: experiments inside a freely falling lab cannot measure any effects of gravity.

We close with a case study illustrating this point. Alice stands on the ground and throws a ball straight up in the air at an initial velocity of 20 m/s while Bob, skydiving above, freely falls toward the ball (neglect air resistance in all of this). Alice will observe Bob plummeting toward her, and she will observe the ball going up and then coming back down—both accelerated by gravity. What does Bob observe for the ball's motion?

For concreteness, start the experiment at $t = 0$ when the ball leaves Alice's hand, and imagine that Bob has $v = -100$ m/s in the Earth frame at this instant (it is standard to consider downward velocities negative and upward velocities positive). The story says that at $t = 0$ the ball has $v = 20$ m/s in the Earth frame, so the distance between the ball and Bob is closing at 120 m/s (20 m/s due to the ball rising and 100 m/s due to Bob falling). At $t = 1$ s, in the Earth frame the ball has slowed to 10 m/s upward (rounding the 9.8 m/s^2 of gravity to 10) while Bob is now moving downward at 110 m/s. The distance between the ball and Bob thus continues to close at 120 m/s (10 m/s due to the ball rising and 110 m/s due Bob falling). At $t = 2$ s, in the Earth frame the ball has slowed to 0 m/s (it reaches the top of its trajectory and turns around at this instant) while Bob has increased his downward speed to 120 m/s. Thus the distance between the ball and Bob *still* shrinks at 120 m/s. If Bob aimed a radar speed gun at the ball, it would register a constant 120 m/s oncoming speed! Please run the numbers for $t = 3$ s yourself to convince yourself that Bob's speed gun would continue to read 120 m/s even as the ball completes the downward part of its trajectory in the Earth frame.

In Bob's frame the ball does *not* follow the usual up-and-down trajectory. It simply approaches him at a constant velocity of 120 m/s. Figure 13.7 shows graphically that the ball has a constant velocity of 120 m/s relative to Bob. Bob does not see the ball accelerating, so he concludes there is no force on it. In other words, gravity effectively disappeared inside the freely falling "laboratory" we can draw around Bob and the ball. Of course, a long enough free fall always ends in a crash, so the point is not that gravity truly disappears. Rather, "the effect of gravity

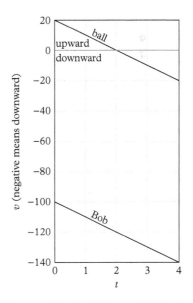

Figure 13.7 *Earth-frame velocities of a ball launched upward from the surface, and of a skydiver (Bob). Positive velocities (above the dotted line) are upward, and negative velocities (below the dotted line) are downward. In this frame the ball goes up and then down, but* relative to Bob *the ball maintains a constant velocity of +120 m/s.*

disappears" means that we may apply the rules of inertial observers in special relativity to experiments conducted entirely inside the freely falling lab. In Chapter 14 we will see how this thinking tool unlocks a new understanding of gravity.

Check your understanding. (a) Fill in the details for $t = 3$ s and convince yourself that Bob observes the ball moving at constant velocity rather than reversing direction. *(b)* Does a freely falling laboratory (such as one we can imagine drawing around Bob) pass the coffee-sloshing test for acceleration?

Box 13.3 Equivalence principle applications

A NASA aircraft trains astronauts for weightlessness by diving steeply in an approximation of free fall. The aircraft can maintain this dive for less than a minute, after which it climbs steeply to gain altitude for another dive, and so on. This explains the affectionate nickname for this aircraft: the Vomit Comet. Experiments there can test the performance of people and devices in weightless conditions without the risk of really being in space. For example, fires on the space station will burn differently than on Earth. Fires on Earth maintain themselves by heating the surrounding air, which then rises and helps fresh oxygen come in from farther away. Without gravity, hot air does not have buoyancy and fires would evolve quite differently. The Vomit Comet can test this to understand space station safety without risking space station safety.

The equivalence principle also implies that we can use acceleration to generate "gravity" where there is none. Our special relativity thought experiments often involved instant acceleration from one frame to another; we see now that hypothetical engines capable of this would cause crushing "gravity" for the crew. Accelerations of 9.8 m/s^2 would simulate Earth gravity. At this rate, accelerating to near c to take advantage of time dilation would take an entire year—plus another year to slow down and stop at the destination. At this rate, a voyage to the nearest stars would require a minimum of about two years of proper time no matter how large the time dilation at full cruising speed. An upside of this scheme is avoiding the decline in crew physical fitness that accompanies long periods of weightlessness. For longer voyages, fuel can be saved by turning the engines off once a large γ factor has been reached, because time dilation ensures that the resulting period of weightlessness will be short in terms of proper time. Of course, this plan still requires a *lot* of energy as pointed out in Section 12.5.

In practice, most spacecraft coast most of the time, so maintaining physical fitness in the face of weightlessness is a pervasive issue for astronauts. One solution would be to spin (at least part of) the spacecraft. As explained in Section 16.1, moving in a circle of a given radius requires constant acceleration, and this is just what is required to mimic gravity. Once the device is spun up, no energy is required to maintain it except to replace frictional losses. This solution is therefore *much* cheaper than constantly accelerating with engines, and is often seen in "realistic" science fiction (e.g., *2001: A Space Odyssey* and *The Martian*). However, to date no such craft has actually been built. For now, astronauts on the International Space Station maintain fitness by running on a treadmill—but to have "weight" on the treadmill they strap themselves into bungee cords that press them against the treadmill. This sounds low-tech, but actually a great deal of engineering goes into protecting the rest of the space station from the pounding and the resulting vibrations.

CHAPTER SUMMARY

- Gravity appears to be the same as uniform acceleration. We use this as a thinking tool to deduce how gravity affects light and clocks.
- The path of light must be bent by gravity (unless it is initially vertical).
- Gravity makes time run more slowly further down.
- The effects of gravity disappear in freely falling laboratories.

☰ FURTHER READING

Was Einstein Right? by Clifford M. Will is a suspenseful account of the race to test the equivalence principle (and other aspects of relativity) ever more stringently in the latter half of the twentieth century.

◢ CHECK YOUR UNDERSTANDING: EXPLANATIONS

13.1 The bowling ball has (a) more inertia; and (b) more force of gravity exerted by Earth on it. But both of these are greater by the same factor so (c) both have the same acceleration toward Earth.

13.2 The pattern is that, the faster the horizontal speed, the less *distance* the projectile falls, because there is less *time* before hitting the wall. This is why the trajectories differ despite having identical vertical accelerations. This is reminiscent of the dropped vs. fired bullet in Chapter 3; both fell with the same acceleration, but the bullet hit Earth very far away due to its large horizontal speed.

13.3 (a) The effect increases with acceleration. (b) No, a small vertical separation would yield only a small effect. In summary, the effect increases with both acceleration and vertical separation.

13.4 (a) A tiny bit bluer. (b) A tiny bit redder.

13.5 (a) After 3 s, the ball now travels *downward* at 10 m/s relative to the ground. Bob now travels downward at 130 m/s relative to the ground, so Bob's speed gun continues to read 120 m/s when pointed at the ball. (b) Yes, because coffee and cup accelerate at exactly the same rate.

? EXERCISES

13.1 In what way(s) is it true that "heavier objects fall faster" and in what way(s) is that statement misleading?

13.2 You push two library carts with the same acceleration: an empty one with your left hand and a fully loaded one with your right hand. Describe the forces exerted by each of your arms. Does gravity arrange forces like that in the case of vertical acceleration?

13.3 Aliens kidnap you and put you aboard a windowless rocket accelerating at 9.8 m/s². *(a)* How, if at all, could you detect that you are not still on Earth? Explain your reasoning. *(b)* Sketch the rocket to show

how the direction you think is "down" relates to the direction the engines push. *(c)* The aliens now turn off their engines and coast, but at very high speed. How do you feel the change?

13.4   You fall out of an airplane side-by-side with a refrigerator. Are you able to lift the refrigerator over your head before you open your parachute? Explain.

13.5   Two baseball players jump out of an airplane and play catch while falling, and you jump along with them. Neglect air resistance. *(a)* Sketch the two players and the ball's path as seen by you (*not* as seen by someone on Earth), if the player on the left throws the ball as he does on Earth, with some upward velocity as well as horizontal velocity. *(b)* On the same diagram, sketch the ball's path if the player on the left throws the ball with only horizontal velocity.

13.6   A scientist jumps out of an airplane, accidentally knocking a flashlight out of the airplane at the same time. *(a)* Before opening her parachute (and neglecting any effects of air resistance), what does she observe happens to the frequency of light from the flashlight as she falls? Why? *(b)* Would your answer change if the flashlight were on the ground pointing up? Why? If your answer changes, describe the different behavior.

13.7   Two magnets are in a vertical glass tube. The upper magnet is glued to the top. The bottom magnet is free to move and is attracted to the upper magnet, but not enough to overcome its weight and actually move. Describe what, if anything, happens when you drop this entire apparatus. Explain your reasoning.

13.8   Sketch an experimental setup in which we would observe a gravitational blueshift.

13.9   Section 13.5 states that the feeling of weight really comes from the floor pushing up on you. What would a bathroom scale read for your weight if you and the scale were in free fall? Explain your reasoning.

➕ PROBLEMS

13.1   Constant moderate acceleration has the benefit of providing long-term artificial gravity for spacecraft as discussed in Box 13.3. What is the corresponding problem associated with a briefer period of high acceleration designed to quickly get the spacecraft to cruising speed?

13.2   Consider a downward-pointing light source on Earth. If the light hits a mirror on the ground and bounces back up to its original height, what is its frequency compared to its original frequency?

13.3   Astronomers observe a range of gravitational redshifts around a celestial object with strong gravity, such as a neutron star, because the light-emitting atoms are scattered at a range of heights on and above the star. Where does the most-redshifted light come from?

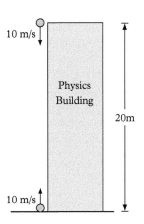

Figure 13.8 *Two balls on a collision course each begin with a speed of 10 m/s.*

13.4   Two students conducting a physics experiment need to make two balls collide at a relative velocity as high as possible. They decide that one will go to the roof of a 20 m high building, the other will stay on the ground, and they will simultaneously throw the balls at each other (see Figure 13.8. They can each throw at 10 m/s. Neglect air resistance. *(a)* They predict that the balls will collide at a relative velocity >20 m/s, because the downward-going ball picks up speed on its way down. Is their prediction correct? What is the relative velocity of the collision, and why? Explain in terms of concepts rather than just raw computation. *(b)* Make a plot of vertical velocity of the upper ball versus time. Right under it, make a plot of vertical acceleration vs. time. Make sure they are aligned so that the relationship between velocity and acceleration is clear. *(c)* On each of the plots you just made, add another line representing the lower ball, and be sure to mark it clearly.

13.5   *(a)* If a clock on board a satellite in low Earth orbit gains a few microseconds per day compared to Earthbound clocks as claimed in Box 13.2, *roughly* how much time does an astronaut gain by remaining on a space station for one year? *(b)* Would the astronaut notice the effect on a daily basis? Explain why or why not. *(c)* Could the astronaut notice the cumulative effect after one year? Explain why or why not.

13.6   Imagine throwing your physics professor into a region of strong gravity such as a supermassive planet while you remained at a safe distance. Making a mighty effort, he stands on the surface of the planet while you attempt to answer these questions. *(a)* How would the light from his red laser pointer appear to you? (Assume he points it straight at you.) *(b)* If you pointed a red laser at him, how would that light appear to him? *(c)* Would you see him scream as if in slow motion, fast forward, or normal speed? Explain your reasoning. *(d)* Would he see you laugh as if in slow motion, fast forward, or normal speed? Explain your reasoning.

13.7   Do the following either as an actual experiment or as a thought experiment. Connect two D batteries or similar weights with a weak spring, such as the kind on a keychain or an old-fashioned telephone cord. Hold one battery so the other hangs down and the spring is extended. *(a)* What does the spring do when you drop the whole assembly? Explain why. *(b)* Why was a *weak* spring specified? *(c)* Relate this to the sinking feeling in your stomach when in free fall. *Hint:* model your stomach as a mass connected to your rib cage.

13.8   As a publicity stunt, Science World, a science museum in Vancouver, placed scales in elevators with the message "You weigh less on the way down." Is this message correct or incorrect, or somewhere in between? Provide a complete explanation of when you weigh less and why.

13.9   Imagine we build an actual freely falling laboratory around Bob in the story of Section 13.5. Now, turn on air resistance. Will Bob still feel weightless once the lab reaches terminal velocity? Explain your reasoning.

14 Gravity Reframed

We saw in Section 13.5 that the effects of gravity disappear in freely falling laboratories. Within such a laboratory, a freely falling particle appears to have no forces on it—it is an inertial particle, following the rules of special relativity. We therefore expect it to follow a worldline of maximum proper time (Section 10.4). This chapter develops new thinking tools for identifying paths of maximum proper time in Earth's frame, where clocks have altitude-dependent tick rates. The reward: we will find that altitude-dependent time *by itself* explains all the trajectories we associate with everyday gravity. This explains how "the force of gravity" manages to accelerate all particles equally. Gravity is not a force at all; it is slow time.

14.1 Maximizing proper time

How do we figure out the maximum-proper-time path between two events when the clock tick rate is altitude-dependent? Figure 14.1 shows one attempt to visualize the situation: a spacetime diagram shaded to remind us of the altitude dependence. Darker shading corresponds to faster tick rates, so the bottom of the laboratory (the planet's surface) is at the right where the shading is lightest and time runs most slowly. Two events A and B are marked, and our job is to determine the maximum-proper-time path between A and B. When presented with a similar question in Section 10.4, we used a powerful thinking tool: choose a new frame that puts the two events at the same position. Unfortunately, we cannot use that tool here because the shading is fixed to the planet-based coordinate system.

To search for another thinking tool, let us state the problem in more detail. Imagine a particle—with a clock attached—moving from A to B. As long as it moves slowly, time dilation is negligible, and its tick rate at any instant must be the same as the tick rate of the stationary clocks around it. The moving clock can maximize its ticks by avoiding regions where stationary clocks tick slowly—but it can't avoid those regions forever because it has to get to event B. Try to pencil in a path based on this logic. Once you have done so, ask yourself whether any part of this path has a very high speed. If it does, you now have to consider that time dilation makes high-speed segments contribute less to the accumulated proper time. See if you can modify your path to reduce the maximum speed a

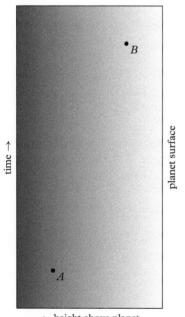

Figure 14.1 *Spacetime diagram with shading to illustrate the tick rate of stationary clocks: darker corresponds to faster rates. What is the maximum-proper-time path from event* A *to* B*?*

time →

planet surface

← height above planet

The Elements of Relativity. David M. Wittman, Oxford University Press (2018).
© David M. Wittman 2018. DOI 10.1093/oso/9780199658633.001.0001

bit without sacrificing the basic idea of keeping most of the worldline in quickly ticking regions.

This problem sounds very abstract, but you perform similar optimization procedures in your own life, probably without realizing it. Imagine being on the beach when someone in the water needs help (Figure 14.2). What is the fastest path? You will find yourself, without even thinking about it, following the solid line in Figure 14.2: running along the beach most of the way so you have less distance to swim. Because swimming is slower than running, the optimal path is one that has less swimming than the straight-line path—even if the total distance is a bit longer. Now imagine you have a friend who understands this concept and takes it to an extreme: he runs further along the beach to reduce swimming to an absolute minimum (dotted path in Figure 14.2). He arrives later than you! Why? Because your swimming distance was only slightly longer than his, while your running distance was *substantially* shorter. You found the optimal path by steering between two types of penalty: the slow speed of the shortest-distance path and the long distance of the highest-speed path. Any deviation from your path would increase the time required to reach the swimmer.

With this in mind, return to the problem of events A and B (Figure 14.3). Although our objective is now to *maximize* the proper time, the reasoning is analogous. The constant-velocity worldline from A to B may be a first guess, but we can improve it by hanging out longer in quickly ticking regions before zooming over to event B. But pushing this strategy too far, as in the dotted path in Figure 14.3, carries its own cost in terms of time dilation. The optimum path must be between these extremes.

So, we found a path of maximum proper time; what does it mean? The particle that follows this path is inertial. We can verify that this object has no forces on it by, say, attaching springs and seeing that they do not compress during the journey. If seeing an inertial particle accelerate surprises you, remember that this always happens in accelerated coordinate systems (Section 4.6, Section 13.2). Our planet-based coordinate system with its nonuniform progression of time is an accelerated coordinate system.

Figure 14.3 represents an object being dropped: it starts at rest and then gradually picks up speed. In other words, *it accelerates downward merely to maximize its proper time*; we need not invoke a "force" to explain this acceleration. This is an important conclusion so let us check it quantitatively. If you make a video of an object dropped from rest until it hits the floor, you will find that after half the time the ball is only ¼ of the way to the floor. [Why? Say $v = 0$ at the start and $v = v_1$ after half the time, for a first-half *average* velocity of $\frac{v_1}{2}$. Now the second half starts with $v = v_1$ and (given the constant acceleration observed in Earthbound labs) ends with $v = 2v_1$ for an average of $\frac{3v_1}{2}$. The second-half speed is triple the first-half speed, so the second-half distance is also triple the first-half distance.] Check Figure 14.3 now to make sure that this is true for the middle path and *not* true for the other paths. Now that you know exactly how to recognize a worldline representing an object falling in an Earthbound lab, you can see that such a path

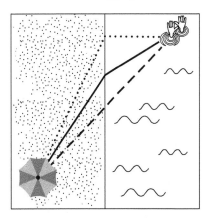

Figure 14.2 *The optimum path from the lifeguard to the swimmer is the solid line; the straight-line path is slower because of the additional swimming, while the minimum-swimming path involves too much total distance.*

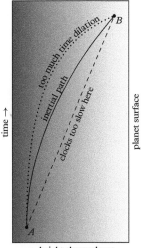

Figure 14.3 *In spacetime the maximum proper time path—the inertial path—from* A *to* B *stays mostly in regions where clocks tick frequently (heavy shading) while also avoiding too much time dilation en route to* B. *The time scale has been compressed to fit on the page, so a 45° tilt is slower than* c.

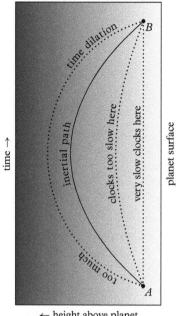

Figure 14.4 *As for Figure 14.3, but events A and B are now at the same altitude.*

does steer a middle ground between avoiding slow-time regions and avoiding too much time dilation. This can, of course, be proven mathematically in more detail.

To further convince you that freely falling objects maximize proper time, we will consider two other event configurations. First, imagine a ball, thrown upward with some initial velocity, that comes to rest on a high shelf. This is the opposite of our previous configuration, so we can simply flip Figure 14.3 vertically to see the answer. In this case, the ball continually slows from its initial speed, but it still spends the requisite large fraction of its time in regions where clocks tick quickly.

Second, imagine events A and B being at the same altitude (Figure 14.4). What is the maximum-proper-time path between these events? We may be tempted to draw a straight line, but the straight-line path accumulates few ticks because it stays entirely in a slow-time region. More ticks are accumulated along a path that deviates upward to more heavily shaded regions before falling back to event *B*. In other words, if a ball is to travel inertially from your hand at one moment to the same hand at a later moment, it *must go up rather than stay in your hand.* Staying in your hand is a path of less-than-maximum proper time and therefore *not* an inertial path. This actually matches what we would measure if we attached force sensors to the ball—these sensors would register a force exerted on the ball when supported by your hand at a constant position, but not when freely falling.

This suggests that we should revise our language when speaking about gravity. The ball accelerates relative to the planet-based coordinate system, but not because of the "force" of gravity; rather, because time runs more slowly at lower altitudes. Gravity is really a slowing of time rather than a force.

This is a new way of thinking so expect to wrestle with it a bit before making your peace with it. If you think you get it right away, you probably need to pay more attention! You should be asking yourself tough questions:

- *The argument of Figure 14.4 seems to imply that the ball "wants" to go up. This is the* opposite *of gravity, so what is going on?* This impression arises from the way we posed the problem. Normally we are given initial conditions (a ball leaves your hand at event *A* with a specified velocity) and asked to predict the end of the story. Here, rather than an initial velocity we were given the end of the story—the ball arrives inertially at the starting location some time later (event *B*)—and asked to fill in the middle. In everyday terms the challenge is to make a ball freely fall from your hand and be in the same hand again at the end of, say, two full seconds. Given this challenge, *you* would start the ball with an appropriate initial upward velocity. So there is no antigravity—the initial upward velocity is just a necessary start to any story that connects events *A* and *B* inertially. Section 14.2 deals with more typical situations where we know the initial velocity.

- *I can see how the discussion of freely falling objects may imply that gravity isn't a force, but what about stationary objects? When I step on the bathroom scale, it definitely registers a force due to gravity.* There is nothing wrong with continuing to use the force model for everyday situations, just as we continue to say "the Sun rises" even after discovering that this is due to Earth's rotation rather than the Sun's motion. But for a deeper understanding of gravity, we should recognize that you are pressed against the scale because you are living in an accelerated coordinate system. If you were in a rocket in interstellar space accelerating at 9.8 m/s^2, the scale would register exactly the same number as it does on Earth. Although your bathroom is not accelerating relative to the surface of the Earth, it *is* accelerating relative to inertial frames.

- *In the accelerating-rocket picture it is clear why inertial objects accelerate toward the rocket floor—because the rocket floor is really accelerating up toward the inertial objects. But the surface of Earth does not accelerate up toward inertial objects in the same way.* This implicitly assumes a reference point somewhere on Earth. If we think of the inertial objects as attached to an inertial coordinate grid and take *that* as our reference then, yes, the surface of Earth does accelerate into the grid.

- *What if we use a distant star as our reference—surely Earth's surface is not accelerating toward that star?* This does seem like a conundrum, but only if we assume that clocks continue to tick faster and faster as we go farther and farther left outside the bounds of Figure 14.4. The next several chapters develop the thinking tools needed to understand how the simplified picture here can be matched smoothly onto the rest of the universe.

- *Do the laws of physics prevent a particle from taking the dotted paths in Figure 14.4?* Not at all. But these paths are not inertial; some force must be applied somewhere. For example, the stationary path results when your hand applies a force supporting the ball throughout. At the other extreme, a path that goes up and down at high speed is entirely possible, but involves a force exerted by the ceiling when the ball bounces off it.

In summary, differential clock rates *alone* account for the trajectories of objects in Earthbound laboratories, without reference to a force. A model of gravity based on time rather than force also explains why the object following the curved inertial worldline *feels* no force while objects following straight worldlines *do* feel forces. And compared to the force model, the time model is particularly good at explaining why *all* free particles, regardless of mass or composition, accelerate downward at the same rate. The time model is conceptually very promising, so Section 14.2 develops it a bit more quantitatively.

Check your understanding. Consider what would happen in Figure 14.4 if the experimenter throws the ball with greater upward velocity than shown by the solid curve. Sketch the path of this ball (starting from the instant it leaves the experimenter's hand) on the diagram. Does it intersect event B? *Hint:* can it follow a noninertial path?

Think about it

"The freely falling object feels no force" can be stated more objectively. We can attach sensors to its surface, we can look at whether an attached spring stretches or compresses, we can look at whether coffee presses against its cup, and so on. All such tests (in laboratories on Earth) reveal a force on stationary objects but not freely falling ones.

14.2 Metrics and the geodesic equation

How can we make the argument of Section 14.1 quantitative and show that slow time by itself explains the effects of gravity on Earth? There are mathematical tools for finding the path that minimizes or maximizes something, but first we must specify mathematically what the "something" is. For example, in the lifeguard story if we specify the time "cost" for running one meter and the time cost for swimming one meter, we can use these tools to minimize the total cost and thus find the optimum place to enter the water.

Specifying the "cost" or "value" of displacements in each direction is exactly what a metric does, so the metric is the place to start when searching for maximum-proper-time paths. Before we get to the specifics, let us note a few differences between the lifeguard story and freely falling particles in spacetime. First, we want to *maximize* proper time, so think of the metric as specifying the proper-time "reward" (rather than cost) for a given displacement. Second, the lifeguard on the beach is free to move in any direction but particles in spacetime *must move forward in time*. The range of spacetime trajectories is further limited by the fact that a particle must advance *at least* one second of coordinate time per second of proper time, and that its "speed through time" (γ) is tied to its speed through space. Third, for the lifeguard, the metric (the cost per meter of travel) changes only once, suddenly, at the sand/water boundary. This results in a one-time change of course. Clock rates in gravity smoothly increase the higher you go, so the proper-time reward smoothly increases with altitude, and we can expect smooth changes in course (i.e., smooth acceleration). Fourth, the mathematics will be more complicated because we need to find an optimum path in space *and* time. Specifying a path through space alone (such as "the ball goes up 5 m and then goes back down 5 m") is not specific enough.

A mathematical tool called the calculus of variations allows physicists to do all this. It yields an equation called the **geodesic equation** that tells us how to find the maximum-proper-time path given a metric. According to the geodesic equation, these paths accelerate only to the extent that the metric varies with position; acceleration will be small where the metric barely changes with position, and large where it changes steeply. This fits with the conceptual picture we developed in Section 14.1, where a nearly uniform metric would correspond to nearly uniform shading in Figures 14.3 and 14.4. In this case clocks on the left side of the diagram would tick only *slightly* faster so the proper-time reward for deviating from a straight worldline would be small—particles would accelerate less. The geodesic equation also reveals that a particle's speed through time is a factor in how much it accelerates toward regions where the $(\Delta t)^2$ part of the metric is smaller, and its speed through space is a factor in how it responds to variations in the $(\Delta x)^2$ part of the metric. Thus, particles at everyday low speed are insensitive to variations, if any exist, in the $(\Delta x)^2$ part of the metric.

Test-drive your understanding by checking what happens in regions of space-time with no (or too-weak-to-measure) gravity. These regions are described by the Minkowski metric (Section 11.2), $c^2(\Delta\tau)^2 = c^2(\Delta t)^2 - (\Delta x)^2$. This metric is the same everywhere—nothing in it depends on where or when a displacement occurs. Therefore, a free particle moving through such a region will exhibit *zero* acceleration, and this indeed matches our experience.

Next, consider a situation where the metric does vary: a room on Earth. Let us use our spatial coordinate, x, to indicate height in the room, with $x = 0$ at the floor and increasing toward the ceiling. Section 13.4 showed that stationary clocks with height x tick faster according to the expression $1 + \frac{ax}{c^2}$. This statement in equation form is:

$$\Delta\tau = (1 + \frac{ax}{c^2})(\Delta t). \qquad (14.1)$$

This is a version of the metric restricted to low-speed objects, for which the Δx term may be dropped because $\Delta x \approx 0$. The factor multiplying Δt is called the *time coefficient*, and it is no longer one as it was in the gravity-free zones where we learned special relativity. The coefficient $(1 + \frac{ax}{c^2})$ does equal one at a particular position, $x = 0$ (the floor), but above the floor the coefficient is larger than one. Clocks on the floor thus record the same time as coordinate time ($\Delta\tau = \Delta t$) but progressively higher clocks record progressively more proper time. The size of this effect is represented by the parameter a: the altitude-dependence of clock rates is steep if a is large, and nonexistent if $a = 0$. In Chapter 15, we will learn how to handle different accelerations in different places, but for now, a is assumed to be a fixed number throughout our laboratory.

Now, the geodesic equation says that two factors determine how quickly particles accelerate toward regions of slower time, where the time coefficient is smaller. The first factor is how steeply the time coefficient changes with position. This is represented by a, and is uniform throughout the room. The second factor is the particle's speed through time, γ, which is exactly one for our stationary particle. Therefore, the geodesic equation predicts that any initially stationary free particle in this room will accelerate toward smaller x with an acceleration exactly equal to a—just what is observed!

After a bit of acceleration the particle is no longer stationary—Δx is no longer zero from one instant to the next—so, in principle, we have to consider how the Δx part of the metric affects the geodesic equation as well as the effect of an infinitesimally larger γ. However, for everyday particles Δx is always much, much smaller than Δt—these two components of the displacement are comparable only for speeds close to c. Similarly, at everyday speeds the speed through time, γ, is still extremely close to one. Therefore, a full mathematical analysis of particles at everyday speeds agrees with our stationary-particle analysis to extremely high precision.

Thus, the geodesic equation confirms the conceptual analysis of Section 14.1— that free particles accelerate toward regions of slow time—with two improvements:

Confusion alert

When reading equations, avoid confusing Δx with a "bare" x, or Δt with t. Δx and Δt describe a displacement such as "1 millimeter up and 1 nanosecond later" and they *always* appear in a metric equation (the very purpose of which is to specify the "value" of these displacements). In contrast, x and t describe where and when the displacement happened; if they do not appear, then the "value" of a given Δx and Δt is independent of position and time.

Confusion alert

Clocks on the floor read coordinate time only in this example; there are many other options for setting up coordinate systems where higher clocks tick more quickly, and we will use some of them later.

- the geodesic equation proves it in general without having to specify starting and ending events along the trajectory.
- the geodesic equation shows that the *amount* of particle acceleration due to altitude-dependent clock rates in a laboratory matches exactly the acceleration that the laboratory would need in order to *cause* those rates in the first place.

In a way, the second statement is obvious: if a rocket moves eastward with acceleration *a*, then of course a coordinate system attached to the rocket measures free particles to accelerate westward by the same amount. The nonobvious part is that clock rates are inextricably involved rather than some curious side effect. The tick rates of stationary clocks determine the accelerations of free particles regardless of their mass, color, or composition—and when particles accelerate the same way regardless of their intrinsic properties we call that gravity. Thus, we now have a metric theory of gravity; that is one that unifies gravity with special relativity. The central role of time in this model requires us to dispense with the idea of "gravitational force." Imagine that humans had noticed high-altitude clocks ticking faster before they noticed that free objects accelerate toward the surface of the Earth: they could have *predicted* the latter from the former. Models of gravity based on "force" would never have arisen.

Is the metric model of gravity better than the force model? Yes, but not (yet) in the sense that it more accurately predicts particle trajectories. Rather, it cleanly accounts for some otherwise thorny points:

- At a given location, gravity accelerates all particles uniformly regardless of their mass, composition, and so on. This is almost inexplicable in the force model of gravity. The force model can claim that each object generates a gravitational force proportional to its inertia, but the metric model offers a far more satisfying explanation.
- Higher-altitude clocks on Earth do tick faster than lower-altitude clocks, by exactly the amount needed to explain observed particle accelerations. The fact that objects fall toward Earth is already completely explained by clock tick rates, with no room for an additional force-based explanation.
- All attempts to *measure* force (with force gauges, the coffee test, etc.) confirm that there is zero force on freely falling objects.
- The metric model subsumes gravity into relativity; we no longer need a separate theory for gravity. Scientists place a high value on models that provide a unified explanation for a vast array of observations.

Check your understanding. (a) A ball falling straight down maximizes its proper time by accelerating straight toward a region where its clock is going to tick more slowly. Explain how this apparently contradictory statement can be true. You may pick any point along its trajectory as an "endpoint" to clarify your thinking. *(b)* Explain how the maximum proper time principle explains the trajectory of a ball launched from the ground onto a balcony.

14.3 Graphical model

Can we illustrate slow time on a spacetime diagram with a grid? Not if we want to be accurate in all details, but a rough mental picture is still useful.

Figure 14.5 shows a spacetime grid that models some aspects of time running more slowly further down. It is reminiscent of the accelerated frames shown in Figure 4.10. Relative to an inertial particle, the floor and ceiling of your room (as well as any height markers in between) are accelerating upward (to the left in this diagram), so fixed positions are shown with curved worldlines. This reflects how maintaining a fixed position on Earth requires an upward force (a support), just as I require the support of a chair to continue typing this section. Inertial particles follow straight lines, so they accelerate downward *relative to this coordinate system*. All this echoes Figure 4.10, but in Figure 14.5 the lines of fixed time are drawn as always perpendicular to those of fixed position. To do so, lines of fixed time spread out as they approach the floor, which gives the sense that time is running more slowly there.

Figure 14.5 includes two inertial worldlines. The worldline near the floor corresponds to a ball thrown upward from the floor. The worldline is straight, but its relationship to the coordinate system keeps evolving: initially moving away from the floor, it gradually becomes parallel to the floor's worldline (i.e., it moves only through time and not through space at its highest point), and eventually returns to the floor. Throughout, this particle feels no force because at every event it keeps moving "straight ahead." The other inertial worldline is initially stationary on the ceiling; perhaps it represents a ceiling tile that has just lost its support. It initially moves only through time, but as the ceiling accelerates away from it the tile begins moving through space as well. It continues accelerating relative to this coordinate system despite feeling no force.

Trajectories are especially sensitive to the behavior of the time coordinate because particles must move forward through time. Given how time runs here, an initial motion forward through time alone *becomes* a motion through space as well (for free particles). This is why when you drop an object, it begins to move: its forward motion through time becomes a motion through space as well. Everyday objects move through space as well, but to a *much* smaller extent than they move through time (using *c* as the standard for equal motion through space and time). To a good approximation, everyday objects move only through time, and (according to the geodesic equation) that is why the time coefficient of the metric completely determines their acceleration.

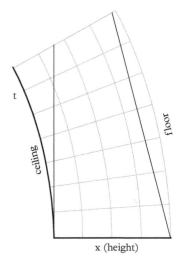

Figure 14.5 *Your room forms a coordinate system that accelerates upward (to the left on this diagram) relative to inertial particles, which follow straight worldlines. The worldline to the left shows that a particle falls from the ceiling because the coordinate system turns this particle's initial motion through time alone into motion through space. Fixed positions have curved worldlines, so a particle can remain stationary only if supported by a force.*

Think about it

Light moves equally through space and time, so it does not fit into the class of everyday particles described here. Chapter 18 examines this in detail.

This chapter has given us a new model of gravity in the simple situation where the acceleration is the same everywhere in the region studied. But gravity is not equally strong everywhere in the universe, even if it is equally strong for all particles at a given event. In the next several chapters we will come to understand how gravity varies throughout the universe and how that can be incorporated into the metric model of gravity.

Check your understanding. On Figure 14.5, sketch the worldline of a ceiling tile leaving the ceiling with a nonzero initial downward velocity. Does it accelerate downward?

CHAPTER SUMMARY

- When tick rates of stationary clocks vary with position, free particles accelerate toward regions of slower time. This maximizes proper time by allowing a particle to spend *most* of its time in regions with more frequent ticks.

- Clocks closer to Earth do tick less frequently, so free particles accelerate toward Earth—even as they feel no force in doing so. Therefore, slow time by itself explains what we call gravity, without reference to force.

- This model of gravity unifies gravity with special relativity. Upgrading the Minkowski metric to match observed tick rates near Earth requires a slightly more complicated time coefficient—but this small change is enough to explain all the everyday effects of gravity.

- The geodesic equation allows us to calculate paths of maximum proper time quantitatively. This confirms that variations in the time part of the metric cause inertial motions through time to acquire a component of motion through space as well—toward regions of slower time.

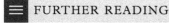 FURTHER READING

Figure 14.5 was inspired in part by a slow-time visualization activity in *Relativity Visualized* by Lewis Carroll Epstein. That book contains several useful activities for modeling aspects of gravity using only pencil, paper, and scissors.

◢ CHECK YOUR UNDERSTANDING: EXPLANATIONS

14.1 Assuming the ball does not bounce off a ceiling, it has no engines or other mechanism that would allow it to travel a noninertial path. Its inertial path would therefore be a magnified version of the solid curve in the figure: it would rise to a higher altitude and take more time before returning to its initial low altitude. Students who imagine the ball bouncing off a ceiling can also be correct, if they draw a sharp kink in the worldline there.

14.2 (a) This one is difficult! The key is to think in terms of *spacetime* rather than space. The path described *is* a correctly curved worldline on a spacetime diagram. If you are still unsatisfied, read Section 14.3; this question was designed to whet your appetite for that section. (b) Imagine a clock inside the ball; this clock ticks most frequently at higher altitudes. To accumulate the most ticks overall, the ball should spend most of its time near the altitude of the balcony. It does this by decreasing its speed after launch.

14.3 Being inertial, this worldline should maintain a fixed slope on the paper. As a result it accelerates relative to the coordinate system (toward the floor grid line).

？ EXERCISES

14.1 Explain why, in Figures 14.3 and 14.4, a stationary path does not maximize proper time.

14.2 Draw a diagram like Figure 14.3 showing the trajectory of a ball thrown upward and landing on a balcony. Explain how that trajectory maximizes its proper time.

14.3 Consider a bouncy ball dropped from a height and bouncing several times. *(a)* draw its path on a diagram like Figure 14.4. *(b)* Overall, does it maximize proper time? Explain how you could determine this *either* from the spacetime diagram alone or from other physical reasoning (and make sure the two types of reasoning agree on the answer). *(c)* Identify any *segment(s)* of the journey along which proper time is maximized.

14.4 Explain in your own words why accelerating *toward* a region of slow time is a way of maximizing proper time.

14.5 Why can we no longer think of gravity as a force?

＋ PROBLEMS

14.1 Figure 14.4 shows only one spatial dimension, the one indicating height. *(a)* Explain why an inertial particle would not accelerate in the other two spatial dimensions (east-west and north-south). *(b)* Explain why the inertial trajectory in that figure *looks* like the path a ball takes through *two spatial dimensions* such as east-west and up-down. You may find it useful to review Galileo's thought experiments on cannonball trajectories (Chapter 3).

14.2 A bicyclist rolls from event A to event B along a straight, flat road at constant speed. *(a)* In what sense, if any, did she take the shortest path between the two events? *(b)* In what sense, if any, did she take the *longest* path between the two events?

14.3 When you ride an elevator down, are you taking a path of maximum proper time? Explain your reasoning.

14.4   Does a helium balloon take a path of maximum proper time after you release it? Explain your reasoning.

14.5   An object hangs from the ceiling by a string, but at a certain time the string breaks. On a copy of Figure 14.5, draw the worldline of this object starting before the break and ending some time after the break.

14.6   Figure 14.6 shows a hypothetical region of spacetime, with lighter shading corresponding to slower time. A particle travels inertially from event A to event B. *(a)* Draw a plausible worldline that maximizes proper time from A to B, and explain why it is plausible. Pay attention to the symmetry of the situation. *(b)* Where is the particle speed lowest and where is it highest? *(c)* Where is the particle acceleration lowest and where is it highest? *(d)* Suggest a physical situation that could correspond to this spacetime diagram.

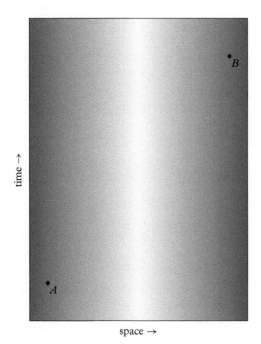

Figure 14.6 *Hypothetical slow-time region for Problems 14.6 and 14.7.*

14.7   Figure 14.6 shows a hypothetical region of spacetime, with lighter shading corresponding to slower time. A stationary particle begins to move inertially at event A. *(a)* Draw a plausible inertial worldline that follows from these initial conditions, and explain why it is plausible. Pay attention to the symmetry of the situation. *(b)* Where is the particle speed lowest and where is it highest? *(c)* Where is the particle acceleration lowest and where is it highest? *(d)* A second inertial particle leaves event A with an initial velocity to the right. Draw its worldline in a way that is consistent with the accelerations evident in your previous worldline.

14.8   Figure 14.7 shows a hypothetical region of spacetime, with lighter shading corresponding to slower time. A particle travels inertially from event A to event B. *(a)* Draw a plausible worldline that maximizes proper time from A to B, and explain why it is plausible. Pay attention to the symmetry of the situation. *(b)* Where is the particle speed lowest and where is it highest? *(c)* Where is the particle acceleration lowest and where is it highest? *(d)* Suggest a physical situation that could correspond to this spacetime diagram. *(e)* If an inertial particle leaves A with zero initial velocity, where must it go?

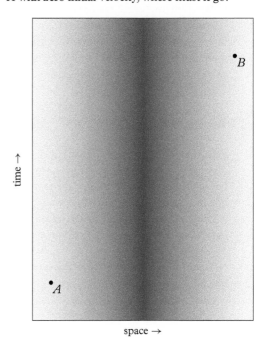

Figure 14.7 *Hypothetical slow-time region for Problems 14.8.*

Potential

15

At any given event, gravity accelerates all particles equally. This does not imply that gravity is the same at all places and events—on the contrary, we know that gravity is very strong in some places and very weak in others. In this chapter we learn a powerful thinking tool to help us deal with these variations.

15.1 Definition of potential

In Section 13.4 we showed that if the gravitational acceleration is a in a room on Earth and the height of the room is Δh, time runs faster on the ceiling than on the floor by a ratio $1 + \frac{a\Delta h}{c^2}$. How do we extend this analysis to higher floors in our building, especially if the acceleration due to gravity becomes smaller as we move up?

Imagine a set of laboratories, each experiencing slightly different accelerations, stacked on top of each other. (Because the direction considered "vertical" is based on the acceleration itself, so too are terms such as "on top" and "upward.") Let a_1 and Δh_1 be the acceleration and height of Lab 1, a_2 and Δh_2 be the acceleration and height of Lab 2, and so on. Then the ceiling-to-floor clock tick ratio in Lab 1 is $1 + \frac{a_1\Delta h_1}{c^2}$ and in Lab 2 it is $1 + \frac{a_2\Delta h_2}{c^2}$. The ceiling of Lab 1 coincides with the floor of Lab 2, so the clock tick ratio from the ceiling of Lab 2 to the floor of Lab 1 is $(1 + \frac{a_1\Delta h_1}{c^2})(1 + \frac{a_2\Delta h_2}{c^2}) \approx 1 + \frac{a_1\Delta h_1 + a_2\Delta h_2}{c^2}$. By the same reasoning, the clock tick ratio from the ceiling of Lab 3 to the floor of Lab 1 is $1 + \frac{a_1\Delta h_1 + a_2\Delta h_2 + a_3\Delta h_3}{c^2}$, the ratio from the ceiling of Lab 4 to the floor of Lab 1 is $1 + \frac{a_1\Delta h_1 + a_2\Delta h_2 + a_3\Delta h_3 + a_4\Delta h_4}{c^2}$, and so on.

The term $a_1\Delta h_1 + a_2\Delta h_2 + a_3\Delta h_3 + a_4\Delta h_4 + \ldots$ could get very tedious to write out for a tall stack of labs, so physicists created a name for it: the **gravitational potential**, or simply, potential, denoted with the Greek letter Φ, pronounced *phi*. By adding up many small steps we are *integrating* the acceleration over the vertical displacement. Think of potential as a generalized version of "acceleration times vertical displacement" that works even in the case where the acceleration changes with height. Figure 15.1 sketches how potential varies with height in a scenario where the acceleration diminishes with distance from Earth. The key thinking tool here is that *the slope of the potential at any point is the acceleration due to gravity at that point.* For example, the steepness of the curve in Figure 15.1 near the Earth indicates a large acceleration there, and the flattening of the curve farther away

---

Confusion alert

Gravity is said to provide uniform acceleration because it treats all particles at a given event equally. This does not imply that gravity is the same everywhere.

Think about it

For those who like math: the approximation to the left works as follows. Writing out $(1 + \frac{a_1\Delta h_1}{c^2})(1 + \frac{a_2\Delta h_2}{c^2})$ yields $1 + \frac{a_1\Delta h_1}{c^2} + \frac{a_2\Delta h_2}{c^2} + \frac{a_1\Delta h_1 a_2\Delta h_2}{c^4}$. We collect the second and third terms to make $\frac{a_1\Delta h1 + a_2\Delta h_2}{c^2}$, and we drop the last term because it is negligible: $\frac{a_1\Delta h_1}{c^2}$ and $\frac{a_2\Delta h_2}{c^2}$ are each much smaller than 1, so their product is far smaller than any other term in this sum. This approximation becomes exact if we make each Δh infinitesimally small.

The Elements of Relativity. David M. Wittman, Oxford University Press (2018).
© David M. Wittman 2018. DOI 10.1093/oso/9780199658633.001.0001

indicates less acceleration there. We can see that acceleration is the slope of the potential *by definition* as follows. Slope is defined as "rise over run"—in this case rise in potential divided by the change in height over which that rise is measured. The increase in potential from, say, the floor of Lab 3 to its ceiling is $a_3 \Delta h_3$, and the height over which that increase is measured is Δh_3. The slope is therefore $\frac{a_3 \Delta h_3}{\Delta h_3} = a_3$.

The potential is so useful because it provides both local and global information. Acceleration is local: at any location it is what it is, without any need to consider other locations. But to compare clock tick rates, say, between two stationary observers Alice and Bob, we need to compute the integrated effect of the accelerations over each little step on a path from Alice to Bob; this is what I mean by global information. If you hand me a map of the acceleration at each point in space with the locations of Alice and Bob marked, I can do that integration but it would be tedious—and I would have to do *another* integration for each additional character to be compared to Alice. But if you hand me a map of the potential, I can instantly compare the clock tick rates across widely separated locations—I get a global view. This convenience comes at little cost, because if I do need to know the acceleration at any point (local information) I can quickly deduce it from the *slope* of the potential at that point.

We will explore clock tick rates in more detail in Section 15.2. In the remainder of this section, we look at two more classical applications of the concept of "acceleration times vertical displacement": one local and one global.

For a local application, we will use Figure 15.1 to help us visualize the distinction between mass and weight. Mass (*m*) is an inherent property of an object (Chapter 12), whereas weight is the force (*ma*) required to support the object against a gravitational acceleration *a*. Thinking of the potential curve in Figure 15.1 as a metaphorical hill, we can see that a 1 kg mass placed near Earth needs substantial support—called its weight—to maintain its position on the steep (high-acceleration) slope. In a region of space where the potential is less steep, the same mass needs less support to maintain its position and thus has less weight. We make little distinction between mass and weight in our everyday lives because in the tiny region of Figure 15.1 where we live (the shaded band), the slope does not vary appreciably.

For a global application of potential, we consider the energy required to lift a satellite into orbital position. When lifting something a small distance over which gravity can be considered constant, the energy used is defined as the upward force of our push (equal to the weight) times the upward displacement. If we push a satellite upward for thousands of km, we now have to account for the gradual decline in lifting force required as we go to higher altitudes. That sounds complicated, but the potential is the perfect tool here: it has already integrated *acceleration times displacement* over this path so we need only multiply by mass to obtain force times displacement (energy). Thus, the energy required to push the mass up from point *A* to point *B* is just *m* times the difference in potential between these two points. By conservation of energy, this is equal to the kinetic

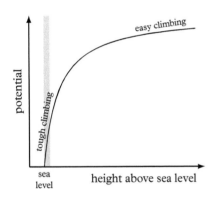

Figure 15.1 *Conceptual view of gravitational potential as a function of distance from the surface of the Earth. The slope of the potential at any point indicates the acceleration due to gravity at the point. We live in a small region (shaded band) where the slope does not vary appreciably.*

Think about it

Astronauts aboard the space station (or in any low Earth orbit) are so close to Earth that the slope of the potential there is about the same as on the ground. They are weightless not because the potential is flat there but because they are freely falling (Chapter 17).

energy gained by the object if it falls from B to A. This helps us recognize yet another form of energy: gravitational potential energy.

Check your understanding. Imagine an initially stationary object far from Earth in Figure 15.1. Which way will it begin to move? After it moves a little, how will the acceleration in its new position compare to the acceleration in its old position? Describe what happens as more time elapses.

15.2 The potential traces slow time

We care about the potential because it traces clock tick rate. Figure 15.2 reminds us of this by replacing the "potential" label in Figure 15.1 with a "clock tick rate" label. We can determine how frequently one clock ticks only by comparison with another, so the potential has meaning only as a *comparison* between two places. Because only *differences* in potential are important, we can choose a zero point or origin that is most convenient for us. If we go "uphill" far enough, we eventually get very far from Earth where the potential is very close to flat. Earth's potential is essentially a sinkhole in the vast, flat plateau that is most of space. If you were to climb up to this plateau, you would be able to coast freely there as if on an air hockey table (but beware of other sinkholes corresponding to other planets and stars). This vast plateau provides a reference throughout the universe, so it is convenient to define the potential to be zero there. Under this convention, the potential is negative everywhere around Earth and climbs up to zero only where it meets the plateau. This is a useful mental picture, but the potential is not actually a physical surface of any kind. When we use phrases such as "fall down the potential" we literally mean "fall toward regions of slower time" and when we say "far down the potential" we mean "in a region of much slower time" rather than any kind of distance in kilometers.

To quantify the connection between potential and slow time, we need only remind ourselves that (Section 15.1) the ratio of clock tick rates between two locations at different heights is $1 + \frac{a_1 \Delta h_1 + a_2 \Delta h_2 + a_3 \Delta h_3 + \dots}{c^2}$ and that the unwieldy term in the numerator there was *defined* as the potential. Therefore, the ratio of clock tick rates is $1 + \frac{\Phi}{c^2}$. Let us more carefully define which clocks are which in this ratio, while dropping the idea that the reference clocks should be on the floor. According to the convention we set up in the previous paragraph, $\Phi = 0$ in distant space, and that would also be a convenient location for reference clocks that tick according to coordinate time t. Indeed, wherever $\Phi = 0$ the expression $1 + \frac{\Phi}{c^2}$ evaluates to exactly one, so we can adopt these clocks as defining our time coordinate. Clocks elsewhere tick at varying rates, and each clock has its own proper time τ. We quantify this by writing $\frac{\Delta \tau}{\Delta t} = 1 + \frac{\Phi}{c^2}$ or

$$\Delta \tau = (1 + \frac{\Phi}{c^2})(\Delta t) \qquad (15.1)$$

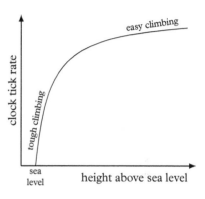

Figure 15.2 *As for Figure 15.1, but with a reminder that the direction of increasing potential is by definition the direction of increasing clock tick rate.*

Confusion alert

Under the convention adopted here, the potential happens to be zero where its slope—acceleration—is also zero. Do not let this obscure the important conceptual distinction between potential and acceleration: there is zero acceleration wherever the potential is *flat*.

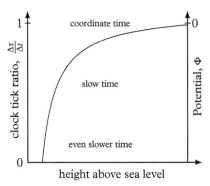

Figure 15.3 *As for Figure 15.2, but with parallel vertical axes showing that* $\frac{\Delta\tau}{\Delta t} = 1$ *where* $\Phi = 0$.

Test-drive this expression: deep in Earth's potential "sinkhole", $\Phi < 0$ so the term in parentheses is less than one and fewer ticks of proper time are recorded, just as we expect. Figure 15.3 shows this graphically by clarifying that $\frac{\Delta\tau}{\Delta t} = 1$ and $\Phi = 0$ at the top of the potential hill. Note that at any position the *depth* of the potential there, not the slope, determines how slowly time proceeds there.

Equation 15.1 has some parallels to the time dilation equation for moving clocks in the absence of gravity, $\Delta\tau = \sqrt{1 - v^2/c^2}(\Delta t)$. In both cases, $\Delta\tau$ is always less than (or at most equal to) Δt. Equation 15.1, though, expresses a gravitational form of time dilation that is experienced *even at rest* in low-potential regions. Beware that gravity makes possible many effects that are never seen in special relativity with its purely inertial grids. Gravity makes us use noninertial grids because "stationary" clocks actually accelerate upward relative to free particles.

You can measure how slow your time is using light that "fell" from a region where $\Phi = 0$, because the frequency of the light acts as a clock (gravitational redshift; Section 13.4). Furthermore, any atom can serve as a clock (Section 9.3) so there is no need to send lab equipment up to regions where $\Phi = 0$. If Bob lives in interstellar space where $\Phi = 0$ and Alice is on Earth ($\Phi_A < 0$) she can compare light from one of her hydrogen atoms to light that "fell" from a hydrogen atom in free space near Bob. Considering each wavecrest as a tick from a clock, she finds that her atoms "tick" at a slower rate—let us grossly exaggerate the strength of Earth's gravity and say half the rate. This means that during any experiment Alice's elapsed proper time $\Delta\tau$ is half the coordinate time Δt kept by interstellar clocks: $\frac{\Delta\tau}{\Delta t} = 0.5$ so Alice ages at half the rate of Bob. If we invert this to read $\frac{\Delta t}{\Delta\tau} = 2$ it looks much like the twin "paradox" with $\gamma = 2$ but with no motion. Bob sees Alice's life unfold at half speed and sees Bob's life unfold at double speed. Each character ages in sync with their local clocks, so neither experiences anything amiss locally, but they *can* measure these relative clock rates and thus infer how Φ varies throughout the universe.

In this example, Alice observes a *blue*shift from Bob because wavecrests from Bob arrive more frequently, and higher-frequency light is perceived by humans as bluer. Conversely, Bob observes Alice's light as redshifted. We can complement the clock-tick analysis by saying that light must "spend" energy to climb the potential hill, and light can do that only by losing frequency—it cannot slow down. This is a reminder that energy and time are inextricably tied together. We focus on the time aspect of the potential because that applies most directly to light. But do not forget that for a particle with mass, the depth of the potential at its location measures how much energy (per unit mass) you would spend pushing that particle up to the top of the potential hill.

Now that we understand how the potential relates to slow time, we can unify gravity and relativity more tightly than we did in Section 14.2. There, the metric $\Delta\tau = (1 + \frac{ah}{c^2})(\Delta t)$ (ignoring any Δx part of the metric while focusing on stationary or slowly moving particles) correctly predicted a free-particle acceleration of a without reference to any "force" of gravity. This was valued because it explained why freely falling particles indeed experience no force, and because it used a rule *we already knew* (free particles maximize their proper time) to explain many observations that previously required a separate and complicated analysis of gravitational "force." Equation 15.1, $\Delta\tau = (1 + \frac{\Phi}{c^2})(\Delta t)$, generalizes that discovery to *any* pattern of gravitational acceleration. The Chapter 14 version of the metric predicted acceleration a because the slope of its time coefficient is $\frac{a}{c^2}$. (To see this, note that increasing the height h by one unit increases the value of $1 + \frac{ah}{c^2}$ by exactly $\frac{a}{c^2}$.) In our more general expression, the slope of Φ at any place is *by definition* the acceleration there. Thus, all the Chapter 14 reasoning about trajectories and the geodesic equation carries over to the general case of position-dependent acceleration. Our metric $\Delta\tau = (1 + \frac{\Phi}{c^2})(\Delta t)$ *cannot fail* to predict the correct acceleration for initially stationary (and slowly moving) free particles.

We can thus be confident that the metric model of gravity is flexible enough to model variations in gravity throughout the universe. We emphasize the unity of gravity and relativity by squaring Equation 15.1 and including again the Δx part of the metric:

$$c^2(\Delta\tau)^2 = c^2(1 + \frac{\Phi}{c^2})^2(\Delta t)^2 - (\Delta x)^2 \qquad (15.2)$$

As we approach regions of space less and less affected by gravity, Φ smoothly approaches zero so the $(1 + \frac{\Phi}{c^2})$ factor in this metric smoothly returns to one—matching the familiar Minkowski metric of Chapter 11. Thus, this model smoothly blends gravity and special relativity.

These successes by no means ensure that our model is correct. We have so far considered only particles moving at a sufficiently small fraction of c that, to a good approximation, they move through time and not through space. We should also test the model with light, which moves as much through space as through time. Trajectories of light, however, are best measured across the entire solar system. Hence, Chapter 16 explores more quantitatively how gravity varies from place to place.

Check your understanding. (a) Rank the following locations from highest potential to lowest: your home, the top of Mount Everest, the shore of the Dead Sea, the International Space Station. *(b)* What information would you need to determine the potential difference between Everest and the space station more quantitatively?

Think about it

The Minkowski metric is not exactly wrong in regions affected by gravity—it describes *freely falling* frames. The modified metric here reflects how this principle works out back in a coordinate system attached to a source of gravity.

15.3 Visualizing the potential

The potential is a metaphorical hill extending upward from your location on Earth. It quantifies the energy expended in the climb up to any other point in space, as well as (via its slope) the difficulty of ascending each additional meter. Section 15.1 asks you to imagine that the climb becomes easier ("the hill flattens") if you climb far from Earth, and we will study this more quantitatively in Chapter 16. But we can learn one more thing before leaving Earth.

Think about it

The meaning of "accelerates" here is not necessarily that your velocity is changing, but that it *would* change if you were not supported by the floor.

Think about it

A complete coordinate system definition will also define a time coordinate and the equivalent of longitude and latitude coordinates, but we need not worry about those details here.

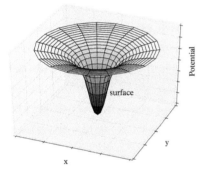

Figure 15.4 *Gravitational potential in and around the Earth. The "down" direction indicates the depth of the potential and is not a physical direction.*

Gravity on Earth accelerates you toward the center of the Earth, regardless of where you are. So, even though the *size* of the acceleration is roughly constant over the surface of the Earth, the *direction* is not at all constant, unless we choose our coordinate system carefully. If we continue to choose a square coordinate system where your floor defines $x = 0$ with x increasing upward from there, then objects in your room fall toward a smaller x, but on the opposite side of Earth objects are falling toward a larger x. As always, we should take advantage of the available symmetry to define a coordinate system that facilitates clear thinking. Let us define an r coordinate that is zero at the center of the Earth and increases with distance from the center. Then objects will *always* fall toward a smaller r.

To visualize this, we can take the "hill" pictured in Figure 15.1 and wrap it around Earth as shown in Figure 15.4. We have not explored quantitatively how it flattens far from Earth so focus on the qualitative meaning of this diagram. This surface does not really exist in space; it is a visualization of the acceleration you would experience in various places. The symmetry indicates that the acceleration always points toward the center of the Earth, and depends only on distance from the center. If, hypothetically, experiments determined that acceleration does *not* decrease with distance from Earth, the same visualization process would yield a downward-pointing cone with the same slope everywhere. This has the same symmetry, and differs only in the prescription for weakening with distance.

This funnel picture was developed long before special relativity, but more as a convenience than an alternative model. The funnel is a great way to visualize gravity, so much so that science museums often display large funnels around which coins can "orbit" before entering a donation box. But sometimes this visualization is mistakenly offered as an *explanation*: objects fall because they "want" to go to regions of lower potential. This just moves the question of why they "want" to fall to a more abstract context without answering it. With relativity the march of time does it automatically: the slope of the potential toward regions of slower time bends motion through time (which all particles share) into motion through space.

This chapter concludes with a speculation to whet your appetite for black holes (Chapter 20). Because of the c^2 in the denominator, $\frac{\Phi}{c^2}$ is, in most cases, extremely small. This makes the clock tick ratio $1 + \frac{\Phi}{c^2}$ very close to one. To see strikingly large effects, we need to look at regions that are *very* deep down in a potential, well beyond what Earth can provide. What if we could find a place so deep down, with such a highly negative value of Φ, that the clock tick ratio $1 + \frac{\Phi}{c^2}$ dives to *zero* as in Figure 15.5? Does time stop there, or does something even weirder happen? What if the potential went even deeper than that—would the clock tick rate get *negative*, and what would that mean? The thinking tools in this chapter set the stage for asking, and answering, these questions in Chapter 20.

Check your understanding. Given the potential pictured in Figure 15.4, imagine an object launched vertically from the planet with one big push, after which it glides. What does "vertically" mean on this figure? Sketch the trajectory of the object. Where is its speed maximum? Minimum?

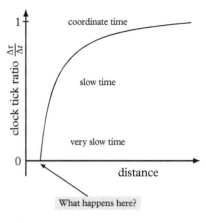

Figure 15.5 *If the potential gets deep enough the clock tick rate should plunge to zero. Can this really happen? We will find out in Chapter 20.*

CHAPTER SUMMARY

- Gravitational potential Φ is a thinking tool that allows us to see local *and* global properties of gravity.

- The potential is defined such that the acceleration at any specific location is the *slope* of the potential there. This is local: it tells us about one location at a time.

- If an object falls from one point to another (possibly very distant) point, the *difference* in the potential of those two points, times the mass of the object, is the kinetic energy gained (and the potential energy lost) by the object. This is global: the cumulative effect of many different accelerations experienced along the path of the object can be read at a glance as the change in potential between two widely separated points.

- By convention we define Φ as being zero far from any source of gravity, where clocks keep coordinate time. Elsewhere, local (stationary) clocks tick more slowly by the factor $1 + \frac{\Phi}{c^2}$. Thus, the metric (for stationary or slow particles) is $\Delta\tau = (1 + \frac{\Phi}{c^2})(\Delta t)$.

- Free particles maximize proper time by accelerating toward regions of lower Φ, with an acceleration given by the slope of Φ. With this formalism in place, the metric model of gravity is flexible enough to model variations of gravity from place to place. But the model is not yet complete because we have not looked at particles that move substantially through space as well as time.

CHECK YOUR UNDERSTANDING: EXPLANATIONS

15.1 It will begin moving toward lower potential. There, it experiences a steeper slope so it accelerates even more toward lower potential. Eventually, the object will fall all the way down and reach a very high speed.

15.2 (a) Space station, Everest, home, Dead Sea. This is really just an altitude ranking because potential by definition increases in the "up" direction. (b)

To know *how much* greater the potential is at, say, the space station versus Mount Everest, we would need to know the acceleration due to gravity at each point along the path from Everest to the station, and add up (integrate) acceleration times vertical displacement along this path.

15.3 First, note that in this visualization particles always stay on the potential surface drawn. Given that, "vertically" means in the direction of steepest potential increase. The velocity is maximum at the point of launch and minimum at the highest point reached.

? EXERCISES

15.1 *(a)* What does the slope of the potential represent? *(b)* What does the value of the potential (the height of the potential curve on a graph such as Figure 15.1 represent?

15.2 Which zones in Figure 15.6 have the strongest acceleration? Which have the least acceleration?

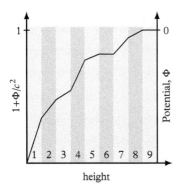

Figure 15.6 *A potential that varies with height in some complicated way. It has been broken into numbered zones for Exercise 15.2.*

15.3 Figure 15.7 presents four hypothetical potentials. *(a)* How does the acceleration of each depend on height? *(b)* What is the observable difference between $P1$ and $P2$? *(c)* What is the observable difference between $P1$ and $P3$? *(d)* Describe what you would experience in potential $P4$.

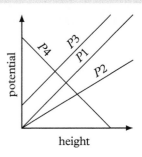

Figure 15.7 *Hypothetical   potentials   for Exercise 15.3.*

15.4 Referring to Figure 15.1, draw a qualitatively correct graph of acceleration versus height.

15.5 Referring to Figure 15.1, draw a qualitatively correct graph of the weight of a 1-kg mass as a function of height.

15.6 Referring to Figure 15.1, draw a qualitatively correct graph of clock tick rates versus height.

15.7 A laser pointer points up from the surface of the planet in Figure 15.1. Draw a qualitatively correct graph of the frequency or energy of the light as a function of height.

15.8 Where in Figure 15.4 would a clock tick most frequently? Least frequently?

15.9 Which has experienced more years since Earth was formed: Earth's core or Earth's crust? If the size of this effect is roughly one part in a billion, by how many years do they differ?

15.10 A laser pointer in space points toward the planet in Figure 15.4. Is the light redder, bluer, or the same color when it hits the planet?

15.11 A balance scale determines the *mass* of an object by comparing its *weight* with that of a standard mass. *(a)* Given that mass is defined as $\frac{F}{a}$ and is independent of weight, why is this considered an acceptable way to determine mass? *(b)* Can you think of any conditions under which the balance would *not* be an acceptable way to determine mass?

✛ PROBLEMS

15.1 A limited-access highway running through a city is often raised above or lowered below the city streets. *(a)* Which choice is more energy-efficient? Consider how cars must accelerate when entering and decelerate when leaving the highway, and how gravity can help with these tasks. *(b)* Explain how one choice allows vehicles to "borrow" gravitational potential energy and return it later. How does this relate to part (a)?

Figure 15.8 *Path of a swinging bowling ball suspended on a wire.*

15.2 Consider a bowling ball hanging from the ceiling on a wire, swinging back and forth as shown in Figure 15.8. The ball follows the path marked by the dotted line. *(a)* Where is the kinetic energy greatest? *(b)* Where is the potential energy greatest? *(c)* Where is the total (potential plus kinetic) energy greatest? The pendulum eventually slows and stops. *(d)* Where does it stop? *(e)* Is the total energy the same as it was initially? If so, how do you explain the zero kinetic energy at this position? If not, what happened to the missing energy?

15.3 When you open a box of cereal, the largest pieces of cereal tend to be on top, with the crumbs at the bottom. *(a)* Explain this in terms of minimizing the potential energy of the cereal. *(b)* The usual explanation of this phenomenon is that when the box is jostled during transport, small pieces are more likely to find a gap they can fall into. Explain how these two explanations complement each other.

15.4 Two identical clocks are manufactured on the surface of the planet in Figure 15.4 and set to read zero at the start of the following experiment. One clock stays on the surface, and the other is slowly taken up to the top of the potential well in Figure 15.4 and then brought back down. *(a)* Which clock, if either, reads a later time? *(b)* Why did the problem specify "slowly"? *(c)* The top of the potential well is in which direction from the surface of the planet?

15.5 Imagine that the planet in the potential well in Figure 15.4 is Earth, and observers there communicate with aliens on a planet with a much deeper potential well. Each uses identically manufactured green lasers to beam a message to the other planet. In the following, consider higher-frequency light to be blue and lower-frequency light to be red, even though in practice the effect is not large enough to see with the human eye. *(a)* What color is the alien light when it is received on Earth? *(b)* What color is the Earth light when it is received on the alien planet? *(c)* What color is the Earth light when it is traveling through interstellar space?

15.6 *(a)* Compute the energy required to send a 1-kg mass from the surface of the Earth to the altitude of the International space station (380 km), assuming gravity provides uniform downward acceleration of 9.8 m/s^2 throughout the trip (Section 17.1 will show that this is not terribly far from the truth). *(b)* What is the cost of this much energy? Use 3×10^7 joules per dollar as the standard cost of electricity. *(c)* Do some research to find out why the actual per-kilogram cost of reaching space is *much* higher than this. Explain what alternative method could bring the cost down to nearly the value you calculated here.

15.7   Equation 15.1 omits the space part of the metric. Square this equation so you can write in the space part as well. Explain how this term requires moving clocks in a region of slow time to experience even less proper time than stationary ones.

15.8   *(a)* Can photon trajectories be predicted using Equation 15.1? Explain why or why not. *(b)* What additional information can be brought to bear on photon trajectories? *Hint:* find the metric with space term included (Problem 15.7) and use the fact that the left side is always zero for photons to simplify it.

15.9   Sketch the potential as a function of position for the hypothetical slow-time patterns in *(a)* Figure 14.6 and *(b)* Figure 14.7.

Newtonian Gravity

16

We now study how gravity behaves beyond the surface of the Earth. We will follow Newton's footsteps, and thus put the principles of special relativity aside for this chapter. After learning about gravity and orbits from the Newtonian perspective in this chapter and the next, in subsequent chapters we will examine how gravity and relativity can be fully unified.

16.1 Invisible string

One of Newton's biggest breakthroughs was the idea that the laws of physics deduced from experiments on Earth could be applied also to heavenly objects. This idea is simple to state and is accepted without question today, but represented a bold departure from pre-Newtonian thinkers. So let us approach the motion of the Moon from the perspective of Newton's first law of motion: an object will maintain constant velocity unless acted upon by a force (Chapter 2). The Moon moves in a circle, which is clearly *not* a constant-velocity path—its direction keeps changing. Therefore, the Moon is acted on by a force.

We may be able to deduce more about this force by describing the acceleration more precisely. How much acceleration, and in what direction, is required to maintain a circular motion? Consider two examples: a person making a ball on a string perform circular motion, and a car driving in circles (assume the speed is constant in each case). The string can exert a force on the ball in only one direction: along the string, toward the hand. The acceleration must be in the same direction as the force: toward the center of the circle. If this surprises you, consider riding in the circling car: you feel flung toward the outside of the circle. To keep you moving in a circle, the car door pushes back on you—toward the center. *Circular motion at constant speed requires a constant centrally directed acceleration.*

The size of this acceleration is $\frac{v^2}{r}$, where v is the speed and r is the distance from the center of the circle. Test-drive this expression: the inverse relationship with r makes sense because you feel less acceleration in a circling car if the car follows a larger circle. To make sense of the v^2, imagine driving a half-circle at speed v_1 in time t_1: the total change in velocity is $2v_1$ because you completely reversed direction. Your acceleration is therefore $2\frac{v_1}{t_1}$. Now drive the same half-circle at double the speed: your velocity now changes by $4v_1$ in half the time. Your

> **Confusion alert**
>
> Circular motion maintains speed while constantly changing the velocity. Change in velocity is conceptually more important because it is linked to force through Newton's second law, $F = ma$: a is rate of change of velocity, not of speed.

The Elements of Relativity. David M. Wittman, Oxford University Press (2018).
© David M. Wittman 2018. DOI 10.1093/oso/9780199658633.001.0001

acceleration is therefore $\frac{4v_1}{\frac{1}{2}t_1} = 8\frac{v_1}{t_1}$. Acceleration *quadrupled* when speed doubled, so a is proportional to v^2.

So, what keeps the Moon moving in a circle around Earth? If we saw a ball performing circular motion around a person, we would instantly look for the string that tethers the ball. If the string were cut (and there were no other means of exerting force on the ball), by Newton's first law the ball must fly off in a straight line. So, what tethers the Moon and keeps it in a circular path? There *must* be some centrally directed acceleration playing the role of string despite the lack of a material connection. We already know of an acceleration that points to the center of the Moon's orbit: gravity, the common acceleration of all Earthbound objects toward the center of the Earth. Could gravity—the everyday phenomenon that makes things fall—explain why the Moon goes in a circle?

Let us run some numbers to see if this hypothesis can actually work. The size of the acceleration at the surface of the Earth is easy to determine experimentally by dropping things: we find 9.8 m/s^2. The size of the Moon's acceleration must be calculated from $\frac{v^2}{r}$ (v is not measured directly but is distance divided by time: the circumference of the Moon's orbit divided by its orbital period, one month). Because the radius r of the Moon's orbit is so large (400,000 km or 4×10^8 m), its acceleration turns out to be quite low: only 0.0028 m/s^2. This is about 3600 times smaller than the acceleration due to gravity at the Earth's surface, so our hypothesis remains plausible only if we can argue that Earth's gravity weakens greatly at large distances.

In fact, in Newton's day there were *multiple* reasons to believe that gravity weakens at large distances:

- Jupiter had four moons known at the time, orbiting at a range of distances, so perhaps Jupiter and its moons can be used as a model. Their accelerations can be calculated just as for our Moon, and they definitely weaken with distance from Jupiter. More specifically, doubling the distance from Jupiter weakens the acceleration by a factor of *four*. In other words, the acceleration in the vicinity of Jupiter is inversely proportional to the square of the distance from Jupiter: $a \propto \frac{1}{r^2}$.

- The orbits of planets around the Sun can be used as a similar model. Measuring the accelerations of planets at various distances from the Sun reveals a similar $1/r^2$ pattern. Properties of the planets themselves—such as size or mass—seem to have no effect on this pattern; *only* distance from the Sun seems to matter.

- Even stuck on the surface of the Earth, we are able to find places slightly closer to or farther from its center. In 1672 it was discovered that the

Think about it

A common misconception is that astronauts orbiting the Earth (e.g., aboard the International Space Station) feel weightless because there is no gravity there. If this were true, the astronauts and Space Station would fly off in a straight line rather than continue to orbit Earth.

Think about it

How can acceleration be $\frac{v^2}{r}$ *and* be proportional to $\frac{1}{r^2}$? The first expression describes *any* type of circular motion, while the second describes gravity. If the circular motion is *due to* gravity, then both must be true. Chapter 17 shows that this works mathematically if v itself decreases with r.

acceleration in French Guiana is less (by a few tenths of a percent) than in France, and the general pattern is that acceleration decreases at higher altitude (farther from the center of the Earth) and closer to the equator. Newton showed that this matches the $1/r^2$ pattern because French Guiana is farther from the center of the Earth than is France (Earth bulges at the equator due to its rotation).

- Newton himself favored deductive reasoning over inductive reasoning, so he focused on the following argument. The orbits of planets and moons are not perfectly circular, as we have been assuming for simplicity. The distance between Earth and Sun, for example, varies throughout the year—but Earth starts each new year at exactly the same position and velocity as it started the previous year, so that each year's orbit retraces the previous one in an ever-repeating ellipse called a **closed orbit**. Newton showed that closed orbits occur only if the size of the centrally directed acceleration varies as $1/r^2$. This is called the **inverse-square law**.

The inverse-square law weakening of gravity thus has very strong empirical support. We can also understand it theoretically in terms of basic geometry (Box 16.1). Armed with this knowledge of how gravity weakens with distance, we can account for the Moon's acceleration around Earth. Putting the origin of our coordinate system, $r = 0$, at the center of Earth because of symmetry, we must remember that when we stand on the surface of the Earth we are at $r \approx 6400$ km (rounding to the nearest hundred km). The distance of the Moon is about 60 times as large as this. According to the proposed $1/r^2$ law the Moon should then experience an acceleration $\frac{1}{3600}$ as large as we do at the Earth's surface. This matches the data nicely.

So gravity—the everyday force that makes apples fall from trees—provides the "invisible string" that keeps the Moon from continuing in a straight line forever. Newton not only made this bold conceptual leap, he provided exhaustive mathematical proofs showing that the inverse-square law explains phenomena as diverse as orbits (Chapter 17), the variation of gravity over the surface of the Earth, and tides (Section 16.7).

Check your understanding. The International Space Station orbits roughly 400 km above the surface (not much more than the distance from London to Paris). Remembering that $r \approx 6400$ km at the surface where $a \approx 10$ m/s^2, estimate qualitatively what the inverse-square law implies for the acceleration of the Space Station and its astronauts.

Box 16.1 Inverse-square law

 To make geometric sense of the inverse-square law, imagine a situation where you can see only one source of light, perhaps a campfire at night. The apparent brightness or intensity of the light drops rapidly if you move away from the campfire. This box quantifies that effect.

In the diagram below we focus on the light cast by the fire in your direction. At $r = 1$ m from the fire you measure some intensity I_0. This corresponds to a certain number of photons per second entering your eye; imagine that each square in the diagram represents the area A of the pupil of your eye and that a certain number of photons per second fly through that area. Those photons continue to fan out from the fire and are spread out over an area $4A$ when they get to a distance of 2 m. Thus, doubling your distance dilutes the rate of photons entering your eye—which has a fixed area—by a factor of *four*: the intensity I at $r = 2$ is $\frac{I_0}{4}$. The same photons become spread out over an area $9A$ by the time they reach a distance of 3 m, so I at $r = 3$ is $\frac{I_0}{9}$. In summary, the intensity is proportional to the inverse of the square of the distance.

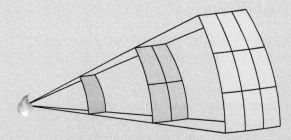

This is just geometry. The *only* assumption we made about the nature of the photons is that they fly straight away from the fire and do not stop. So the law is not followed exactly if the camp is surrounded by trees, or if photons die after a short lifetime.

Gravity follows identical behavior, so it seems likely that gravitational "intensity" is also transmitted by messenger particles flying directly away from sources of gravity (masses). These theoretical particles—called gravitons—have not yet been directly detected. From the inverse-square behavior of gravity we can conclude that gravitons, like photons, do not decay. However, physicists continue to search for any tiny departure from the inverse-square law, because any such departure could imply that gravitons *do* decay and could thereby shed further light on the nature of gravity.

16.2 Fields and test masses

We can make a map of the gravitational acceleration at every point in space around Earth, as in Figure 16.1. We can build this map empirically by releasing a **test mass** at any given position and recording its acceleration—any convenient mass can serve as a test mass because gravity accelerates all objects at a given position equally. Or, we can use the inverse-square law to build a *model* of this map, by drawing arrows pointing toward the center of the Earth, with lengths proportional to $1/r^2$. Figure 16.1 is a model, but it rests on a great deal of empirical support: the accelerations of thousands of orbiting satellites have always been consistent with the $1/r^2$ law. In any case, the concept of assigning a number or an arrow to each point in space is called a **field**, hence the term gravitational field.

Once we understand the arrow map, we can think of better ways to represent the same information. Arrows very far from Earth are too short to read, and those very close to Earth would be so long that they are omitted for clarity in the top panel of Figure 16.1. In other words, the range of arrow lengths is simply too extreme to render clearly in one map. The lower panel of Figure 16.1 renders the "arrow length" (size of acceleration) as a color instead, so we can simultaneously see the size of the acceleration at every location. Rendering it as a color eliminates information about the direction, but this is no real loss because we know it is always centrally directed. The lower panel of Figure 16.1 should remind you of the potential; it is as if we are looking "down" into the potential "funnel." This is no accident: acceleration is the slope of the potential.

We must return to one subtlety of test masses: they cannot be *too* massive. To understand why, we must first establish that each and every gram of mass participates equally in gravity. Certainly, it is true that (neglecting air resistance) if you break a falling object into parts gravity will affect each part equally. By extension, each gram of the Earth must also participate in gravity; we cannot identify any specific part as "the gravity generator." So, Earth's gravitational field must be the sum of the fields of its parts (fields add just like forces). Therefore, any rock or other test mass on its own must generate its own field, many trillions of times weaker than Earth's field. Such a weak field can safely be ignored when mapping out Earth's field, but an extremely massive test mass (Venus, say) would itself change the field we are trying to test. Test masses must be much less massive than the object whose field you are testing, which is called the **source** of the field.

We use m to indicate the test mass and M to indicate the source mass. Whether an object qualifies as a test mass depends on the context. For example, the Moon has not much more than 1% of the mass of the Earth, so the Moon may be considered a test mass probing Earth's field. But when considering Earth and Sun, Earth is the test mass because it has only 1/333,000 of the Sun's mass. When two important masses are roughly equal we need additional thinking tools, and we briefly touch on that case in Chapter 17.

The equal participation of each and every gram of mass implies that test masses do pull on source masses as well as vice versa. This is also required by Newton's third law (Chapter 2). If so, why do we make a distinction between test and source masses and why does Earth apparently not accelerate upward to meet a falling ball? The answer is that Newton's third law requires the ball to exert the same size force on Earth as Earth exerts on the ball. If we call this force F, the upward

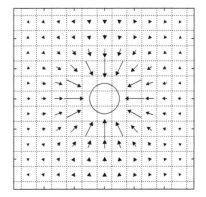

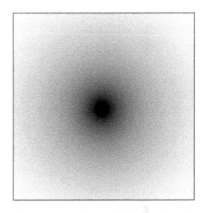

Figure 16.1 *Maps of the acceleration field around Earth.* Top: *arrows represent the size and direction of the acceleration; those nearest Earth are omitted for clarity (they would be very long).* Bottom: *shading represents the size of the acceleration, with the direction understood to be toward the center. This view should remind you of the potential; it is as if we are looking "down" into the potential "funnel."*

acceleration of Earth is $\frac{F}{M_{\text{Earth}}}$ and the downward acceleration of the ball is $\frac{F}{M_{\text{ball}}}$. Because M_{Earth} is approximately 6×10^{24} times greater than M_{ball} (assuming a 1 kg ball), the acceleration of Earth is approximately 6×10^{24} times smaller than the acceleration of the ball; in other words, immeasurably small.

Figure 16.2 illustrates this argument in detail. If two 1-kg masses float in space, by symmetry each must exert equally tiny gravitational forces F on the other (call this F), and each must have equally tiny accelerations a. If we now add two more 1-kg masses to the mass on the left, they each exert a force F on the mass on the right for a total of $3F$. Therefore, the leftward acceleration of m_{right} triples to $3a$. The mass on the right in turn pulls on each 1-kg mass with force F, so they *each* accelerate to the right with acceleration a. Now imagine gluing the three 1-kg masses into a single body: the motion of each part of the glued body is described by the acceleration a, so the motion of the *entire body* is also described by the acceleration a. In summary, the *force* on the glued body is triple the single-body force but the *acceleration* is unaltered because the mass also tripled. Extending this argument until the mass on the left represents Earth (6×10^{24} kg), we see that Earth has some tiny acceleration a toward a 1-kg body, while the 1-kg body accelerates toward Earth by the noticeable amount $6 \times 10^{24}a$. The accelerations are so lopsided that one mass can reasonably be called a *source* of gravity and the other is relegated to the status of *test mass*.

Check your understanding. Describe how the upper and lower panels of Figure 16.1 relate to each other. What are the strengths and weaknesses of each way of representing the gravitational field?

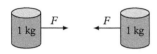

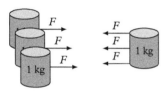

Figure 16.2 *Massive bodies accelerate less even though each bit of mass participates equally in gravity.* Top: *the gravitational forces* F *are are symmetric and each body has acceleration* a. Bottom: *each body on the left has the same relationship with the body on the right, but* collectively *they exert force* 3F *on it so its acceleration is* 3a. *The mass on the right still exerts force* F *on each of the bodies on the left, causing each to accelerate by* a. *Thus, a single 3 kg body would also accelerate by* a.

16.3 Newton's law of universal gravitation

We are well on our way to being able to predict the gravitational force between *any* two masses. We know the inverse-square dependence on distance, and that the force is proportional to m (so that all test masses fall with the same acceleration). By the reasoning of Section 16.2, the force must *also* be proportional to M. So we can immediately write, as a guess, that

$$F_{\text{grav}} \stackrel{?}{=} \frac{Mm}{r^2} \tag{16.1}$$

where the question mark indicates that we are not certain yet.

Test-drive this equation: it predicts that the gravity of two 1-kg masses placed 1 m apart ($M = 1$, $m = 1$, and $r = 1$) will provide an acceleration of 1 m/s², or that two 100-kg people placed 1 m apart ($M = 100$, $m = 100$, and $r = 1$) will accelerate toward each other at 10,000 m/s²! Gravity is *much* weaker than this, so much so that everyday objects have completely negligible gravitational attraction

Think about it

Gravity is weak: the entire mass of the Earth pulls down on your hair but a small electric charge (from, say, by rubbing a balloon) is sufficient to make your hair stand up against this force! Gravity nevertheless rules the universe because positive and negative electric charges tend to cancel each other out, while gravity has only one kind of "charge" (mass).

to each other. The $\frac{Mm}{r^2}$ part of our guess is based on solid reasoning, so our guess must be incomplete rather than wrong. The simplest way to keep our reasoning intact while capturing the empirical weakness of gravity is to include an overall calibration factor that must be a small number.

Therefore, we write

$$F_{\text{grav}} = \frac{GMm}{r^2} \qquad (16.2)$$

where G is called the **gravitational constant** (sometimes also called Newton's constant). We will assume it is a constant of nature with the same value anywhere in the universe, at all times, unless at some point we find evidence to the contrary (to date, no such evidence has been found). Equation 16.2, called **Newton's law of universal gravitation**, gives the magnitude of the force; the direction always points along the line separating the two masses. It is often useful to rewrite this as an acceleration law:

$$a_{\text{grav}} = \frac{GM}{r^2}. \qquad (16.3)$$

This form is most useful when we are describing the field around M, for which the mass of the test mass is irrelevant. Students should practice thinking of gravity as an *acceleration field* rather than a force field until that thinking becomes second nature. Also, beware that r is the distance between mass centers and has nothing to do with the size (radius) of either mass.

Measuring G requires some careful lab work: eliminating friction and other sources of force such as static electricity, so that the tiny gravitational force of two laboratory masses on each other can be revealed. Figure 16.3 shows the classic experiment to measure G. This setup takes advantage of symmetry: there are *two* test masses so they can be balanced and hung from a wire, which minimizes friction. The attraction between m and M then causes the wire to twist measurably; the greater the force, the greater the twist. The exact relation between the amount of twist and the force required to perform that twist can be carefully calibrated even before the masses are set up. The most precise current estimate of G is 6.6741×10^{-11} m^3/(kg s^2).

Once G is known, the mass of any M such as Earth can be inferred from the law of gravitation: simply rearrange Equation 16.3 to read

$$M = \frac{a_{\text{grav}} r^2}{G}. \qquad (16.4)$$

Again, beware that r is the distance between (the center of) M and the test particle that accelerates with a_{grav}. To infer Earth's mass, plug in $a_{\text{grav}} = 9.8$ m/s^2 and (because our experiments at Earth's surface are 6400 km from Earth's center) $r = 6.4$ million meters; the result is $M_{\text{Earth}} \approx 6 \times 10^{24}$ kg. For Jupiter, the acceleration and orbital radius of any of its moons tell us that planet has about 300 times the mass of the Earth. Similarly, the acceleration and orbital radius of any planet tells

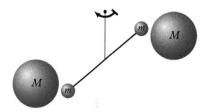

Figure 16.3 *The classic experiment to measure* G, *originally developed by Henry Cavendish (1731–1810).*

us that the mass of the Sun is about 1000 times larger yet. Today these masses are known to a precision of about one-hundredth of one percent, and astronomers continue to use this law to measure the masses of stars and galaxies.

Check your understanding. The mass of the Moon is about 1/80 that of the Earth. *(a)* If you place yourself 6400 km from the center of the Moon (the same distance as you now are from the center of the Earth), how would your acceleration compare to the 9.8 m/s^2 you are accustomed to on the surface of Earth? *(b)* If you stood on the surface of the Moon, your distance from the center of the Moon would be 1600 km rather than 6400 km. How does this change your acceleration compared to part (a)?

16.4 Gravity in and around spheres*

If each gram (each atom, really) participates in gravity, how can we treat the entire Earth as a single source mass? The inverse-square law (Equation 16.3) in principle describes the acceleration caused by a mass located exactly at $r = 0$; that is with no spatial extent. Such hypothetical masses are called point masses, and we can think of a ball such as the Earth as being composed of innumerable point masses. The net force of the Earth on a test mass is the sum of the forces exerted by the point masses composing Earth; this is known as the **principle of superposition** and applies to accelerations as well. To represent the acceleration field surrounding Earth, in principle we should sum up the accelerations caused by all its atoms. This calculation can be quite laborious because the atoms are spread out at different positions and thus exert pulls of differing size in differing directions. But Newton found clever ways to use symmetry to find the net gravity of a spherical ball—a good approximation for planets and stars—at positions inside as well as outside the ball.

Think of a solid ball as being composed of a nested set of spherical shells as in Figure 16.4. If we were to pull out a shell and analyze it individually, it would be very thin and completely empty inside. It turns out that the gravitational acceleration field in and around a thin spherical shell is relatively easy to compute, so we will first find the field in and around one shell, then sum up these fields to find the gravitational acceleration field in and around the spherical ball.

The gravitational acceleration field of a thin spherical shell. The shell is rotationally symmetric so the acceleration field can depend only on r and must point toward the center if it points anywhere. So, we can choose to think about *any* point at distance r from the center and our conclusion must hold for all other points at this distance. The next step is to split the problem into two distinct cases: finding the field at points inside the shell, and at points outside the shell. We will start with the easier of the two cases: inside the shell.

Figure 16.4 *A solid ball can be thought of as a set of nested spherical shells, like an onion.*

Inside the shell. Consider a test mass at point P in Figure 16.5. Focus first on the forces exerted by the parts of the shell nearest to and farthest from P, lying at distance r_n and r_f, respectively. The cones in Figure 16.5 highlight what the test mass "sees" in opposite directions. The black parts of the shell (the "caps") thus represent source masses pulling on the test mass in opposing directions. The mass of the near cap, M_n, is proportional to r_n^2 because the cap is a circle with a radius proportional to r_n. The acceleration due to M_n is, by the inverse-square law, proportional to $\frac{M_n}{r_n^2}$. With M_n itself proportional to r_n^2, the acceleration is then proportional to r_n^2/r_n^2, which means that r_n cancels out and *the effect of the cap does not depend on r_n at all.* The same argument applies to the far cap, which has a greater mass but also a greater distance. The acceleration exerted by a cap is independent of its distance! The near and far caps therefore exert equal and opposite accelerations, leaving no net acceleration. Now, picture the cones as opposing flashlight beams illuminating areas of the shell "seen" by the test mass, and sweep these beams all around the shell: the same argument applies no matter where you point the opposing beams. Therefore, the net effect of the *entire* shell reduces to zero net acceleration at this point. Because this point could be *any* point—we did not choose any special value of r—the acceleration is zero *everywhere* inside the shell.

Outside the shell. This is a more difficult case, so we will look at the outline of the argument rather than the details of the proof. Along the lines of the previous argument, Newton considered the net effect of all the bits of mass "seen" by a test mass placed a distance r from the center of a shell of radius R (where $r > R$ so the test mass is outside the shell; Figure 16.6). Assuming the inverse-square law describes the effect of each bit of mass, Newton used geometry (e.g., similar triangles) to show that the net effect of all the bits of mass in the shell is proportional to $\frac{R^2}{r^2}$. Now, notice that R^2 is proportional to the *mass* of the shell, because the mass is confined to a thin surface of area proportional to R^2. Therefore, the net effect of the shell is proportional to $\frac{M}{r^2}$. This is identical to the effect of a point mass prescribed by Equation 16.3. Therefore, *the net effect outside a shell is the same as if we had placed all the mass in a point at the center.* This greatly simplifies the reasoning that follows.

From shells to a solid ball. Outside a solid ball, we sum the effects of all the nested shells that form the ball. Each shell behaves as if all its mass were at its center and all the shells are concentric, so the ball *also* behaves as if all its mass were at the center. This is why we apply Equation 16.3 to an entire planet or star as if it were a single unit, and also explains why distance to the *center* of the planet or star is always the relevant distance.

To see what happens *inside* the ball, return to Figure 16.4 and imagine standing on, say, the fifth shell from the center. The shells outside your position have no effect because you are inside them, so we need only consider the cumulative effect of the five shells interior to your position. The effect of each shell is the same as if all its mass were concentrated at the central point. The cumulative effect of the

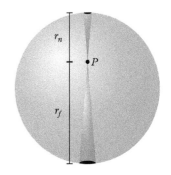

Figure 16.5 *The balance of forces exerted by opposite parts of a spherical shell on a test mass inside the shell. The near (top) cap is less massive than the opposing cap, but this is exactly balanced by the inverse-square effect of the top cap's smaller distance from the test mass. The net effect on the test mass is therefore zero. This argument applies to test masses at* any *location inside the shell.*

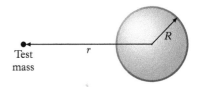

Figure 16.6 *The geometric setup examined by Newton to find the net effect of a mass shell on a test mass outside the shell.*

Think about it

Few planets or stars are perfectly spherical, but most are quite close. For precision work such as computing spacecraft orbits around Jupiter (which is notably oblate because it spins rapidly), scientists *do* have to perform more detailed calculations. This idea can also be turned on its head: some missions infer the mass distribution of a planet by measuring the very small variations in acceleration along its orbit.

interior shells is therefore the same as if *all* the interior mass were concentrated at the central point. The final result is that Equation 16.3 is directly applicable with the caveat that M must now be the mass *interior to* your position. Testing this proposition at the extreme, if you were at the center there would be zero mass interior to you so zero acceleration is predicted. This matches an argument from symmetry: at the center you have equal amounts of mass pulling in all directions, so the net effect must be zero. If you are near (but not exactly at) the center, there is very little interior mass, so the acceleration is very weak despite the fact that r is small.

Check your understanding. If the Sun collapes to a very small radius while keeping all its mass, how (if at all) would its gravitational effect on Earth change? Justify your reasoning.

16.5 Gravitational potential revisited

If there is no gravitational force or acceleration at the center of the Earth, then do all the effects of gravity disappear there? If clocks run slowly further down, we would expect clocks to be *slowest* here, yet there is no force or acceleration there. How do we reconcile these two pictures?

The potential is the best thinking tool for these situations. Recall (Section 15.1) that the potential difference between two points is a cumulative measure of acceleration times displacement in height, obtained by summing over steps so small that the acceleration within each may be considered constant; this makes the slope of the potential at any point equal to the acceleration there. There is no gravitational acceleration inside an empty spherical shell so the potential must be flat there (Figure 16.7). But if you move from inside to outside the shell, you immediately feel an acceleration pulling you back toward the shell. This means the entire interior of the shell is at the bottom of the potential and must have slow-running clocks despite the lack of acceleration there. Similar arguments apply to the inside of a solid ball: despite the zero acceleration at the center, it is actually at the bottom of the potential, and that means clocks tick least frequently there.

Quantitatively, what is the potential Φ outside a point mass, shell, or ball? We need an expression whose slope is equal to the known acceleration $\frac{GM}{r^2}$. Those who are familiar with calculus will see that this implies

$$\Phi = -\frac{GM}{r}. \tag{16.5}$$

A graph of this expression (Figure 16.8) yields the familiar funnel-shaped potential explored in Chapter 15. The shaded area in Figure 16.8 represents the

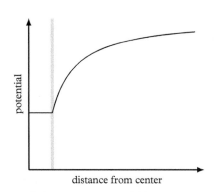

Figure 16.7 *The potential inside a spherical shell must be flat, because there is no acceleration there. Outside the shell (shaded band), the potential is the same as if the mass of the shell were in a point at its center.*

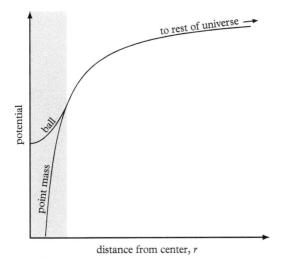

Figure 16.8 *Gravitational potential of a uniform ball versus a point of the same mass. Outside the ball the potentials are identical. Inside (shaded area), the potential falls much less rapidly than for the corresponding point mass, because only the mass interior to a given r contributes to the acceleration there. This yields zero acceleration (flat potential) at the very center.*

surface of a ball mass such as Earth, so focus first on the region outside this line. The acceleration (the slope of the potential) is quite steep on the initial climb away from the surface, but the climb becomes substantially easier once we are several more Earth radii away from the center. Despite this flattening, the potential on this figure still falls a bit short of the idealized value of zero it would reach infinitely far from Earth. In practice, we can never get *infinitely* far from Earth but a few hundred Earth radii would be far enough to put us very, very close to zero potential. The potential *inside* the Earth departs from the point-mass curve because mass exterior to a given r contributes nothing to the acceleration at that r. If we were to tunnel to $r = 0$, we would find zero acceleration there (Section 16.4) so the potential must be flat there.

The potential is a thinking tool, not just a visualization tool. For answering many questions, "how far down the potential" (global information) is more important than the acceleration you experience locally. Consider the opening question of this section: does time run slowly at the center of the Earth, where there is no acceleration? To find the answer, recall that at any given r, the *depth* of the potential (rather than its slope) determines the clock rate and the energy required to escape to interstellar space. Because the potential is deepest at the center of the planet, time runs most slowly there, exactly where gravity provides *zero* acceleration. Similarly, when we study black holes in Chapter 20 we will see that what makes them unique is the depth of their potential, rather than the acceleration they cause. This depth effect suggests that potential is closely related to a concept you may have encountered previously: escape velocity. Box 16.2 explains the relationship, but feel free to skip that box if you feel no need to think in terms of escape velocity.

Box 16.2 Escape velocity

One measure of the depth of the potential at a given point is the **escape velocity** from that point. We already know that a particle falling down the potential gains kinetic energy (per unit mass) equal to the drop in the potential it experiences, so think about this idea in reverse. If we are deep in the potential and launch a particle with kinetic energy per unit mass equal to our potential depth, the particle can coast all the way up to the top of the potential (but with zero kinetic energy left at the top). To find the required launch speed, equate $\frac{1}{2}v^2$ (kinetic energy per unit mass) to $\frac{GM}{r}$ (the potential "climb" from r to infinity). Rearranging $\frac{1}{2}v^2 = \frac{GM}{r}$ yields $v_{esc} = \sqrt{\frac{2GM}{r}}$. This is the escape velocity. (The direction of the escape velocity does not matter, so it should really be known as escape speed, but the name has stuck.) Particles with v less than this may climb much of the way up the potential, but are doomed to fall back in—unless they get assistance from engines or the like. You may read about escape velocity in other texts and resources, but the potential is more fundamental because v_{esc} works only for particles with mass while potential applies also to clocks and massless particles. Escape velocity is also a misleading way to think about rockets because they do the *opposite* of launching at high speed and coasting; they start with zero speed and escape by continually firing engines to push slowly upward. If, nevertheless, you feel the need to compute v_{esc}, it is just the square root of -2 times the potential.

If multiple source masses are present, the potential at a given point is the sum of the potentials due to each source mass. Figure 16.9 captures this idea in our solar system. Study this figure closely, as it contains many lessons. For example, more massive planets have deeper potentials. Where the potential has a peak between planets, its slope at that point is zero; this reflects equal and opposite tugs from nearby planets resulting in zero acceleration at that one location. Ignoring the localized potentials (informally called gravity wells here) due to the planets, the general trend is a steep rise from lower left fading into a shallower rise to upper right; this is the potential of the Sun. The steep potential outside a planet becomes shallow *inside* the planet (shaded regions), and is flat at its center. One unrealistic aspect has been introduced into this figure for the sake of clarity: the distances between the planets (and between Earth and Moon) would be much, much larger if the drawing were to scale. Between Jupiter and Saturn, for example, a large region of nearly flat potential has been eliminated to make room for more interesting features. Off the right edge of this figure is the truly vast plateau of flat potential that is the space between stars.

The potential is a valuable thinking tool, but keep in mind that it contains no new information compared to the acceleration field. You can always convert an acceleration field to a potential and vice versa. The virtue of the potential is that it tabulates the *cumulative* effect of all the accelerations experienced on the journey from one location to another. This gives us a global view that facilitates comparison of widely separated points.

Check your understanding. Consider Figure 16.9. *(a)* Imagine hanging out at the potential peak between Mars and Jupiter. If you remain exactly at the peak there

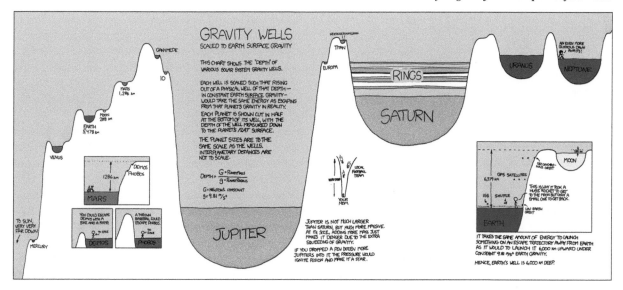

Figure 16.9 *This xkcd comic shows gravitational potential in the solar system, with the vast space between the planets suppressed for clarity. Source: xkcd.com/681.*

is no acceleration, but what happens if you drift slightly away from the peak? *(b)* In the inset of Figure 16.9, the Moon is drawn as sitting in a region of exactly flat potential. Can this be correct? *Hint:* consider the acceleration implied by a flat potential, and the path the Moon would follow if it had this acceleration.

16.6 Surface gravity and compact objects

Figure 16.10 shows the potentials of two planets with identical radii (represented by the shaded band), one twice as massive as the other. We will measure masses and radii in Earth units so we can call the $M = 1$ mass "Earth." At any exterior value of r the potential of the $M = 2$ mass is twice as steep *and* twice as deep as Earth's potential. "Twice as steep" means that a test mass experiences twice the acceleration at that value of r. As a result, the test mass requires twice the support if we wish to prevent it from falling—it has twice the weight. "Twice as deep" means that a test mass falling from infinitely far away (or, in practice, any extreme distance) gains twice the kinetic energy by falling to that value of r. (Relativity adds further nuance to the doubling of potential depth: stationary clocks at a given distance from the $M = 2$ planet lag coordinate time by twice as much as stationary clocks at the same distance from Earth, which doubles the gravitational redshift seen by distant observers.) Now we add a twist: imagine that the $M = 2$ planet actually has triple the radius of Earth. Find $r = 3$ on Figure 16.10 and cover the

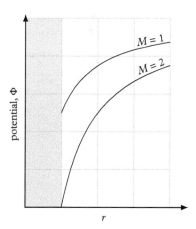

Figure 16.10 *Comparison of the gravitational potential of two different masses.*

Think about it

Some of the most massive stars are also some of the fluffiest; their surfaces are so far from their centers that surface gas can drift away at the slightest provocation despite their high mass.

Confusion alert

Students often equate *smaller* with *lower mass*. Do not fall into this trap! Unless otherwise specified, *smaller* means *smaller in radius*, with no preconception as to the mass. Stars and planets exhibit an enormous range of densities, so assume no connection between mass and radius.

part of the $M = 2$ potential curve to the left of this; because the same mass is spread over a larger volume we can no longer access this region. The potential at the surface of the $M = 2$ planet is now *higher* than at Earth's surface, so a rock falling from distant space will have *less* kinetic energy on impact. Furthermore, the acceleration (the slope of the potential) at the surface of this planet is less than on Earth, so a given object will weigh less on this planet. We say that this planet has weaker **surface gravity** than Earth.

By the same token, if an Earth-mass planet were made of solid lead it would occupy less volume so its surface would lie at smaller r. So in Figure 16.10, pencil in how the $M = 1$ potential would continue downward: it becomes deeper and steeper as it reveals more of a point-mass potential. Gravity is stronger at the surface of this smaller Earth-mass planet. Squeezing the mass of the Earth into a smaller radius would make Earth's "gravitational engine" more powerful. In other words, *regions of strong gravity are not necessarily regions around high-mass objects*. High-mass objects that are puffed up, such as supergiant stars, do not expose any regions of steep or deep potential (and even in the region "hidden" inside the star the potential flattens out and never becomes very deep). And a low-mass object *can* expose regions of steep and deep potential if its surface is close to its center.

The densest stars—made entirely of neutrons—pack the mass of the Sun into a radius of about 10 km. A neutron star provides a powerful gravitational engine in the sense that gas falling onto its surface releases a great deal of energy. But if the Sun became a neutron star (preserving its mass in the process), *nothing would happen to Earth's orbit*. Newton's law of gravity (Equation 16.3) dictates that the acceleration of Earth depends *only* on the mass of the Sun and Earth's distance from the Sun. If the Sun were to shrink in size while retaining its mass, it would expose new regions of stronger gravity *without changing the potential exterior to its original radius*.

A perennial source of confusion in this regard is the use of the symbol r to indicate distance from the origin of a coordinate system; students often misinterpret this as the radius of the source mass ball. We will denote a fixed distance such as the radius of a ball with R or r_{surf} and reserve r for variable distances, such as that of a test mass we can place anywhere. With that convention in place, Equation 16.3 tells us that the acceleration on the surface of a sphere (also known as its surface gravity) is $\frac{GM}{r_{surf}^2}$ and the potential there is $-\frac{GM}{r_{surf}}$.

Particles falling onto objects with large $\frac{M}{r_{surf}}$ ratios can release a great deal of energy. Such objects are called **compact objects** and are intensively studied by astronomers. We will return to compact objects later in the book; the key point here is the relationship between size (radius), mass, and potential. Compacting a given mass has a dramatic effect on the potential in the newly exposed region, as shown in Figure 16.11. For this reason, a low-mass ball may actually expose small regions of steeper and deeper potential than a less compact high-mass ball, even as the latter has a greater effect on distant regions. The ratio $\frac{M}{r_{surf}}$ determines the depth of the potential at the surface of the ball.

Check your understanding. How will the Sun's surface gravity and potential depth at the surface change (compared to the current situation) when it becomes: *(a)* a red giant of the same mass but about 100 times its current radius; *(b)* a white dwarf with half its current mass and about 1/100 its current radius?

16.7 Tides

Another triumph of Newtonian gravity is that it explains the rise and fall of the sea known as **tides**. Tides result from small differences in gravitational acceleration from one place to another, so physicists now use *tide, tidal effect,* or *tidal acceleration* to describe *any* difference in gravitational acceleration from one place to another. The left panel of Figure 16.12 shows Earth placed in the acceleration field created by the Moon, which is off to the left. If we subtract the acceleration arrow at Earth's center from each of the arrows in that panel, we see the acceleration field *as measured in a frame attached to the center of the Earth* (right panel). Note that we are using Earth only as a test body here; we are not showing its own gravitational field.

This field stretches Earth along the Earth-Moon axis, and pinches Earth in the perpendicular directions. As Earth rotates (the small rotation arrows in Figure 16.12 suggest that we looking down from the north pole), your location alternately rides up and down on this stretching and pinching pattern, so you experience two high tides and two low tides daily. However, by itself this does not explain why oceans rise and fall *relative to the land*. For that, focus on the red arrows. They show that accelerations *along* the surface move sea water toward the Earth-Moon axis where it can pile up relative to the land, which does not flow.

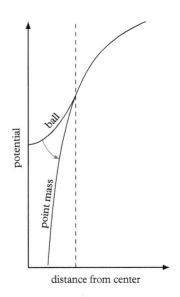

Figure 16.11 *If the ball in this example were to be compacted down to a point mass, the potential in the formerly interior region would deepen dramatically, without affecting the original exterior region.*

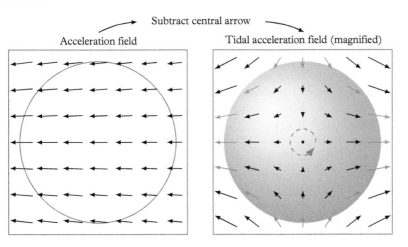

Figure 16.12 Left: *Earth in Moon's acceleration field.* Right: *the average acceleration is subtracted to highlight variations (magnified for clarity), which we call the tidal acceleration field. The dashed circle shows Earth's rotation; picture yourself rotating through this entire pattern daily. The red tidal acceleration arrows highlight how sea water is pushed toward, and piles up along, the Earth-Moon axis.*

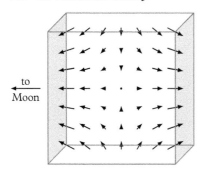

to
Moon

Figure 16.13 *The four white sides of this box have small inward-pointing arrows (arrows are omitted from sides facing us and opposite us for clarity). The two shaded sides have double-length arrows that point outward, leaving zero net arrow flow into or out of the box.*

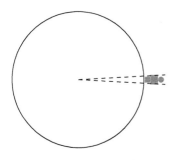

Figure 16.14 *Gravity squeezes your shoulders together because each shoulder is attracted to Earth's center. This lateral squeezing is tiny in practice, but provides a frame-independent signature of gravity–it happens even in a freely falling laboratory, where other effects of gravity disappear.*

There are many more fascinating consequences of tides in the Earth-Moon system—such as Earth's rotation slowing down—and in many other areas of astronomy; see *Further Reading*. Here, we focus on the consequences for relativity. The right panel of Figure 16.12 (repeated as Figure 16.13) shows a freely falling frame, because Earth is freely falling in the Moon's gravitational field. The equivalence principle claims that in a freely falling frame the effects of gravity disappear, but this is really true only at the center of the frame. The surroundings still contain an imprint of the gravitational field—the tidal stretching and pinching we just discussed. If you are weightless you can use this tidal test to determine whether your frame is freely falling in a gravitational field, or truly gravity-free. You can also use the test if you *do* have weight (i.e., there is an overall acceleration of your frame): just subtract off the average acceleration vector in your area to help expose any variations. Figure 16.14 shows another aspect of this test: in a rocket all parts of your body are accelerated equally but with gravity your shoulders accelerate in slightly different directions resulting in a slight inward squeeze.

This tidal test is important because it gives us a frame-independent way of describing gravity, and relativity trains us to look for frame-independent properties as the most physically meaningful ones. A remarkable frame-independent property of the acceleration pattern in Figure 16.13 is that the net convergence of arrows into the box is zero. By net convergence, I mean the total length of inward-pointing arrows minus the length of outward-pointing arrows; for each arrow oblique to the box boundary, count only its inward or outward component. The balance between inward and outward is not evident in Figure 16.13 at first glance, because the arrows exiting left and right are twice as long as the arrows entering at top and bottom. However, short arrows *also* enter the white sides facing toward and away from us, although I have not attempted to draw them. This makes four sides with entering arrows and two with double-length exiting arrows, for zero net convergence of arrows. Zero net convergence must also apply to the original field in the left panel of Figure 16.13, because adding back the *uniform* part of the acceleration field has no effect on the convergence.

Now that you have the basic idea, try drawing smaller boxes, spheres, or other surfaces in Figure 16.13: the net convergence is *always* zero. (If it appears not to be zero for your surface, try accounting for the fact that *each point* has an associated arrow even if not shown in the drawing.) This turns out to be a unique feature of the inverse-square law: in regions without a source mass, the net convergence of the acceleration field is zero. Does this change if we stuff the box in Figure 16.13 with mass? Yes, in this case acceleration arrows all around will point into the box. In fact, the net length of entering arrows would increase in proportion to the amount of mass we put in the box. Because this rule applies to arbitrarily small boxes or volumes of any shape, it is best stated as: *the net convergence of the acceleration field at each point is proportional to the density of mass at that point.* This statement turns out to be mathematically equivalent to Newton's law of gravity.

We will resume this thread in Chapter 18, when we go further in combining gravity and relativity. But first we need one more practical skill: understanding how we get information about the gravitational field over large volumes of space without actually going there. In Chapter 17 we look at how orbits give us this information.

Check your understanding. Imagine a model of gravity in which acceleration always points toward the source mass, but never weakens with distance. How would each panel of Figure 16.12 change? Would there still be zero net convergence of arrows in each panel?

CHAPTER SUMMARY

- Newton's universal law of gravitation: a point mass M causes an acceleration $a_{grav} = \frac{GM}{r^2}$ toward the mass.

- Principle of superposition: the accelerations exerted by multiple point masses add. A series of symmetry and superposition arguments shows that the acceleration outside a ball of mass M is also $a_{grav} = \frac{GM}{r^2}$.

- The difference in gravitational potential $-\frac{GM}{r}$ between two points represents the cumulative effect of gravity along a path from one point to the other. The potential determines how slowly clocks run and the resultant gravitational redshift.

- Compact objects allow test masses to fall to (or orbit at) very small r, where the potential is very steep and deep.

- A tidal acceleration is a *difference* in gravitational acceleration from one location to another. This is a frame-independent sign of gravity.

☰ FURTHER READING

Any introductory astronomy textbook will offer a thorough explanation of tides on Earth (including the effect of the Sun as well as the Moon) and examples of tidal effects in other contexts. Tidal heating of Jupiter's moons, for example, is potentially life-enabling because it causes water to be liquid on moons that would otherwise be frozen through.

Gravity From the Ground Up by Bernard Schutz explains how tides act to slow Earth's rotation—making the day longer—and have already slowed the Moon's rotation to the point that it is "locked" to the Moon's four-week orbit.

This has a further effect on the Earth-Moon orbit, pumping the Moon up to a higher orbit at a current rate of about 1 cm per year. Most astronomy texts cover some or many of these effects as well, but Schutz excels at tying them all together.

Back on Earth, for a lively introduction to ocean tides (and an explanation of why lakes do not have tides) see the *PBS Spacetime* video *What Physics Teachers Get Wrong About Tides*, available at https://www.youtube.com/watch?v=pwChk4S99i4.

◢ CHECK YOUR UNDERSTANDING: EXPLANATIONS

16.1 The ratio of acceleration at the Space Station to that at the surface should be the ratio of $1/6800^2$ to $1/6400^2$ (because the Space Station is 6800 km from the center of the Earth). This is the same as $\frac{6400^2}{6800^2}$ or about 90%. Acceleration at the surface is about 10 m/s^2 so acceleration at the Space Station is about 9 m/s^2.

16.2 The shading in the right panel is darker where arrows in the left panel are longer. The right panel therefore encodes the *size* of the acceleration but the *direction* is more subtle: the acceleration always points toward the direction of darker shading. The arrow map is very explicit about the direction, but arrows become confusing when too many are placed and when the arrows would be very large. In contrast, the shaded map easily portrays the acceleration at every single point.

16.3 (a) You would have about $1/80$ the acceleration. (b) You would now have a much larger acceleration than in part (a)—16 times larger if we are now at $1/4$ the distance. *Bonus:* combining the fact that the Moon's surface gravity would be $1/80$ of Earth's based on mass alone, but 16 times larger based on surface radius alone, we find that the Moon's surface gravity is $16/80$ or $1/5$ that of Earth.

16.4 Nothing would change, because neither the Sun's mass (M) nor Earth's distance from the Sun (r) would change in this scenario.

16.5 (a) Drifting away from the peak means that you will start "sliding down the slope." Very near the peak, the slope is shallow so you will not accelerate rapidly, but this will pick up as you get further down the slope. The peak is not a stable place to hang out, but as long as you remain *near* the peak you only need minimal engine power to push yourself back toward the peak. (b) A potential provides zero acceleration where it is flat, so the Moon cannot orbit if it sits in a region of flat potential. There should actually be a slight slope up away from Earth.

16.6 (a) Potential depth decreases by a factor of 100; surface gravity decreases by a factor of 10,000 because an atom on the surface is now $100^2 = 10,000$ times further from the center. (b) Potential depth increases by a factor of 50 and surface gravity by a factor of 5000 (would have been 100 and 10,000 but for the mass loss).

16.7 The acceleration field would have arrows pointing in the same directions as shown in the left panel of Figure 16.12, but they would all be the same length. Subtracting off the average would then reveal essentially *no* residual arrows pointing along the Earth-Moon axis (the left-right direction). The top-bottom trend would remain, though, causing a net inflow of arrows into the box.

▢ EXERCISES

16.1 The 1672 measurement that gravitational acceleration is smaller in French Guiana than in France came from a pendulum clock. The period of a pendulum depends only on its length and the acceleration due to gravity; because the period and length are easily measured to good precision, the acceleration can be determined to good precision. In fact, the Wikipedia entry for *pendulum* states that "moving a pendulum clock to the top of a tall building can cause it to lose measurable time from the reduction in gravity." How is this distinct from the fact that higher clocks tick more frequently in an accelerated laboratory?

16.2 Explain the distinction between source masses and test masses in your own words.

16.3 In your own words, what is a field? What is the gravitational field?

16.4 Why does G have units of m^3/(kg s^2)? *Hint:* review the units of all the quantities on both sides of Equation 16.3.

16.5 If you drilled a deep shaft in the Earth so that objects could fall to extremely small r, would they experience extremely large accelerations? Explain your reasoning.

16.6 The Sun is about 300,000 times more massive than Earth, and at some point in the future will become a white dwarf with roughly the same radius as Earth. *(a)* Compare (quantitatively) the surface gravity of the white dwarf Sun to that of Earth. *(b)* The Sun's current radius is about 100 times Earth's radius. Compare the Sun's current surface gravity to that of Earth.

16.7 Figure 16.8 ends at $r = 25000$ km, about four times the radius of the planet. *(a)* Compare the depth of the potential at $r = 25000$ km to the depth of the potential at the surface of the planet. *Hint:* measuring the depths with a ruler may help you recognize a pattern. *(b)* Compare the acceleration at $r = 25000$ km to the acceleration at the surface of the planet. *(c)* What would you have to do to measure the accelerations from the graph?

16.8 The radius of the Moon is about ¼ that of Earth. *(a)* If the mass of the two bodies were equal, how would the Moon's surface gravity compare to Earth's? *(b)* Now factor in the mass of the Moon: about 1/80 that of Earth. How does the surface gravity of the Moon compare to that of Earth?

16.9 There must be a point between Earth and Moon where a spacecraft would find the Moon's gravitational pull to be equal and opposite to that of Earth. Is this point halfway between Earth and Moon, or somewhere else (if so, where)? Explain your reasoning.

16.10 In Figure 16.8, at what point is escape velocity largest? Where is it smallest? Consider only points at or outside the surface of the planet.

16.11 Imagine you dig a tunnel to the center of the planet in Figure 16.8, and you transport the diggings to deep space. Where is more work done: transporting diggings from the center to the surface, or from the surface to deep space?

16.12 Explain why the potential at a given point is the sum of the potentials due to each source mass. *Hint:* assume the superposition principle is true for forces. Relate that to superpositions of accelerations and then to the potential.

16.13 Do clocks tick slowly inside a hollow spherical shell of mass?

16.14 If Planet A has stronger surface gravity than Planet B, does it necessarily follow that Planet A has a lower potential at its surface? Explain your reasoning.

16.15 Draw a version of the left panel of Figure 16.12 for the hypothetical case where Earth is much closer to the Moon. Is the difference between one arrow and the next smaller, larger, or the same as before? Use this to make a prediction about the resulting size of tides on Earth.

16.16 Look at the right panel of Figure 16.12 and imagine a small body of water such as a lake. Describe how much the acceleration field varies over this body, and use that to make a prediction regarding the sizes of tides in lakes.

+ PROBLEMS

16.1 *(Requires a bit of algebra.)* The Sun experiences an acceleration of 1.8×10^{-10} m/s^2 toward the center of our Milky Way galaxy (this is inferred from v^2/r rather than measured directly) and is about 2.6×10^{20} m from the center of the galaxy. How much mass does the Milky Way contain interior to the Sun's orbit?

16.2 Why are planets and stars approximately spherical? *Hint:* consider the gravitational forces on outlying parts of a (hypothetical) nonspherical planet.

16.3 Imagine that Earth is coin-shaped. Draw a coin seen edge-on and draw arrows all around indicating the direction and size of the acceleration due to gravity at various points. What would people on this planet experience as they explore its surface?

16.4 Use arrows to sketch the gravitational field around a hypothetical cigar-shaped planet. *Hint:* points off the long axis are further from the center than are points just off the short axis. It may also help to think of the cigar as two or more spheres touching each other.

16.5 In proving that the effect of a spherical shell of mass on a test mass outside the shell is the same as that of a point mass, why is it sufficient to demonstrate that the effect is proportional to $\frac{M}{r^2}$? A skeptic could offer other expressions, such as $\frac{2GM}{r^2}$, that are proportional to $\frac{M}{r^2}$ but do not match that of a point mass. How would you answer such a skeptic? *Hint:* consider what must happen very far from the sphere.

16.6 *(Requires algebra.)* Consider a solid ball with uniform density, radius R, and total mass M. *(a)* You tunnel from the surface to the center and measure the acceleration at various points from $r = 0$ to $r = R$. How does the acceleration depend on r? *(b)* To illustrate part (a) with a specific example, what is the acceleration at $r = R/2$ if it is 9.8 m/s^2 at the surface? *(c)* Earth is denser in the core than at the surface. Would you expect the acceleration at $r = R/2$ inside Earth to be lower, the same, or higher than your answer to part (b)?

16.7 Consider a spherical ball M with radius R surrounded by a thin, hollow spherical shell of mass M and radius $3R$ (Figure 16.15). The space in between is completely empty. *(a)* Copy Figure 16.15 to your paper, and add arrows showing the direction and relative size of the gravitational acceleration at the surface of the ball, just inside the shell, just outside the shell, and at $r = 6R$. *(b)* Graph the qualitative behavior of the potential from $r = 0$ to $r = 6R$. *(c)* Where do clocks tick most slowly? *(d)* Where is escape velocity largest?

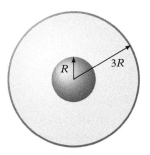

Figure 16.15 *Spherical mass inside a thin shell of mass, with empty space in between.*

16.8 Consider two bodies of differing mass but equal $\frac{M}{r_{\text{surf}}}$ ratios. *(a)* Which (if either) has stronger surface gravity (acceleration): the one with larger M or the one with smaller M? *(b)* Which (if either) has the steeper potential at the surface and which has the deeper potential at the surface?

16.9 *(a)* Compute the energy required to send a 1-kg mass from the surface of the Earth to the altitude of the International Space Station (380 km) in two ways: first, assuming gravity provides a constant 9.8 m/s^2 acceleration (Problem 15.6), and second, using the correct expression for the potential. Compare the two results and explain why one is larger but not substantially larger. *(b)* Do the same two calculations and comparison for geosynchronous orbit (about 36,000 km above Earth's surface). Why do you now see a substantial difference? *(c)* What fraction of the energy required to reach 36,000 km is used in the first 380 km?

16.10 *(a)* Compare the Sun's surface gravity (that is, the acceleration due to gravity at its surface) when the Sun is a red giant compared to its current surface gravity. If it is larger or smaller, describe how much larger or smaller. *(b)* Use the result of part (a) to explain why red giant stars often lose their outer layers of gas.

16.11 In *The Little Prince* Antoine de Saint-Exupéry shows asteroid B-612 as having a tiny radius (perhaps 2 m) but surface gravity roughly as strong as Earth. *(a)* Find the mass of this asteroid. *(b)*

Describe the tidal effects you would feel if you lived on this asteroid.

16.12 Escape velocity from the surface of Earth is 11 km/s (7 miles per second). Consider a rocket launched from the surface that escapes: it reaches such a large distance from Earth that it experiences negligible acceleration, and it coasts at at 5 km/s after engines have stopped firing. When, if ever, did the rocket actually travel at escape velocity? Explain your reasoning.

16.13 A shaft is bored from the surface of Earth all the way through the center to the surface on the other side, and an apple is dropped through the shaft (neglect air resistance, as air has been removed from the shaft). Define an x coordinate along this shaft, with the center of Earth at $x = 0$, the apple dropper at $x = -6400$ km, and the far end of the shaft at $x = +6400$ km. *(a)* Sketch the velocity of the apple as a function of distance. *(b)* Sketch the acceleration as a function of distance. Make no calculations, but make sure your plots are conceptually accurate; consider carefully, for example, how the velocity at the far end must relate to the velocity at the near end. *(c)* What will the apple do after completing a one-way trip through the Earth?

16.14 Which causes a larger tidal effect on you: a person standing next to you, or the planet Jupiter?

16.15 Research the Gravity Recovery and Climate Experiment. Explain how it uses the principles described in the chapter, and list a few of its discoveries. Has anything similar been used to map other planets?

16.16 Sketch the acceleration $\frac{GM}{r^2}$ as a function of r and use this sketch to illustrate how strongly the *difference* in acceleration across a test body varies from small values of r to large values of r. Those who know calculus may be able to compute the r-dependence quantitatively as well.

16.17 Repeat the *Check Your Understanding* exercise at the end of Section 16.7 with two more hypothetical force laws: $a \propto \frac{1}{r}$ and $a \propto \frac{1}{r^3}$.

16.18 Use calculus to show that the horizontal arrows in the right panel of Figure 16.12 must be twice the size of the corresponding vertical arrows.

17 Orbits

Orbits are ubiquitous in the universe: moons orbit planets, planets orbit stars, stars orbit around the center of the Milky Way galaxy, and so on. Any theory of gravity will have to explain the properties of all these orbits. Conversely, we can use our understanding of gravity to infer the masses and other properties of these cosmic systems. This chapter consists of three introductory sections that pave the way for developing the metric theory of gravity (general relativity) and four optional sections that provide starting points for further explorations of the cosmos that do not require general relativity.

17.1 Circular orbits

The International Space Station orbits about 400 km above the surface of Earth, and a popular misconception holds that gravity is quite weak so far up. However, Earth's gravity is defined by its *center*, and at Earth's surface we are already $r \approx 6400$ km from the center. Thus, an additional 400 km has only a modest effect: Equation 16.3 predicts 8.7 m/s^2 acceleration there, nearly as much as the 9.8 m/s^2 acceleration at the surface. We can also appreciate this acceleration *without reference to gravity* by noting that *any* circular motion requires an acceleration of v^2/r (Section 16.1). The space station circles Earth at $v \approx 8000$ m/s so it would rapidly fly off in a straight line without a substantial gravitational acceleration to pull it back.

How does the space station accelerate toward Earth's center without ever losing altitude? Just as a ball thrown in the air can be traveling upward while accelerating downward, the trajectory of the space station depends on the interplay between its acceleration and its *initial velocity*. Figure 17.1 shows how the initial velocity can be arranged so that this acceleration changes the *direction* but not the size of the velocity, thereby maintaining a constant distance from Earth. *All* types of circular motion, whether related to gravity or not, are made possible by this type of balance. As with a car driving in circles, at each instant the nudge toward the center is just what is needed to keep the velocity vector pointing along the circle. So the Space Station keeps falling toward Earth, but the distance never decreases. Douglas Adams surely had this in mind when he wrote in *Life, the Universe, and Everything* that "There is an art, [the Encyclopaedia Galactica] says, or rather, a knack to flying. The knack lies in learning how to throw yourself at the ground and miss."

Figure 17.1 *The changes in velocity required to keep an object moving along a circle are always directed toward the center. A judicious amount of centrally directed acceleration therefore prevents the distance of the orbiting body from either growing or shrinking.*

The Elements of Relativity. David M. Wittman, Oxford University Press (2018).
© David M. Wittman 2018. DOI 10.1093/oso/9780199658633.001.0001

Newton's cannonball provides another useful mental picture (Figure 17.2). Consider cannonballs launched parallel to (and, say, 1 m above) the ground at a range of speeds. A low-speed cannonball does not travel far before falling the 1 m and hitting the ground. A moderate-speed cannonball travels farther before hitting the ground, but is still unremarkable. At very high speed, the cannonball travels so far before hitting the ground that the curvature of the Earth must be considered. The ground slopes away and gives the cannonball extra distance before hitting the ground. At high enough speed (and in the absence of air resistance), the cannonball could travel all the way around the Earth this way; once around, it will continue the same pattern and keep circling. If launched at a much higher speed yet, the cannonball will follow a straighter line and leave the Earth before gravity can bend its path sufficiently. In *any* of these cases, the cannonball feels weightless while it is in free fall. Weightlessness while circling Earth is simply weightlessness in free fall. When an astronaut drops a hammer, astronaut and hammer accelerate equally so the astronaut sees (and feels) the hammer just floating there.

What speed is just right for circling the Earth? We need to match two types of centrally directed acceleration: the $\frac{v^2}{r}$ that is required for *any* type of circular motion, and the $\frac{GM}{r^2}$ that is provided by gravity. So

$$\frac{v^2_{\text{circ}}}{r} = \frac{GM}{r^2}$$
$$v^2_{\text{circ}} = \frac{GM}{r}$$
$$v_{\text{circ}} = \sqrt{\frac{GM}{r}} \qquad (17.1)$$

This is the fundamental equation for circular orbits. It implies that smaller orbits require *larger* speeds to maintain a circular orbit around a given mass M.

A useful model for orbits is the gravity wishing well or coin funnel often seen in science museums. This model works precisely because it is shaped like the gravitational potential outside a spherical source mass (Figure 17.3). Coins started at low speeds simply drop into the center, and coins started at very high speed tend to escape from the system. Forming a circular orbit requires a specific initial speed *and direction* so coin funnels include on-ramps that start the coin orbiting at that speed. (When satellites are launched from Earth, rockets push the payload up to the desired altitude *and* give it the right horizontal velocity.) Unlike real orbits, the coin's orbit gradually decays due to rolling resistance, so we can watch a single coin pass through a series of nearly-circular orbits that demonstrate the higher speed of smaller orbits. Notice also that the relationship between orbital speed and distance does *not* depend on the mass of the coin, just as real orbits do not depend on the test particle mass m. A given coin funnel represents the gravitational potential around a given planet or star; M is fixed. To represent greater M a funnel would have to provide a faster orbit at any given radius—it would have to be steeper at all radii and therefore deeper as well.

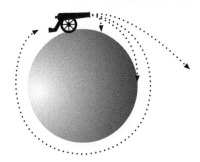

Figure 17.2 *Newton's cannonball thought experiment: cannonballs fired at progressively higher speeds will travel progressively farther around the Earth. Absent air resistance, a cannonball with just the right speed will make it all the way around. This orbit will repeat forever in the absence of air resistance.*

Think about it

Note that v_{circ} is the square root of the absolute value of the potential, making it equal to the escape velocity divided by $\sqrt{2}$ (Section 16.5).

Figure 17.3 *A coin funnel wishing well is a good model for orbits, because it is shaped just like the potential. The weaknesses of this model are that coins are confined to a surface and that they spiral in due to rolling resistance; neither is true for real orbits.*

To cast this idea in equation form, simply rearrange Equation 17.1 to read

$$M = \frac{v_{circ}^2 r}{G}.$$

(17.2)

Astronomers use this to infer the mass M of any celestial object around which an orbiting test mass can be observed. Conveniently, we need not determine the test mass m because it is not relevant. This makes the task relatively easy: we need only measure the velocity of a test mass and its orbital radius r. High orbital speeds at moderate to large values of r are a sure sign that M is large.

Centuries ago, astronomers measured velocity not through the Doppler effect, but by counting the days required to complete a full orbit. This time is called the period, P, and is related to v_{circ} by the fact that velocity is distance divided by time. A distance of $2\pi r$ around the orbit in a time P implies $v_{circ} = \frac{2\pi r}{P}$ or $P = \frac{2\pi r}{v_{circ}}$. Plugging $v_{circ} = \frac{2\pi r}{P}$ into Equation 17.1 yields

$$\frac{2\pi r}{P} = \sqrt{\frac{GM}{r}}$$
$$\frac{4\pi^2 r^2}{P^2} = \frac{GM}{r}$$
$$\frac{4\pi^2 r^3}{GM} = P^2$$

(17.3)

Equation 17.3 has the same physical content as Equation 17.1 but can be applied in cases where we do not measure v directly. It is known as **Kepler's third law** after Johannes Kepler (1571–1630).

Box 17.2 summarizes all three of Kepler's laws of planetary motion, but the third law is most important for our purposes because it relates orbital properties to the amount of source mass being orbited. Kepler actually worked out only a rudimentary form of the law, $r^3 \propto P^2$, but this was already a stunning accomplishment. The apparent motions of the planets are quite complicated because we see them from Earth, which itself is moving. Nevertheless, Kepler was able to work out models of the orbits that explained the apparent motions if and only if Venus orbits the Sun at 0.72 times the Earth-Sun distance, if Mars orbits at 1.52 times the Earth-Sun distance, Jupiter at 5.20, and so on. This was truly a scale model of the solar system because the actual distances were unknown. Nevertheless, it enabled Kepler to identify the proportionality $r^3 \propto P^2$.

That proportionality already implies that larger orbits must have slower speeds—Saturn, for example, must have about ⅓ Earth's speed because it takes about thirty years to go around an orbit only about ten times as large as Earth's ($10^3 \approx 30^2$). This in turn is all that is required for us to picture a potential that is steep near the Sun and flatter farther away as in Figure 17.3. However, it took Newton to tie all these variables to the *mass* of the Sun through Equations 17.1–17.3.

Check your understanding. List three ways in which the coin funnel accurately reflects trajectories of particles in a gravitational potential (be sure to include infalling or escaping motion as well as circular motion). In what way does the coin funnel fail to provide an accurate model?

Box 17.1 Orbits: putting it all together

Understanding orbits is one of the great stories of physics, so it is worth reviewing how far we have come over several chapters:

- By the definitions of velocity and acceleration, *any* circular motion requires a *centrally directed* acceleration of size $\frac{v^2}{r}$.

- Gravity on Earth accelerates all objects by an equal amount, and is always directed toward the center of the Earth.

- The centrally directed acceleration keeping the Moon on its path around Earth could therefore be the same centrally directed acceleration we experience on Earth's surface (gravity)—but the small size of the Moon's acceleration is a hint that gravity weakens with distance from the center of the Earth. A variety of additional observations and geometrical reasoning lead us to conclude that the acceleration due to gravity declines as $\frac{1}{r^2}$.

- Because a force is an interaction between two objects, the equation describing the force must include the source mass M in the same way it includes m. This led us to $F_{grav} = \frac{GMm}{r^2}$, where G is a constant of nature determined by measuring the (very small) gravitational force between 1-kg masses placed 1 m apart. The gravitational field around a massive object can therefore be described by $a_{grav} = \frac{GM}{r^2}$.

- If circular motion is due to gravity, we equate the circular acceleration to the gravitational acceleration: $\frac{v_{circ}^2}{r} = \frac{GM}{r^2}$ leads to $v_{circ} = \sqrt{\frac{GM}{r}}$. This is the fundamental equation of circular orbits. Smaller orbits must be faster than larger orbits around the same mass, and orbits of a given size must be faster around a large mass than around a small mass.

- Rearranging $v_{circ} = \sqrt{\frac{GM}{r}}$ yields $M = \frac{v_{circ}^2 r}{G}$. Armed with the value of G and a measurement of orbital size and velocity (or period) we can determine the mass of the body being orbited. For example, Earth's mass is determined using the Moon's orbital v and r, and the mass of the Sun is determined using the v and r of any planet.

17.2 Elliptical orbits

It seems like an amazing coincidence that each moon and planet has *exactly* the speed required to maintain a circular orbit at a fixed distance from its parent body, so let us now admit that real orbits are not exactly circular. Imagine a planet with too much speed to maintain a circular orbit at its initial distance r_0 as in Figure 17.4. Gravity still bends the path, but not enough to maintain a circle. The

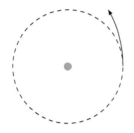

Figure 17.4 *A planet with too much initial speed cannot follow a circle, but still follows a curved path.*

Figure 17.5 *Elliptical orbits result from the interplay of kinetic and potential energy. The test mass moves rapidly when close to the source mass because it has gained kinetic energy by falling down the potential hill.*

Think about it

Coin funnel orbits exhibit precession, but due to rolling resistance, which does not apply to planets.

Confusion alert

Here *a* refers to semi-major axis rather than acceleration.

---

planet thus gains some distance from the Sun. This means moving up the potential hill, so the planet that started too fast begins to slow. This allows gravity to bend the path more effectively and bring it toward the Sun. The planet then regains speed by falling back down the potential as it completes the loop. The end result is an elliptical orbit as shown in Figure 17.5. Think of an elliptical orbit as a circular orbit plus an excursion up and down the potential hill.

You can observe this behavior directly in a coin funnel if the coin's initial speed differs noticeably from the required circular speed. The coin's behavior is much like a pendulum that swings back and forth exchanging kinetic with potential energy, except that this swinging is relative to a circular loop. Newton showed that an inverse-square law (and *only* an inverse-square law) results in *exactly one* swing of the "pendulum" per orbit. This means that the orbiting particle starts each orbit with exactly the same position and velocity as the previous orbit, and therefore repeats the same orbit over and over. Exactly repeating or **closed** orbits are therefore indicative of the inverse-square law. Orbits in solar systems are indeed mostly closed, thus supporting the inverse-square law. However, there is a wrinkle.

Even with the inverse-square law, orbits close exactly only when the potential is provided by a point or spherical source mass. Although the Sun is very close to spherical, 0.1% of the mass of the solar system is in Jupiter. Jupiter (and to a lesser extent the other planets) pulls on the orbit of each planet enough to shift its perihelion, or point of closest approach to the Sun, slightly from one orbit to the next, an effect called **precession**. Still, the inverse-square law is confirmed because for most planets Jupiter's inverse-square effect neatly explains any departure from a completely closed orbit. This does not quite explain all of Mercury's precession, however; a point we will return to in Chapter 18.

In elliptical orbits the distance r to the source mass is constantly changing with time, so we need a more stable definition of the overall size of the orbit. Astronomers take the longest distance across the ellipse (the *major axis*) and cut this in half to define the *semi-major axis*, denoted a. This definition of orbital size is useful because it preserves the relationship between period, source mass, and orbital size given by Equation 17.3. Thus, Kepler's third law is best expressed as

$$\frac{4\pi^2 a^3}{GM} = P^2. \tag{17.4}$$

In fact, one of Kepler's great contributions was realizing that orbits are elliptical and that those ellipses are *not* centered on the Sun. Kepler did not know why, but we can see that the shape of the potential in Figure 17.5 forces more-centered orbits to also be more circular. As a result, only circular orbits can be perfectly centered, and greater departures from circularity require more miscentering.

Much more can be said about orbits (Box 17.2), but do not lose sight of the main concepts: a circular orbit is an unchanging balance between gravity and motion given by $v_{\text{circ}} = \sqrt{\frac{GM}{r}}$. In an elliptical orbit the balance shifts back and forth over the course of an orbit but on average is described by the same principles. We can re-express this relationship in different ways to emphasize period, mass, or distance rather than velocity, but the physical idea is always the same.

Check your understanding. If there were only one planet in the solar system, would its orbit precess?

Box 17.2 Kepler's laws of planetary motion

Kepler spent many years poring over detailed records of planetary motion (as seen from Earth) and found that the following three laws summarize the data.

1. Planets follow ellipses with the Sun off-center. Kepler showed this empirically, but we can understand this in relation to Newton's proof that orbits are closed. To close, a noncircular orbit must move up and down the potential hill exactly once per orbit. This means that the center of the potential *cannot* be in the center of the orbit; if it were, the two ends of the elliptical orbit would be equally far up the hill, implying two hill climbs per orbit. As seen in Figure 17.5, the center of the potential must be closer to the high-speed end of the orbit than to the low-speed end of the orbit because the test mass gains speed as it comes down the potential hill. Kepler's first law quantifies how far the Sun is from the center of the ellipse as a function of the ellipticity of the orbit.

2. Any given planet moves faster when it is closer to the Sun; speed and distance are inversely proportional. Clearly, a particle must pick up speed as it falls down the potential hill, but why must the speed be inversely proportional to the distance? (Note that this is a very different context from Equation 17.1, which prescribes the *constant* speed about which an elliptical orbit oscillates.) The inverse relationship between r and v is another way of saying that the product rv (technically known as angular momentum per unit mass) is conserved. It turns out that this product is conserved by *any* potential that depends only on r; in other words, any spherically symmetric potential. Emmy Noether (1882–1935) showed that there is a deep relationship between symmetries of the potential and conserved quantities. Another example of this connection: time symmetry (i.e., potentials that do not depend on time) results in conservation of energy.

3. The square of the period is proportional to the cube of the semimajor axis. This is Equation 17.4, but Kepler stated it empirically as $a^3 \propto P^2$ (here, a represents the semimajor axis, *not* acceleration). Newton showed that this is true *only* if gravity obeys the inverse-square law; check the reasoning leading up to Equation 17.1 and you will see that it depends on the gravitational acceleration obeying this law. Equation 17.1 in turn is essentially Kepler's third law, as you can see from the reasoning leading to Equation 17.3.

17.3 Symmetry of orbits

In the Newtonian model of gravity, orbits exhibit a symmetry that is important to note here because we will find later that this symmetry can be violated in a metric theory of gravity. Consider the highly elliptical orbit in Figure 17.6; it is symmetric about the dashed line. The left side is a mirror image of the right, and the symmetry goes deeper than a sketch of the path through space. Whatever the *speed* of the planet at point A, for example, it will have the same speed at A_M,

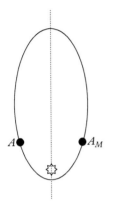

Figure 17.6 *This orbit is symmetric (in space) about the dotted line. It is also symmetric in* time *in the sense that it repeats exactly. In the Newtonian model of gravity, the test mass can come arbitrarily close to the source mass without breaking this symmetry.*

the mirror image of point *A*. This means that if you made a movie of the orbit and then played it backward, the motion of the planet in the time-reversed movie would still be perfectly described by Newton's laws. This is called time symmetry. In the Newtonian model of gravity, orbits have this symmetry *no matter how closely they approach a point mass*. Gravity may be very strong close to a point mass, but in Newton's model that strength gives an approaching particle exactly the kinetic energy it needs to climb back up the potential hill and recede to its original starting distance.

This is the Newtonian picture for pointlike source masses and single orbiting particles. Box 17.3 presents some situations in which this symmetry is broken. You may skip this box if you prefer to focus on the important conceptual point: the Newtonian model of gravity has no way of breaking the symmetry of a single orbiting particle in a spherically symmetric potential. If we idealize the source mass as having an arbitrarily small radius (so collisions do not enter the picture), then orbits can come arbitrarily close to $r = 0$ and come back out unscathed in this model. The metric model will have something different to say on this point.

The remainder of this chapter surveys a variety of optional topics related to Newtonian orbits, and the main thread of this book resumes at the start of Chapter 18.

Check your understanding. (a) A particle is launched vertically from the surface of a planet, slows due to gravity as it climbs, and then falls vertically back to the planet. Neglecting air resistance, is the return trajectory a sort of mirror image of the outbound trajectory? Convince yourself that the speeds at each point match each other. *(b)* Do the same for a particle falling straight onto a source mass and bouncing off frictionlessly.

Box 17.3 Breaking time symmetry of orbits

Orbital symmetry can be broken if we consider physical processes beyond those considered in the main text. First, many source masses (stars and planets) do *not* have very small radii, so collisions can happen and completely disrupt the orbit. Impacts are rare because space is so big, but they can be spectacular. The impact of Comet Shoemaker-Levy 9 on Jupiter in 1994 produced fireballs easily visible through small telescopes on Earth, and blackened the face of the planet for months. Based on analysis of Moon and Earth rocks, our own Moon is thought to have formed when a large body collided with the then-forming Earth.

A second asymmetric process is friction. Although space is generally empty and frictionless, sufficiently dense gas can make a difference. When a star is surrounded by a gas disk the inner parts of the disk naturally circle faster than the outer parts. Collisions between gas atoms then cause a few atoms to gain energy and be ejected while most lose energy and fall into smaller, lower orbits. The net effect is a steady spiral drain of gas toward the center, reminiscent of coins in a coin funnel. Interactions between planets and a gas disk are also thought to be responsible for inspiraling of planets in some solar systems.

Box 17.3 *continued*

A third body can also introduce asymmetry. Picture a space rock called Alice encountering Jupiter for the first time: Alice has kinetic energy even before entering Jupiter's potential, so she must retain this amount of kinetic energy when it *leaves* that potential. Therefore, Alice cannot be captured into orbit around Jupiter—a direct consequence of time symmetry. But capture *is* possible in the presence of a third body. Imagine that Alice and Bob are modest-sized rocks orbiting each other; their velocities have similar sizes but always point in opposite directions. As they near Jupiter, Alice and Bob experience slightly different accelerations along their slightly different paths, and these acceleration differences can increase their velocity differences. Alice may gain enough kinetic energy to be flung out while Bob loses kinetic energy and is captured by the planet (or vice versa). Many of the moons of the outer planets have orbits that suggest they have been captured this way. Similar interactions cause star clusters to evaporate (lose most of their stars), and allow spacecraft to gain kinetic energy from planets in the frame of the Sun (see Section 17.4).

17.4 Slingshot maneuver*

You have probably heard about a space probe slingshotting around a planet to gain speed. But we just established that a particle gains no more kinetic energy falling toward a source mass than it loses climbing away from the same mass. Just as a pendulum cannot swing higher than it started, so the probe cannot be moving faster at the end of this maneuver. How can the probe receive a *net* gain of energy?

The answer becomes clear if we think in different frames (Figure 17.7). Imagine that Earth, moving at 30 km/s relative to the Sun, catches up to a space probe moving at only 25 km/s relative to the Sun. In the Earth frame, the probe approaches Earth at an initial speed of 5 km/s, swings around in a highly elliptical close encounter, and completely reverses its direction of motion, leaving the vicinity with a final speed of 5 km/s, equal to its initial speed. The probe may fire its engines here and there to fine-tune the encounter, but the hard work of reversing the probe's direction is done by Earth's gravity rather than the engines. But what is the point of merely reversing direction? Think back to the Sun frame: the probe now moves 5 km/s *faster* than Earth—and Earth still moves at 30 km/s so the probe moves at 35 km/s in the Sun frame. This is a huge gain for a small fuel expenditure, so such maneuvers are now built into nearly every space mission. In practice, gains are not as large as 10 km/s, but the maneuver still pays off handsomely. The energy must come from somewhere, though: Earth's orbital motion loses a bit of kinetic energy. However, the loss is so small that environmental impact statements need not be filed.

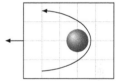

Figure 17.7 *Slingshot maneuver: in the planet frame gravity can reverse the direction of a spacecraft, but not increase its final speed. The fact that the planet moves relative to the Sun means that this maneuver does change the spacecraft speed relative to the Sun.*

Check your understanding. (a) Why is gaining or losing speed *in the Sun frame* the important criterion for these maneuvers? *(b)* For what kind of mission would one want to gain speed in the Sun frame, and for what kind of mission would one want to lose speed?

17.5 Dark matter versus modified gravity*

The planet Uranus was discovered in 1781, and it soon became clear that the planet did not quite follow the orbit predicted by Newton's laws. Two hypotheses were suggested. One—*modified gravity*—was that the true law of gravity departed from Newton's inverse-square model at such great distances from the Sun. The second—*unseen matter*—was that another (as-yet unseen) planet was exerting forces on Uranus. Working on the latter hypothesis, Urbain Le Verrier (1811–77) determined in 1846 roughly where a hidden planet would have to be in order to explain the orbit of Uranus. He initially had difficulty persuading astronomers to even look there, but when they did they discovered Neptune after only *one hour* of searching—a stunning success for the Newtonian model of gravity.

Within another decade astronomers also noticed that Mercury did not orbit quite as expected. In 1859 Le Verrier, naturally, hypothesized an unseen planet called Vulcan close to the Sun, where it would have escaped detection in the glare. But this time modified gravity would have been the correct hypothesis: Chapter 18 explains how Einstein's model of gravity fits Mercury as well as all the other planets.

At that time, little was known about the more distant cosmos. We now know that orbits of stars in their galaxies, and orbits of galaxies in their clusters, are universally faster than expected from $\sqrt{GM_{\text{vis}}/r}$ where M_{vis} is the amount of visible matter (Figure 17.8; see also Section 17.6 for how astronomers know the masses of stars). Either the cosmos contains enormous amounts of unseen matter, or our misunderstanding of gravity is substantial.

Astronomers are now confident in the former interpretation, and we can add some details. Not only does this matter not shine, but it also does not absorb, reflect, or interact with light in any way; this gave rise to the term **dark matter**. Gravity tells us not only how much dark matter there is (about five times as much as normal matter), but also how that dark matter is distributed in galaxies and throughout the universe. Figure 17.8, for example, shows that, compared to normal matter, dark matter must be less concentrated around galactic centers.

This is so astonishing that modified gravity may seem to be a more attractive explanation. However, it has been difficult to find a modified gravity model that works across the wide range of environments in which we see mismatches between

> **Think about it**
> _____
>
> Because dark matter does not reflect or absorb light, in some languages it is called *invisible matter*.

gravity and normal matter. For example, one model posits that the gravitational acceleration cannot fall below a certain minimum amount, no matter how far a test particle is from the center of mass. This works well (by design) to explain the too-high orbital speeds in the outskirts of galaxies (Figure 17.8), which are low-acceleration environments. But this tweak does not help in high-acceleration environments such as galaxy cluster cores, which *also* indicate a need for dark matter to explain the orbits. Additional support for dark matter has come from collisions of galaxy clusters, which yield gravitational fields that are *not at all* centered on the majority of the normal matter (gas in this case). This qualitative mismatch is predicted by the dark matter model (the collision dislodges the gas from the dark matter) but is extremely difficult to explain with modified gravity. As a result, dark matter has become the scientific consensus.

The lack of interaction with light suggests that dark matter is fundamentally composed of different types of particles than is normal matter. In fact dark matter cannot be made of atoms, because astronomers know the average density of atoms in the universe, and that turns out to be substantially lower than the density of dark matter. Dark matter must therefore be some as-yet identified kind of particle, which nicely connects the science of big things to the science of small things: particle physicists have good reasons to believe that there are more types of particles than have yet been discovered, and astronomers believe that at least one of those must make up the dark matter. If individual dark matter particles are not too massive, the Large Hadron Collider (LHC) may have the ability to create them in energetic collisions. Other experimenters are searching for dark matter particles that pass through Earth, and yet others are looking at places where dark matter congregates to see if dark matter particles ever decay or otherwise interact with each other.

This summary barely scratches the surface of dark matter studies, but it may be enough to pique your interest and look at some of the resources in *Further Reading*.

Check your understanding. A census of stars and gas in the Milky Way reveals enough normal matter to cause the Sun to orbit the center of the Milky Way at about 125 km/s. But it actually orbits at about 250 km/s. How much dark matter is there compared to normal matter? Remember that circular velocity goes as the square root of the mass.

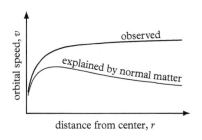

Figure 17.8 *Orbits in a typical galaxy are much faster than predicted by the mass in normal matter like stars and gas. Adding dark matter to the mass model increases the predicted speeds to match the observed speeds. Dark matter is not just more of the same: it makes up an increasing fraction of the mass budget on the outskirts of each galaxy, so it is distributed more widely than normal matter.*

17.6 Masses of stars*

So far, we have neglected the fact that an orbiting body (such as a planet orbiting a star) has some mass, m, that, in principle, sources a weak gravitational field in addition to the much stronger field of the central body. We can also view this issue through the lens of Newton's third law: the planet pulls on the star just as the star pulls on the planet, but the star—thanks to its higher mass M—accelerates so much less that we have approximated the star as fixed. In reality, the star moves slightly

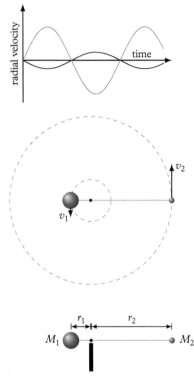

Figure 17.9 *Mass ratio inference for a binary star.* Top: *observed Doppler effect for the two stars over time. One star has quadruple the speed (and acceleration) of the other.* Middle: *the 4:1 speed ratio indicates that one star travels quadruple the distance of the other. The dot marks the center of mass, the point that remains fixed as the system spins.* Bottom: *if the system balances at the dot, the 4:1 distance ratio indicates that* M_2 *has* 1/4 *the mass of* M_1.

so the relationship between orbital size, period, and mass is more complicated. Newton showed that Equation 17.4 should really be written

$$\frac{4\pi^2 a^3}{G(M + m)} = P^2. \tag{17.5}$$

Measurements of the period P and the semi-major axis a thus tell us about $M + m$ rather than M. In the case of a planet orbiting a (much more massive) star, $M + m$ is nearly the same as M, so our previous approximation was justified—but more care is required when analyzing orbits of nearly equal masses such as two stars orbiting each other (binary stars).

Equation 17.5 tells us the total mass, so how do we find each star's portion of this total? Newton's third law tells us that the force of star A on star B must be the same size as the force of star B on star A; and since $F = ma$ we can write $m_A a_A = m_B a_B$ or $\frac{m_A}{m_B} = \frac{a_B}{a_A}$. Furthermore, both stars share a common period so the ratio of the accelerations is also the ratio of the velocities: $\frac{m_A}{m_B} = \frac{v_B}{v_A}$ (see Figure 17.9 for a graphical view of these connections). The velocity ratio, which is measured from each star's Doppler effect, thus tells us the mass ratio. Knowing the mass ratio and the total mass then allows us to assign a mass to *each* star.

Measuring the mass of an isolated star is practically impossible, but astronomers have found a way to transfer the knowledge gained from studies of binary stars to isolated stars. Having studied many stars in binary systems, astronomers found that the color and luminosity of a star depends on its mass. Astronomers use this relationship to infer the mass of an isolated star from its color and luminosity. Binary stars thus serve as the foundation for understanding the masses of all stars.

Check your understanding. If the velocity of Star A in a binary system varies from ±50 km/s throughout its orbit and the velocity of Star B in the same system varies from ±100 km/s, what can you say about the masses of Stars A and B?

17.7 Extrasolar planets*

For many years, astronomers searched for exoplanets by looking for variations in the Doppler effect of potential host stars. Imagine that in Figure 17.9 you could see only M_1 because M_2 is a very faint planet. The host star still moves even if this **reflex motion** is much smaller and the planet cannot be seen at all. Taking Jupiter and the Sun as an example, $\frac{m_{\text{Jupiter}}}{m_{\text{Sun}}} \approx \frac{1}{1000}$, so the Sun moves at 1/1000 Jupiter's rate—small, but not impossible for dedicated alien astronomers to measure through changes in the Sun's Doppler ratio. Regular oscillations in a star's Doppler ratio are a sure sign of an orbiting planet even if (as is typical) the glare of the star prevents us from seeing the nearby planet in a direct image.

Hundreds of exoplanets have been discovered by measuring reflex motions of sunlike stars. For years, the exoplanets that were detected were only the most massive ones, simply because it is easier to detect large reflex motions. They also tended to be the ones closest to their host stars—because that yields more reflex motion but also because it yields a short period, which facilitates the confirmation of a complete orbit. But improvements in technology keep revealing lower-mass planets and planets further from their host stars. Earth-mass planets (1/318 the mass of Jupiter) are now being discovered, including some that are far enough from their host stars to potentially be habitable.

Most discoveries are now made with the transit method, which relies on the planet coming between us and the star once per orbit. This blocks some of the starlight and produces a repeating series of small dips in the apparent brightness of the star (Figure 17.10). This is an efficient way to search many stars with a single set of wide-field time-lapse images; candidates are then followed with the Doppler method for unambiguous confirmation and for estimating the exoplanet mass through the velocity it induces in the star. A planet blocks a tiny fraction of its star's light so transit monitoring is best conducted from space, above the turbulence of our atmosphere. The first such mission, NASA's *Kepler*, alone detected about 3600 planet-like transit signals; more than 1200 of these have been confirmed as of this writing. The *TESS* mission will be even more sensitive to smaller planets, and is scheduled for launch in March 2018.

Known exoplanets now number more than 3500 and span an extraordinary range of properties: gas giants close to their hosts ("hot Jupiters" that can have the density of styrofoam), Earth analogs, planets around binary stars (Figure 17.11), a disintegrating planet, a lava planet, a waterworld, and so on. Even more amazing may be the statistical result that *most stars have planets, and many have multiple planets.* To be clear, *Kepler* found planets around only a fraction of the stars it studied—but that is because the transit method is sensitive only to the small fraction of planetary orbits that actually interrupt our view of the host star. After correcting for this and other known insensitivities, the discoveries indicate that most stars have at least one planet. That in turn may change your perspective on whether we are alone in the universe.

Check your understanding. Earth's mass is about 1/300,000 that of the Sun, and Earth orbits the Sun at about 30 km/s. What is the reflex velocity of the Sun due to Earth?

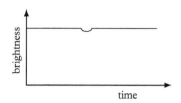

Figure 17.10 *The transit method monitors the brightness of a star and looks for small dips that repeat at regular intervals. Jupiter would block only about 1% of the Sun's light, and Earth would block only about 0.01%.*

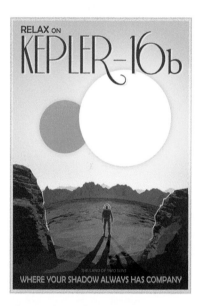

Figure 17.11 *Some of the planets discovered by the Kepler mission orbit binary stars. Courtesy of the Exoplanet Travel Bureau at NASA/JPL-Caltech.*

CHAPTER SUMMARY

- Objects in orbit are in free fall; they are *not* in a region of zero gravity.

- Maintaining a circular orbit requires a speed $v_{\mathrm{circ}} = \sqrt{\frac{GM}{r}}$.

- This further implies that $M = \frac{v_{\mathrm{circ}}^2 r}{G}$ can be used to determine the mass of sources of gravity given observations of the orbits of test masses.

Applications of this law also include inferring the existence of extrasolar planets and dark matter.

- Elliptical orbits are based on the same principles, but—like a pendulum—have surplus kinetic energy at some points and surplus potential energy at others.

- The Newtonian model of gravity predicts that elliptical orbits in the potential of a spherical mass are closed, and that all orbits in such a potential are symmetric.

FURTHER READING

Any introductory astronomy text will cover topics in this chapter in much more detail.

Dark matter: the Astronomy Picture of the Day website (http://apod.nasa.gov/apod/ap060824.html) has a striking picture of a galaxy cluster collision and a short explanation of how this constitutes proof of dark matter, with many links for further exploration. In the context of a single galaxy, take the challenge of trying to match observed velocity data with normal matter alone, at http://wittman.physics.ucdavis.edu/Animations/RotationCurve.

Exoplanets: The Exoplanet Travel Bureau (http://planetquest.jpl.nasa.gov/exoplanettravelbureau) and sister sites provide fun places to begin learning about the amazing variety of exoplanets that have been discovered. Also check out exoplanets.org for the latest numbers on exoplanets have been discovered, as well as the properties of those planets. You can easily generate plots of exoplanet distance from the host star, exoplanet mass, and so on—always with the most recently updated numbers in this rapidly-developing field. The essay *Yes, There Have Been Aliens* by Adam Frank in the *New York Times* (June 10, 2016) is a provocative view on recent exoplanet discoveries.

CHECK YOUR UNDERSTANDING: EXPLANATIONS

17.1   Circular orbits are possible; the speed of a circular orbit varies inversely with the size of the orbit; elliptical orbits have greater speed closer to the center; particles with too much speed will escape completely; and particles with too little (or zero) speed fall sharply toward (or into) the center. The shortcoming of the coin funnel model is that, unlike in space, there is friction that causes all orbits to decay over time.

17.2   No. In the Newtonian model of gravity, orbits in a perfectly spherical potential cannot precess. (The Sun does happen to be incredibly close to spherical.)

17.3   (a,b) A trajectories are symmetric in both cases in the sense that "running the movie backward" looks the same as running it forward. If this is not clear, draw a spacetime diagram of each motion.

17.4   (a) The Sun's potential dominates the solar system, so getting to a different planet (which is at higher or lower potential) requires gaining or losing speed relative to the Sun. (b) Gaining speed in the Sun frame allows a craft to climb the potential further and visit outer planets. Conversely, to visit inner planets a craft should lose speed—it will then gain speed back as it falls toward the Sun, so when it reaches the inner planets its speed will match their high speeds.

17.5 We need to quadruple the mass to get twice the orbital speed, because the latter goes as the square root of the mass. Therefore, normal matter comprises only about one quarter of the total mass. This is a somewhat larger fraction than in the cosmos generally, because dark matter and normal matter are not distributed in the same way throughout the Galaxy.

17.6 Star A has twice the mass because it has half the acceleration. Although the question specified that A has half the *velocity*, this implies half the acceleration because they share a common period.

17.7 $1/300,000$ of 30 km/s yields $1/10,000$ km/s or 0.1 m/s. Amazingly, astronomers are now able to determine velocities of stars down to this level and detect Earth-mass exoplanets. Higher-mass planets are often present in the same system, so the larger reflex motions due to those planets must first be modeled and subtracted out.

? EXERCISES

17.1 What is the circular speed of the Moon as predicted by Equation 17.1? What is the actual circular speed, judging by its orbital circumference and period?

17.2 Deduce the mass of Jupiter by using the orbital period and radius of one of its moons. Repeat for another of its moons. How well do the answers agree?

17.3 Compared to Jupiter, Saturn is only about twice as far from the Sun, yet its period is more than twice as long: about thirty years compared to about twelve years. Explain why.

17.4 Sedna, one of the most distant known objects in our solar system, orbits with a semimajor axis of 500 times the Earth-Sun distance. Estimate the period of Sedna's orbit in years. Is it 500 years, or shorter or longer (and if so, roughly how much shorter or longer)? Explain your reasoning.

17.5 *(a)* What would happen to the Earth's orbit if the Sun suddenly vanished? *(b)* What would happen to the Earth's orbit if the Sun collapsed into a neutron star with its original mass but a much smaller radius of 10 km (the Sun's current radius is 1.4 million km)? *(c)* What would happen to the Earth's orbit if the Sun becomes a red giant with its original mass but a 100 times larger radius? (This is half as large as Earth's orbit; ignore possible nongravitational effects such as Earth encountering friction with the Sun's atmosphere.)

17.6 Imagine that the Sun's potential has the same slope everywhere, regardless of distance from the Sun. For each of Kepler's laws, determine whether the law would still be valid in this situation, and explain why or why not.

17.7 Mercury is in a highly elliptical orbit. In what part of its orbit does it move most rapidly? Most slowly?

17.8 Not all trajectories are ellipses. A particle that starts on the potential "plateau" can fall into the solar system and climb back out. Sketch this trajectory on a flat piece of paper. Is it symmetric?

17.9 If gravity did *not* weaken with distance from the Sun, could planets still have closed circular orbits? Explain your reasoning.

17.10 Does a more massive planet provide for a better slingshot effect?

17.11 If Galaxy A has four times as much mass within $r = 30,000$ light-years as Galaxy B, how does v_{circ} for stars at $r = 30,000$ light-years compare between Galaxy A and Galaxy B? (You may imagine Galaxy B as a hypothetical galaxy with normal matter only and Galaxy A as a real galaxy with dark matter as well.)

17.12 Use the galaxy rotation curve builder at http://wittman.physics.ucdavis.edu/Animations/RotationCurve to match observed velocity data with some combination of normal matter, dark matter, and central point mass. How much of each do you need? How and why will a central point mass (such as a supermassive black hole) always fail to explain the observed high orbital speeds?

17.13 In a binary star system, which star has the smaller orbit?

17.14 If a hypothetical alien astronomer observes our solar system and looks for changes in the Sun's velocity due to planets, the alien will infer the existence of which planet first? How many years must the alien wait to see a complete orbit?

✛ PROBLEMS

17.1 *(Requires algebra.)* The Moon orbits in one month, while the space station orbits much closer to the surface in ninety-six minutes. How far above the surface would an orbit have a period $\left(P = \frac{2\pi r}{v}\right)$ of twenty-four hours? What would such an orbit be useful for?

17.2 The three plots in this problem should be vertically aligned so that the behavior at a given distance can easily be compared. *(a)* Sketch the velocity required to maintain a circular orbit around the Sun, as a function of distance from the center of the Sun. Make your horizontal axis go from zero to ten times the radius of the Sun. For less than one solar radius, an orbit is impossible, so start the curve at one solar radius. *(b)* Do the same for a star that has the same mass as the Sun but a much smaller radius. Think carefully about where to start the curve. *(c)* Do the same for a star that has the same radius as the Sun but much less mass. *(d)* If the size of v_{circ} is an indicator of how strong the gravitational field is, where should an astronomer look to observe strong gravitational fields?

17.3 Imagine gravity does *not* weaken with distance from the Sun. *(a)* Could planets still have closed elliptical orbits? Explain your reasoning. *(b)* What property of elliptical orbits would be the same as with the inverse-square law?

17.4 Imagine a source mass M moving with velocity v through a uniform sea of test masses. (The test masses never actually hit the source mass because its radius is so much smaller than the average distance between test masses.) *(a)* Sketch the situation and draw acceleration arrows on some of the test masses. *(b)* Some time later, the test masses have moved in the directions you drew, but the source mass has moved forward. Sketch this new situation, and explain why the source mass must decelerate. (This is called *dynamical friction.* Can you guess how it depends on M and on v?)

17.5 Research more about the slingshot maneuver. What are the practical limits on the boost in speed, and why? What are the risks? It is nice to get to your destination faster, but this implies that more deceleration is required to stay when the destination is reached. What are some solutions to this problem?

17.6 Clusters of galaxies contain 1000 or more galaxies, each moving on a highly *non* circular orbit. Can the speeds of these galaxies be used to infer the mass of the cluster? Explain why or why not.

17.7 Given a typical mass of about 10^{30} kg per star and a separation of 10^8 km, estimate the orbital speeds of binary stars.

17.8 How large are the changes in velocity that Jupiter induces in the Sun? How long does it take for that velocity to go through a full cycle?

General Relativity and the Schwarzschild Metric

18

In Chapter 14 we saw how the metric can explain trajectories of free particles without reference to gravitational forces, thus unifying gravity and relativity. In Chapter 15 we extended this to the case where gravitational acceleration varies with position. In Chapters 16 and 17 we looked at how acceleration does vary with position in our universe, and how we study that through observations of orbits. But finding a metric that empirically explains orbits takes us only so far. In this chapter, we learn how physicists *predict* the metric in the spacetime surrounding any source mass. In the process, we will see how spacetime can exhibit much more dynamic and complicated behavior than we ever suspected.

18.1 From Newton to Einstein

Newton's inverse-square force model of gravity works extremely well as a scientific theory: it explains a diverse array of phenomena—falling apples, orbits, and tides—with one simple fundamental law. The potential as a thinking tool was developed after Newton, but Section 16.5 showed that we can express Newton's law with the potential $\Phi(r) = -\frac{GM}{r}$. Section 15.2, meanwhile, showed that the potential determines the time coefficient of the metric, $1 + \frac{\Phi}{c^2}$. This explains the altitude-dependent tick rates of stationary clocks *and* explains why freely falling particles accelerate downward without feeling any force: they are simply maximizing their proper time like any other inertial particle. We can thus write a "Newtonian metric model" of gravity by taking the special relativity metric and making sure that $\Delta\tau = \left(1 + \frac{\Phi}{c^2}\right)(\Delta t)$ for stationary clocks:

$$c^2(\Delta\tau)^2 = c^2\left(1 - \frac{GM}{rc^2}\right)^2(\Delta t)^2 - (\Delta r)^2 \qquad (18.1)$$

where I am now using r for the spatial coordinate to reflect the spherical symmetry of gravity around a point or ball source mass. This model suggests that gravity and relativity are unified but still leaves some open questions. This section looks at four of these questions.

Is the Δr term affected by gravity? So far, we have ignored the Δr term on the basis that everyday particles and orbits are so much slower than c that their

The Elements of Relativity. David M. Wittman, Oxford University Press (2018).
© David M. Wittman 2018. DOI 10.1093/oso/9780199658633.001.0001

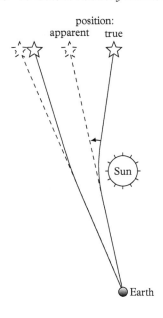

position:
apparent   true

Sun

Earth

Picture from Earth:

Sun

Figure 18.1 *The Sun's gravity deflects the path of light from distant stars (not to scale). The deflection angle is indicated by the arrow, and is larger for starlight passing closer to the Sun. We see the stars in their true (solid) positions when the Sun is not in the picture; we see the stars as being in the dashed positions when the Sun is placed as shown.*

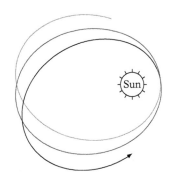

Sun

Figure 18.2 *Precession, highly exaggerated.*

motion through time is overwhelmingly larger than their motion through space. We should test the Δr term with the highest-speed particle we can think of: light. Figures 18.1 sketches the path of light from a distant star as it passes near the Sun. In the Newtonian view, the light accelerates toward the Sun, more so for light passing closer to the Sun. The solid lines in Figures 18.1 show the actual path of the light, and the dashed lines show where each star appears to be as seen by us; the angle between solid and dashes lines is known as the deflection angle, and is greatly exaggerated here. If the acceleration is $a = \frac{GM}{r^2}$ like it is for any other particle, the deflection angle near the Sun's surface will be only 0.00024 degrees. How would we observe this? The true position of any star is well known from pictures taken when the Sun is nowhere near the line of sight to that star. Observing the same star when the Sun is near the line of sight is, however, no easy task because the Sun is so bright. Testing this prediction therefore requires detailed observations during a total solar eclipse, when background stars are briefly visible.

In the metric model, light follows a path of zero proper time. The path of light is determined by applying a modified version of the geodesic equation (to find the zero-proper-time path rather than the maximum-proper-time path followed by particles with mass) to a metric such as Equation 18.1. Working out the path in this model, Einstein found the same deflection angle as the Newtonian prediction. He nevertheless urged astronomers to undertake the measurement to see if the metric theory of gravity worked for high-speed particles. Before a measurement could be made, however, Einstein made a discovery about the Δr term that caused him to revise his prediction substantially. We will return to this drama later; the point here is that a complete theory should make a prediction about the Δr term rather than assume it is unaffected by gravity.

The highest-speed planet may be able to test the Δr term as well. Section 17.2 hinted that Mercury's orbit does not *quite* fit the Newtonian model. Tugs from other planets, primarily Jupiter, cause Mercury's orbit to precess by 0.00148° per year; that is, if you look "down" on the solar system and watch the planets orbiting counterclockwise as in Figure 18.2, Mercury's point of closest approach to the Sun shifts counterclockwise by 0.00148° each year. (*Year* here means 31,557,600 seconds, based on the Earth year; Mercury completes about four orbits in this time.) By the mid-1800s astronomers found that the actual rate of precession was a bit higher, 0.00160° per year, and that this discrepancy was *not* due to uncertainty in their measurements. Searches for hypothetical unseen planets causing this discrepancy (Section 17.5) found nothing. If the space part of the metric is affected by gravity, we would expect Mercury to be the most-affected planet because it has the highest speed through space. Furthermore, if the space nearest the Sun is most affected, then Mercury is best poised to probe that as well. Mercury's anomalous orbit thus provides additional motivation to think about how gravity may affect the space part of the metric.

How can we make this model frame-independent? Despite our motivation to think about how gravity may affect the space part of the metric, we lack appropriate thinking tools. In fact, even the time part of the metric in Equation 18.1 is a bit of an *ad hoc* theory. It works in one frame, attached to the source mass, but it is not really based on any frame-independent principle about how gravity affects spacetime. It would be nice if a unifying frame-independent principle could be used to deduce *both* time and space parts of the metric.

Furthermore, it is not even clear that the metric should be the ultimate focus of our investigation, because the mathematical form of the metric depends very much on the coordinate system we choose. If we choose to cover the same spacetime with a different type of grid, we will get a different metric. Yet those different metrics covering the same spacetime surely must have *something* in common—what is it?

How do energy and momentum affect gravity? The M in Equation 18.1 refers to mass, but mass is just one form of energy. Do all forms of energy act as sources of gravity? (Physicists use *source* as a verb here: do all forms of energy source gravity?) Experimentally, all forms of energy *respond* to gravity, to incredibly high precision (Box 13.2), so we will take it as a foothold idea that all forms source gravity as well. (This foothold idea is also supported by experiment, but the discussion would take us too far astray here; see the first two books listed in *Further Reading* for more discussion on this point.) From now on, look at the $\frac{M}{c^2}$ in Equation 18.1 as including all forms of energy in the source, not just mass.

Momentum is inextricably linked to energy (Section 12.3) so does momentum source gravity? Here is a fascinating thought experiment from physicist Bernard Schutz, involving hypothetical mass *currents*: straight, infinitely long rivers of mass. The upper panel of Figure 18.3 shows a perfectly symmetric situation: two such currents running in opposite directions with a stationary particle halfway between them. By symmetry, the particle has no reason to move up or down, so it remains midway between the two mass currents; this is a frame-independent statement.

Now, think in the frame of the lower current (lower panel of Figure 18.3). Due to length contraction, the upper current has higher density (illustrated by the contracted striping) and the lower current has lower density, than in the symmetric frame. The upper current squeezes much more mass-energy closer to the test particle, so we would expect the particle to accelerate toward the upper source—but *we know from the symmetric frame that it does not*. An observer in the lower frame must conclude that energy is *not* the only criterion determining gravitational attraction; source momentum must also play a role.

The contribution of momentum is actually *repulsive* in Figure 18.3 because it partially cancels the greater attraction of the current with higher energy density. To be clear, any effect of source momentum is in addition to the primary attractive effect of source energy; because a source's momentum can never quite equal its energy, the repulsive effect of a single source can never quite overcome its attractive effect. We have deduced all this from a simple thought experiment involving symmetry. In real life, source masses are not necessarily so simple and symmetric, so we need a theory that lets us work out the metric around *any* configuration of sources. We could then plug that metric into the geodesic equation to predict particle trajectories around those sources and compare to observations.

How can we apply this model to dynamic situations? In the Newtonian model we developed, the potential is fixed to the source mass. If we bump the source mass, the potential throughout all space must shift to reflect the new source position. This instantaneity violates relativity, so let us look at a possible fix. The

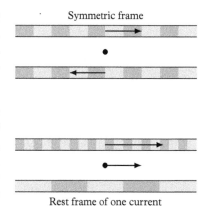

Figure 18.3 *Thought experiment demonstrating that momentum sources gravity. With mass currents flowing equally in opposite directions (upper panel), the test particle in between will not accelerate in any direction. In another frame (lower panel), this lack of acceleration is remarkable considering that one current is denser (due to length contraction) and thus more attractive. We conclude that the momentum of the upper current has an offsetting* repulsive *effect on this particle. Inspired by a figure and physical argument presented by Bernard Schutz in* Gravity from the Ground Up.

Confusion alert

Momentum is repulsive only if the test particle is moving *parallel* to the source mass. A stationary test particle is indifferent to source momentum, and a particle moving in the opposite direction actually feels more attraction. This is a hint that metrics need to be much more flexible than we may have imagined (Section 18.2).

problem really stems from expressing the inverse-square law globally: $a = \frac{GM}{r}^2$ prescribes what happens throughout the universe (at all values of r) based only on where M is now. Can we change this to a local description, one that specifies the local acceleration arrows based on local conditions? Section 16.7 gives us a thinking tool for developing such a local rule: the net convergence of acceleration arrows into any volume you care to draw is proportional to the density of source mass there.

How would this rule work in practice? Say you are given a point mass in an otherwise empty space. Start by drawing a small sphere around the point mass; the local proportionality rule tells us the net length of arrows that should enter this sphere; call this L. This sphere has no exiting arrows, so L determines the total length of entering arrows. Given the spherical symmetry we divide L into equal-length arrows entering evenly around the sphere as in the left panel of Figure 18.4. Next, we ask about immediately adjacent parcels of space. We take advantage of the symmetry by drawing wedges as in the middle panel; wedges in a given shell will be identical. We start with the top wedge: the net length entering must be *zero* because there is no source mass in this wedge. Therefore, arrows entering the top must exactly cancel the arrows exiting the bottom (the lateral sides have no effect because arrows there neither enter nor exit the wedge; they are parallel to the wedge side by design). Now the top and bottom are actually parts of spherical shells, with the top having an area larger than the bottom by a factor $\frac{r_{\text{top}}^2}{r_{\text{bottom}}^2}$. Because there is an arrow *each point in space*, to compensate for this area factor we we need smaller arrow lengths at the top, by the factor $\frac{a_{\text{top}}}{a_{\text{bottom}}} = \frac{r_{\text{bottom}}^2}{r_{\text{top}}^2}$. This gives us the inverse-square relation $a \propto \frac{1}{r^2}$.

By repeating this process for additional wedges throughout space, we build an acceleration field based only on local information. This idea was developed by

Figure 18.4 Left: *the mass in a region of space determines the net length of inflowing acceleration arrows; symmetry here tells us to divide this length evenly around the sphere.* Middle: *Poisson's law says there is no net inflow in cells with no mass, so the top of this cell needs an inflowing arrow size that balances the known flow out of this cell.* Bottom: *because each* point *on each face has an associated arrow, the arrow length at the top face is shorter than at bottom, by the ratio of face areas. Repeating this process yields an inverse-square field throughout space while using only local information.*

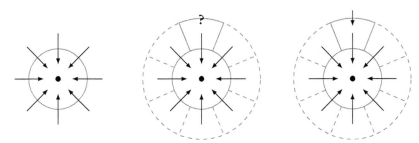

Siméon Denis Poisson (1781–1840) and is now called a *local field equation*. We can imagine that if we bump the source mass we must go through this process again, starting from the new source mass position. In this way, changes in the gravitational field can ripple out toward the farthest reaches of the universe rather than violate relativity by doing so instantaneously. But I have not yet described a *physical* model for this process. Such a model would tell us the speed with which these changes propagate (which is presumably faster than the speed with which we can deduce the changes).

Collectively, these four issues suggest that we need new thinking tools to uncover a deeper theory. If nature is kind, a single unifying theory will give us answers to *all* these questions.

Check your understanding. What are the four issues that suggest we need new thinking tools?

18.2 Elements of general relativity

Einstein adapted Poisson's approach to deal not with acceleration, but with a more physically meaningful and frame-dependent quantity explained next. Furthermore, Einstein added a space*time* perspective to what had been a purely spatial perspective. Because he built time in as just another coordinate, his model follows changes in time just as naturally as Poisson's model tracks variations in space.

Gravity as geometry. The key insight for a frame-independent description of gravity is to compare each inertial worldline *to its neighbors* rather than to the coordinate system. Figure 18.5 compares worldlines of free particles aboard an accelerating rocket to those of free particles near Earth. With the "artificial gravity" aboard a rocket, initially parallel worldlines remain forever parallel—and this will remain true no matter how we reframe this diagram. With real gravity the initially parallel worldlines deviate from parallel—and this too will remain true no matter how we reframe this diagram. By comparing inertial worldlines to each other, Einstein discovered a *geometric* interpretation of gravity.

Mathematicians had already found that initially parallel lines diverge or converge *when the lines are drawn on a curved surface*. For example, longitude lines on Earth start out parallel at the equator, but converge at the poles. By Einstein's time, a whole branch of math was dedicated to understanding curved surfaces. Mathematicians had actually developed the geodesic equation to define the shortest possible line between two points on a curved surface (which is the original definition of geodesic). Informally, we can find the shortest line from Beijing to San Franciso by pulling a string taut along a globe; you will find that this string goes far north, across Alaska. To find the same route mathematically, we would apply the geodesic equation to the metric for a globe. But if the geodesic equation gives the shortest route along Earth's surface, why does it give the

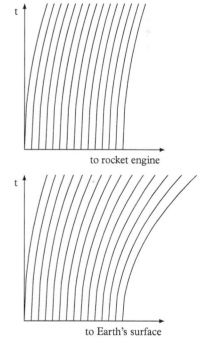

Figure 18.5 *Initially parallel worldlines of free particles remain parallel aboard an accelerating rocket* (top) *but not in a gravitational field* (bottom) *because the acceleration is greater near the floor.*

longest proper time route in spacetime? This is because on Earth, latitude and longitude displacements each contribute positively to the distance traveled; while in spacetime, spatial displacements contribute *negatively* to the proper time. This illustrates that while the geodesic equation works with any metric, the result must be understood in the context of the particular coordinate system.

Mathematicians had proved something else about the geodesic equation that is important for us: the shortest path is also the straightest path. The straightest path would be one in which the traveler's position always moves forward directly along his velocity vector. Pushing the traveler forward on the globe changes the relationship between the velocity vector and the longitude-latitude coordinate system, so to use that coordinate system we have to continually recalculate what "forward" means—even as the traveler does nothing more than follow his nose. The mathematical prescription for propagating a vector this way turns out to be exactly the same prescription as the one for making the shortest path—the geodesic equation. This neatly reunifies the traditional sense of *inertial path*—the straight path taken when a particle's velocity vector remains unmodified by an applied force—with the longest-proper-time sense we have been using since we completed our understanding of special relativity.

This is an important bridge from special relativity to general relativity. We no longer need to ask *why* inertial particles maximize their proper time; they are simply moving "straight ahead" through a curved spacetime. This does *not* mean that there is a higher dimension for spacetime to curve *through*. Mathematically and physically this is just not necessary; "curvature" is simply a statement about initially parallel lines.

Confusion alert

Remember that curved spacetime is not the same as curved worldlines. The top panel of Figure 18.5 shows how inertial worldlines may curve even in a flat spacetime—i.e., one in which initially parallel inertial worldlines remain parallel.

Box 18.1 Curved spacetime and special versus general relativity

In a curved spacetime, widely separated vectors cannot really be compared with each other because bringing them side-by-side for a comparison would change them. For an analogy, think of the Beijing-to-San Francisco air route: the plane leaves Beijing to the northeast, follows a straight path along Earth's curved surface, and finds itself on a southeast path as it glides into San Francisco. This plane's velocity vector originally seemed to be the same as that of a plane leaving San Francisco to the northeast, but the vectors look quite different when we compare them *at the same place*.

In special relativity, comparing vectors was easier because a vector arrow had the same meaning *regardless* of where we drew it. You will therefore need to unlearn some things you took for granted in special relativity. For example, light traveling at c relative to a local observer need not travel at c relative to a *distant* observer. In special relativity, a zero proper time path means $c^2(\Delta t)^2 - (\Delta x)^2 = 0$ so $\frac{\Delta x}{\Delta t} = c$. But in general relativity we have more complicated metrics such as $c^2 A^2 (\Delta t)^2 - B^2 (\Delta x)^2$ (where A and B stand in for complicated expressions involving proximity to source mass). In this case a zero proper time path means $c^2 A^2 (\Delta t)^2 = B^2 (\Delta x)^2$ so $\frac{\Delta x}{\Delta t} = \frac{A}{B} c$. The x and t coordinates are defined far from gravity where $\frac{A}{B} = 1$ so the speed of light there is c. But near a source of gravity, $\frac{A}{B} < 1$ so the speed of light there is $< c$ when measured *in these coordinates defined by distant observers*.

Box 18.1 *continued*

You can reconcile this with your special relativity intuition as follows. Local observers measure the speed of light passing them using local clocks and rulers, whose relationships to the distant coordinates are also affected by A and B; the net result is that local observers find the speed to be c. Distant observers, in contrast, measure the speed indirectly—they send a flash of light to a distant mirror and back (Section 18.6) and find that the round-trip time is greater than the round-trip distance divided by c. Special relativity can be true only locally because each patch of the universe freely falls in a different way. General relativity stitches these patches together into a global model, with relationships between distant patches unlike anything you have seen in special relativity.

The mathematics of curved spaces also tells us to make a distinction between coordinates and geometry. A worldline may curve one way relative to one coordinate system, but curve another way (or not curve at all) relative to another coordinate system. The coordinates are, in a way, superficial—they are painted onto an underlying geometry, and the geometry determines how two worldlines relate to each other. Worldline relationships are easily visualized on a spacetime diagram, but the math required to calculate them is *not* simple. Any given coordinate system has a metric that depends on the underlying geometry *and* on the setup of the coordinate system. Painting a complicated coordinate system onto a flat sheet of paper yields a complicated metric—but the underlying geometry is still flat. As a second example, the accelerating-rocket metric that would describe the top panel of Figure 18.5 differs from the special relativity metric, but the underlying geometry is still flat there as well. The mathematical tools for calculating how parallel lines do or do not deviate in any given metric are quite sophisticated, so we will not pursue them here; just be aware that most problems cannot be solved simply by looking at the metric.

A spacetime framework: Einstein borrowed heavily from established mathematics but added a spacetime perspective, for example realizing that time is a coordinate that enters the metric with a sign opposite to the spatial coordinates. We now dig a bit deeper to show how a spacetime framework naturally accounts for dynamic effects and the effects of momentum.

Dynamic effects. We have talked much about initially parallel inertial worldlines. Let us now begin using the shorter term **geodesics** for inertial worldlines, and **geodesic deviation** for the divergence or convergence of initially parallel geodesics. So far, we have only considered deviation of geodesics that were initially separated by a small amount of space as in Figure 18.5. We gain additional insight by looking at two geodesics initially separated by a small amount of *time*. Given the unchanging gravity at Earth's surface, two test masses dropped one second apart would trace out similar worldlines separated by one second in time. But if the source mass is changing somehow—losing mass, gaining mass, or wiggling its position—the second worldline will be accelerated differently, and these two geodesics *will* deviate from each other. Because the spacetime framework fully includes time as a coordinate, geodesic deviation naturally includes deviation through time as well as space, and so naturally captures the effects of a dynamically evolving source mass. Because c is the exchange rate between space and time coordinates, you will not be surprised to find that changes in the metric propagate away from the changed source at speed c.

We will use *dynamic* to mean that the metric changes with time, due to some change in the source. It is useful to distinguish this from *static* (no changes ever) and *stationary*, which means that the source moves but in a way that is constant with time. For example, most stars and planets spin at a constant rate. In this case the metric of the surrounding spacetime will not change with time. To gain some insight into how the metric *is* affected by this motion, we examine momentum in more detail.

Momentum. We have been a bit simplistic about dividing the metric into time and space terms; there can be mixed terms as well. In a (t, x, y) coordinate system such as may be used to describe Figure 18.3, the full list of mathematically possible terms in the metric is

$$
\begin{aligned}
(\Delta\tau)^2 \; = & A(\Delta t)^2 + B(\Delta x)(\Delta t) + C(\Delta y)(\Delta t) \\
& + D(\Delta x)^2 + E(\Delta x)(\Delta y) \\
& + F(\Delta y)^2
\end{aligned}
\tag{18.2}
$$

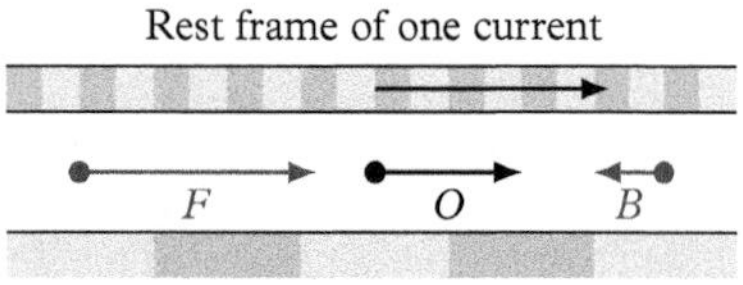

Rest frame of one current

Figure 18.6 *Revisiting the source momentum effect in Figure 18.3. The original particle (O) is equally attracted to both currents, because in O's rest frame the two currents are equally dense. In B's frame contraction makes the upper current much denser than shown, without much effect on the lower current; B thus accelerates upward strongly. Back in* this *frame, source density accounts for only some of that can be attraction; the remainder is attributed to source momentum being* attractive *for test particles moving "upstream."*

and we have simply assumed so far that $B = C = E = 0$. Physics tells us that A must be positive, and D and F negative—but does it tell us that all the other coefficients must be zero? To get a feel for this, consider how a mixed $(\Delta x)(\Delta t)$ term would behave differently from a pure term involving $(\Delta x)^2$ or $(\Delta t)^2$. The sign of a term like $D(\Delta x)^2$ is determined purely by the coefficient D because $(\Delta x)^2$ is always positive regardless of the sign of Δx. But the sign of a term like $B(\Delta x)(\Delta t)$ depends on the sign of Δx. Whatever B is, this term will subtract from the proper time for particles moving in one direction, but add to the proper time for particles moving in the opposite direction! As a result, the acceleration predicted by the geodesic equation will depend on a particle's direction of motion, not only on its position. This is a qualitatively new phenomenon that is *mathematically* possible if the mixed terms in the metric are not zero—but is it *physically* possible?

For an example of such a situation, let us return to the mass currents in Section 18.1. The lower panel of Figure 18.3 is reproduced here as Figure 18.6. We previously concluded that the momentum of the upper current had a repulsive effect that would explain (in this frame) why the particle was not more attracted to it despite it being the denser current. The following argument will make it plausible that this effect depends on the *particle's* velocity as well. The original particle in our thought experiment, now labeled O, has (by construction) the unique velocity that makes it "see" the two currents as having equal densities; in its rest frame the upper current is much less contracted, and the lower current is much more contracted, than shown in the figure. The faster particle F sees the upper current as even less dense, and the lower current as even more dense; thus F falls toward the lower current. But observers in *this* frame merely see a faster particle, leading them to conclude that the momentum of the upper current repels F more than it repels O. So, the repulsive effect of source momentum depends on test particle speed; does it depend on test particle direction as well? Consider particle B in Figure 18.6; it sees the upper current as dramatically denser than shown, because at high relative speed the contraction factor γ increases sharply. But it sees little contraction in the lower current, because γ increases slowly at low relative speeds. The rest frame of B thus argues for a dramatic upward acceleration, more than we would expect from the densities shown in *this* frame. Back in this frame, we attribute the extra attraction to source momentum being *attractive* for a test particle moving "upstream." So, over a time step Δt, a particle's acceleration depends on the size *and* direction of its Δx—and this can be the case only if the metric contains a mixed term such as $(\Delta x)(\Delta t)$.

We can think of the $(\Delta x)(\Delta t)$ coefficient as responding to the transport of energy in the x direction; the $(\Delta y)(\Delta t)$ coefficient responding to the transport of energy in the y direction, and so on. If so, what does the $(\Delta t)^2$ coefficient respond to? Transport of energy in the *time* direction; in other words, energy at rest (mass). The relationships between source properties and metric coefficients can be more complicated in general, but this at least gives a glimpse into the need for a mathematical framework to track all these source terms in a way that enforces spacetime rules such as energy in one frame looking like energy *and* momentum in another. Think of this framework as a generalization of the stretching-triangle picture (Chapter 12) to *sixteen* elements, as each of four interrelated quantities—energy, x momentum, y momentum, and z momentum—can be transported through each of four interrelated coordinates—t, x, y, and z. Although complicated, this sixteen-element framework provides just enough information to determine the sixteen metric coefficients needed to describe a four-dimensional spacetime.

Section 18.3 will look in more detail at one component of the source description by considering a static source mass. But we can be assured that the procedure there will generalize to the other components thanks to the spacetime framework in which Einstein developed these equations.

Check your understanding. Draw Figure 18.6 back in the frame where the two currents have equal and opposite velocities. Take care drawing the velocities of the test particles. *In this frame*, which way does particle B accelerate and why?

> **Think about it**
>
> We had not previously recognized the need for mixed terms because we considered only extremely symmetric sources. In general, we should picture the metric as having all possible mixed terms; if their coefficients happen to be zero we should thank symmetries of the source.

18.3 The Einstein equation

To recap Section 18.2, mathematicians bequeathed to physicists a set of equations for determining geodesics—and therefore geodesic deviation—given a metric; and Einstein adapted this to a spacetime context. However, many of the mathematically possible patterns of geodesic deviation have nothing to do with the patterns

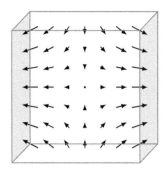

Figure 18.7 *The tidal acceleration field outside a source mass (which is off to the left or right). The net flow of arrows into the box is zero because four sides have inward-pointing arrows (for clarity, arrows are omitted from the near and far sides of the box) while two sides have double-length outward-pointing arrows.*

generated by gravity, so Einstein distilled the equations for geodesic deviation down to just the physically relevant parts as follows.

We saw in Figure 18.5 that geodesics diverge along the direction toward a source mass, but we did not take time to consider what happens in the other two spatial directions. Spacetime diagrams are not good at showing additional spatial dimensions, so Figure 18.7 switches to a fully spatial view of variations in the acceleration field (this figure is a copy of the tidal acceleration field in Figure 16.13). According to Figure 18.7, freely falling particles indeed diverge along the direction to a source mass (as we saw in Figure 18.5), but *converge* along the two transverse directions. As noted in Section 16.7, the combination of four weakly converging sides and two strongly diverging sides yields zero net convergence into the box. But remember, this is a box *outside* a source mass—a box with a mass inside does have geodesics converging on it from all directions. The net convergence of geodesics into a box thus tells us how much mass is inside.

Einstein recognized this pattern and turned it around: given a source mass we can *predict* the net geodesic convergence everywhere, and that in turn tells us what the metric must be. The solar system, for example, can be modeled as 2×10^{30} kg (the Sun) resting at $r = 0$ and zero density elsewhere. Einstein hypothesized that this source density "map" times some constant of proportionality is *also* a map of the net convergence of geodesics at every point in the solar system. Finding the metric in the solar system then consists of finding the metric that yields the right amount of net convergence at each point. Einstein implemented this mathematically by developing a recipe for computing, given a metric, the net geodesic convergence at each point. By setting this proportional to the source density at each point he obtained what is now called the **Einstein equation**.

But Einstein was not able to solve his own equation. The recipe for forecasting the geodesic convergence from a set of metric coefficients is straightforward; the problem is that we need to search through all possible metric coefficients to find the one set that yields the particular convergence pattern we need. This search is guided by considerations of symmetry and so on, but it is nevertheless difficult. Einstein found no exact solution for the solar system, but he did find an approximate solution that confirmed that his model matched the Newtonian model at low speed and far from the Sun. Closer to the Sun and at slightly higher speed, though, his approximate solution *also* predicted that Mercury's orbit would depart from the Newtonian model in exactly the way that had been observed. This triumph confirmed the value of Einstein's model.

Before examining that triumph in detail, we address the broader meaning of the Einstein equation. Net geodesic convergence is rather abstract, so we will visualize it in terms of the *volume* of a cloud of test particles (this is inspired by John Baez and Emory Bunn; see *Further Reading*). Figure 18.8 shows two small clouds full of test particles initially at rest, one surrounding a source mass and the other some distance away. In each case the dashed circle outlines the cloud initially, and the shaded volume represents the same cloud a bit later in time. The mass-containing cloud shrinks as freely falling particles converge on all sides. The mass-free cloud, in contrast, follows the tidal field by stretching toward the mass but compressing (half as much) in each transverse direction, and this leaves its volume unchanged. The Einstein equation can thus be seen as shrinking the volume of a test-particle cloud in proportion to the local density of source mass.

This allows us to build up a complete picture of a spacetime using only local information, just as Poisson's method (Section 18.1) allowed us to build up a complete picture of an acceleration field using only local information. We used the tidal field as a thinking tool to get here, but now we can drop it completely and focus just on these volumes and the local proportionality dictated by the Einstein equation. Starting at a source mass, the Einstein equation dictates the shrinkage of a small surrounding cloud of test particles. Now, consider a neighboring mass-free cloud: it must stretch in one direction to maintain contact with its shrinking neighbor. Yet the Einstein equation dictates that its volume be preserved, so it must be squeezed in the other two dimensions (by half as much in each dimension). Although this region is mass-free, the fact that it changes shape means that geodesics there converge along some directions and diverge along others—in other words, it still has spacetime curvature, and that curvature was deduced using only local information. This idea can be extended outward as far as we wish, until we obtain a complete map of spacetime curvature everywhere.

The genius of this approach is that it works in a space*time* context. If a source moves through space or otherwise changes with time, we simply apply the same rules to a space*time* grid and we automatically capture variations of the metric in time as well as space. By revealing how the metric can change with time, the Einstein equation forever changed our view of spacetime. In place of the eternally fixed grid that had been imagined for centuries, we now picture events playing out on a dynamic and evolving spacetime stage.

The Einstein equation also unifies a remarkably diverse set of phenomena, discussed in more detail in the remainder of this chapter and the following chapters. To appreciate both the unity and diversity, think of the Einstein equation as a set of rules by which energy curves spacetime. These rules are so comprehensive that most particular situations can be studied with a subset of the rules. For example, the study of stationary situations eliminate any action in the time dimension, and if you read extensively about stationary situations you may even forget that the rules apply to the time dimension. Yet those rules are there when we need them; in fact they take center stage in the study of gravitational waves (Section 19.3), where the *static* aspect of the Einstein equation lies dormant. Although a wide array of specialized solutions have been developed, keep in mind that the Einstein equation unifies them all.

Check your understanding. Explain why changes in the volume of a cloud of test particles is a good visual indicator of the net geodesic convergence there.

Figure 18.8 *A cloud of test particles around source mass M shrinks in volume as M makes local geodesics converge. The upper cloud stretches toward M but with no local mass, geodesics have no net convergence; we infer a lateral squeezing that preserves the volume of the upper cloud. Net motion of the upper cloud is not shown. All this behavior is predicted with a single rule: local proportionality between mass and change in volume.*

Think about it

The cloud visualization considers only one frame, in which the setup looks particularly simple. The complexity of the Einstein equation relates to the fact that statements about cloud volume must be true in all frames; in some of those frames, the test-particle behavior looks more complicated and is ascribed to more complicated effects such as source momentum.

18.4 The Schwarzschild solution

The Einstein equation is so complicated that Einstein did not find an exact solution for the metric even in the limited case of a single static mass like the Sun. But with help from Michele Besso (1873–1955; see *Further Reading*) Einstein was able to find an approximate solution and show that orbits of planets would not *quite* be closed. This differs from Newton's model, which predicts that orbits in a spherical potential are closed (Section 16.1). The effect is larger for orbits closer to the source mass, so among the known planets Mercury would be most affected—and in fact Einstein showed that his theory completely accounted for the anomalous precession of Mercury. To Einstein, this confirmed the theory beyond doubt. But the history of science is full of theories that explain already-known phenomena (known as *retrodiction*), so the gold standard of scientific proof is a *prediction* that differs from those of previous theories and is ultimately confirmed by experiment.

To provide a foundation for understanding the predictions of general relativity, we turn to Karl Schwarzschild (1873–1916), a German physicist and astronomer serving on the eastern front in 1915. Schwarzschild read Einstein's papers as they arrived at the front and soon found an exact solution for the metric around a static spherical mass. The modern form of the **Schwarzschild metric** is

$$c^2(\Delta\tau)^2 = c^2\left(1 - \frac{2GM}{c^2 r}\right)(\Delta t)^2 - \left(1 - \frac{2GM}{c^2 r}\right)^{-1}(\Delta r)^2 \qquad (18.3)$$

plus terms involving the equivalent of latitude and longitude, omitted here because they are the same as in space without gravity. Compare this with the "Newtonian" metric model (Equation 18.1, repeated here):

$$c^2(\Delta\tau)^2 = c^2\left(1 - \frac{GM}{rc^2}\right)^2(\Delta t)^2 - (\Delta r)^2 \qquad (18.4)$$

The models disagree on the $(\Delta t)^2$ *and* $(\Delta r)^2$ terms. Modification of the $(\Delta r)^2$ term is a conceptually new feature, to which we devote all of Section 18.5. This section provides a warmup by examining the $(\Delta t)^2$ term, with which we are already somewhat familiar.

In either model, the coefficient multiplying the $(\Delta t)^2$ term is the primary effect of gravity on low-speed particles because their displacements in time are much, much larger than their displacements in space. We can therefore simplify the fearsome-looking Equations 18.3 and 18.4 by letting $\Delta r = 0$; any result of this simplification will apply exactly to stationary particles and approximately to low-speed particles (we followed the same process in Sections 14.2 and 15.2). After dropping the Δr term in each metric, we can simplify more by dividing out the c^2 and taking a square root. The result is

$$\frac{\Delta\tau}{\Delta t} = \begin{cases} 1 - \frac{GM}{c^2 r} & \text{Newtonian} \\ \sqrt{1 - \frac{2GM}{c^2 r}} & \text{Schwarzschild} \end{cases} \qquad (18.5)$$

As a reminder, this expression tells us how much time ($\Delta\tau$) elapses on a stationary clock at r, as a fraction of the elapsed coordinate time Δt. Figure 18.9 compares these two predictions for $M = 2 \times 10^{30}$ kg (one solar mass). If you think of these curves as potential wells the Newtonian curve has the familiar shape while the Schwarzschild curve is steeper and deeper. The difference is substantial only within about 10 km, so experimental confirmation of the Schwarzschild curve requires a star with the mass of the Sun and a radius less than 10 km: a *very* compact object. (For comparison, the radius of the Sun is about 700,000 km, and less than a trillionth of the Sun's mass lies within 10 km of the center.) Compact objects are valuable laboratories because they directly expose a difference between Newtonian gravity and general relativity—and could potentially expose flaws, if they exist, in general relativity.

The slowness of time can be tested with gravitational redshift; the $\frac{\Delta\tau}{\Delta t}$ axis in Figure 18.9 directly indicates the fraction of its emission frequency retained by a photon reaching a distant observer. Stationary clocks at $r = 4$ km (the first small mark on the r axis), for example, experience $\frac{\Delta\tau}{\Delta t} \approx 0.5$ so photons they emit arrive at distant observers with only half their original frequency; distant observers see these clocks ticking at only half the rate of the t coordinate. Humans sent to $r = 4$ km would appear to distant observers to be living in slow motion, and would return having aged less than their counterparts who stayed far away at mission control. In practice, astronomers do not know the value of r at which a photon was emitted, so the astronomical evidence that supports general relativity in this context requires a bit more context to understand (Section 20.3).

Although $\frac{\Delta\tau}{\Delta t}$ corresponds directly to gravitational redshift and the gravity well picture, its inverse $\frac{\Delta t}{\Delta\tau}$ (marked on the right vertical axis in Figure 18.9) connects more easily to special relativity. In Chapter 10 we discussed $\frac{\Delta t}{\Delta\tau}$ as a measure of speed through time: displacement in the time coordinate t per unit time τ on the traveler's clock. Higher speed through time means faster travel into the future. Continuing our $r = 4$ km example, where we read $\frac{\Delta\tau}{\Delta t} \approx 0.5$ in the previous paragraph we can just as well read $\frac{\Delta t}{\Delta\tau} \approx 2$; observers there travel two years through the t coordinate per year on their own watches. Unlike special relativity, the situation is *not* symmetric: *all* observers agree that clocks at $r = 4$ km tick less frequently than distant clocks. In other words, you could spend one year (on your watch) near a compact object and emerge to find that *two* years of coordinate time have passed. This has obvious parallels with the twin paradox and the accelerating-rocket experiment in Section 13.3, but with gravity the acceleration depends strongly on position—and *does not require motion through the spatial coordinates.*

Based on the time coefficient, the Schwarzschild model looks quite similar to the Newtonian model of the solar system. The Schwarzschild model does predict slightly stronger gravitational effects, but this becomes substantial only if we can probe very close to a massive compact object. Experimental tests of the Schwarzschild model may therefore seem quite difficult, but let us not forget about the spatial term of the Schwarzschild metric. This term will become important for particles that travel through comparable amounts of space and time, which suggests that the behavior of light will provide a critical test of the model.

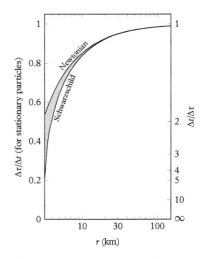

Figure 18.9 *Time runs slowly near a mass, more so for the Schwarzschild model than the Newtonian model. The left vertical axis lists, for a photon emitted at r, the fraction of its emission frequency or energy that reaches a distant observer; the right vertical axis is analogous to the γ factor but applies even to stationary particles. The r values assume a solar-mass compact object.*

Think about it

The fraction of frequency retained goes to *zero* for photons emitted at $r = \frac{2GM}{c^2}$ (≈ 3 km for a solar-mass compact object) in the Schwarzschild metric. Distant observers therefore cannot observe such photons. This is our first hint of black holes.

Think about it

The mathematical expressions for the time coefficients in the two models are clearly different, so how can they have essentially the same numerical values in the solar system? See Problem 18.15 if this intrigues you.

250 18 General Relativity and the Schwarzschild Metric

Check your understanding. Refer to Figure 18.9 to predict the effect of gravity at $r = 100$ km from the center of a compact solar-mass object. *(a)* How slowly does a stationary clock there tick compared to a distant clock? *(b)* How could you observe this? *(c)* Do the two models considered here differ on this prediction?

18.5 Curved space

The conceptually new part of the Schwarzschild metric is that the cofficient on the $(\Delta r)^2$ term is not simply -1 everywhere. To build a mental picture of this part of the metric, we must first understand the particular way in which the Schwarzschild r coordinate of an event is *defined*: it is *not* the distance one would measure by counting meter sticks from the origin to the event. Rather, we take a circle centered on the origin and running through the event, and $2\pi r$ is the circumference of that circle: $r = \frac{C}{2\pi}$. In the absence of gravity this would be the same as measuring directly from the origin, but in the Schwarzschild metric they are *not*, so we need names to clearly distinguish the two ideas. Let us write r_m for meter-stick distance (also called proper distance), and let us think "$\frac{C}{2\pi}$" whenever we see the coordinate called r.

Just as in Section 18.4 we isolated the meaning of the Δt term by considering events with $\Delta r = 0$, we now isolate the meaning of the Δr term by considering events with $\Delta t = 0$. For event pairs with $\Delta t = 0$ (representing simultaneous events at opposite ends of a meter stick or tape measure) the left side of the metric determines rest length or proper distance rather than proper time (Box 11.1). In Box 11.1 we called this L because we focused on the lengths of objects in their rest frame, but in the absence of a concrete object it is usually called proper distance; I will call it meter-stick distance r_m to make it more concrete. Setting $\Delta t = 0$ in Equation 18.3 and simplifying, we find

$$\frac{\Delta r_m}{\Delta r} = \begin{cases} 1 & \text{Newtonian} \\ 1/\sqrt{1 - \frac{2GM}{c^2 r}} & \text{Schwarzschild} \end{cases} \quad (18.6)$$

The Schwarzschild relationship is plotted in Figure 18.10 for a solar-mass object (with the usual assumption that all this mass is packed into an r smaller than any shown on the plot).

Test-drive this figure. The surface of our Sun is at $r \approx 700,000$ km, so we must extrapolate far off the right edge of the figure, but we can imagine that $\frac{\Delta r_m}{\Delta r}$ will be very, very close to one anywhere in the solar system (for Mercury it is about 1.00000003). And $\frac{\Delta r_m}{\Delta r} = 1$ is the usual relationship between meter-stick radius and the radius expected based on the circumference of a circle, so no dramatic effects are predicted for our solar system. But near a compact object Δr_m becomes substantially larger than Δr; in other words, *the meter-stick radius becomes larger than expected based on the circumference.*

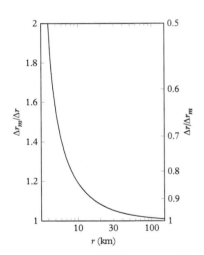

Figure 18.10 *Relationship between change in meter-stick distance from the origin and change in* r *(which is defined as the circumference of a circle divided by 2π) for the Schwarzschild metric. This curve has the same form seen in Figure 18.9 but inverted. The* r *values again assume a solar-mass compact object.*

How is it possible that the meter-stick radius of a circle is larger than the circumference divided by 2π? This happens when space is *curved*. A familiar analogy is the curved surface of Earth (Figure 18.11). Imagine drawing circles in the snow around the north pole of the Earth and measuring their radii (r_m) and circumferences. For small circles (up to, say 100 km in radius) you will find that $C = 2\pi r_m$ within the precision of reasonable measurements. But for progressively larger circles the meter-stick radius r_m becomes progressively *larger* than $C/2\pi$ because r_m is measured along the curved surface of the Earth. Another way to say this is that the circumference is *smaller* than $2\pi r_m$ because the surface curves back on itself.

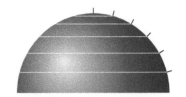

Figure 18.11 *Each of Earth's latitude circles shown here has a circumference smaller than its meter-stick distance from the north pole (measured along the surface by the tickmarks) times 2π. This is the hallmark of a curved surface.*

The sphere analogy can be quite misleading because a sphere has the same curvature everywhere—the pattern shown by circles on Earth would be the same regardless of where we chose to center them. The Schwarzschild metric, in contrast, describes a space that is highly curved near the source mass but negligibly curved—nearly flat—very far away.

We can represent such a space as follows. Start with a large circle centered on the source mass and call it Circle 1; this is the outermost circle in Figure 18.12. Measure inward with a meter stick to find a Circle 2 that is $\Delta r_m = 1$ meter of proper distance inward from Circle 1. In a flat space, this would *also* imply $\Delta r = 1$, meaning the circumference of Circle 2 is $2\pi \Delta r = 2\pi$ meters smaller than Circle 1. However, with smaller Δr as indicated by Figure 18.10, the two circles do not differ so much in circumference; Circle 2 does not shrink as much as flat-space intuition suggests. The only way to fit this largish Circle 2 in the drawing *while keeping it one full meter stick from Circle 1* is to offset it slightly upward on the page. (This is a visual trick; up/down on the page does *not* correspond to any physical direction.) Now we repeat the process: another meter stick inward, and another larger-than expected circumference that must be offset upward even more. This process builds the curved surface in Figure 18.10. Below that, Figure 18.10 shows the same circles separated by only Δr; this does *not* respect the meter-stick separation. *You must imagine particles as confined to the curved surface* if they are to "feel" the unusual relationship between proper (meter-stick) distance and displacement in the coordinate r.

I could have drawn the circles in Figure 18.12 as offset up *or* down on the page; either would demonstrate the same radius-circumference relationship. Following Lewis Carroll Epstein's *Relativity Visualized*, I chose upward offsets to make this drawing as distinct as possible from the funnel we used to represent the potential, which is *an entirely different concept*. For the potential, you are allowed to think that a particle "wants" to fall down to the lower parts of the funnel; this reflects how inertial particles always accelerate toward regions of slower time (see Chapter 14 to review why). But the vertical direction in Figure 18.12 has no such meaning; it is merely a way to draw the radius-circumference relationship given by the spatial part of the metric. Nevertheless, we can use the curved surface in Figure 18.12 to predict two observable consequences.

Think about it

The metric defines *differences* between adjacent circles. Finding the actual meter-stick distance between widely separated circles requires adding up many of these little steps, a task we address in Section 20.2.

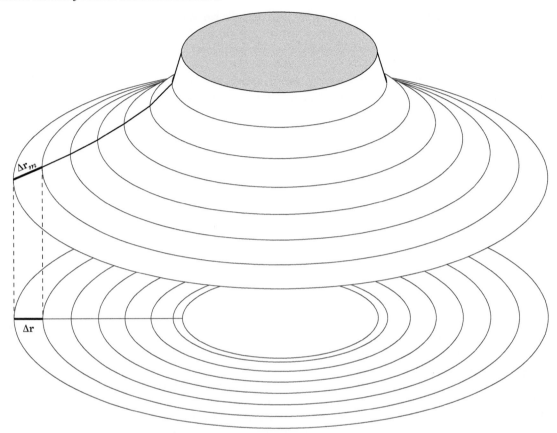

Figure 18.12 *Visualization of spatial curvature around a compact source mass. To draw concentric circles separated by $\Delta r_m = 1$ meter stick of proper distance while making their r coordinates (which define the circumference $2\pi r$) differ by less than this, we displace them vertically. The vertical direction here is not real; it merely enables us to see the relationship between circumference and meter-stick radius.*

First, like the Tardis in *Doctor Who*, Circle 1 encloses more space than one would expect from its perimeter. This is manifest in the distance traveled from A to B in Figure 18.13. To find the shortest path in a curved space, imagine stretching a string from A to B and keeping it taut so it follows the surface; the result is the thick red path in the top panel. (Physicists use the geodesic equation, but the taut string highlights the same concept.) This path is clearly longer than it would be if the space between A and B were flat; Section 18.6 describes how we actually measure the extra distance.

Second, spatial curvature accelerates particles toward the source mass. The trajectory in Figure 18.13 entered the drawing at point A heading a bit south of east and exited at point B heading north of east—a pattern more clear in the bird's-eye view in the lower panel than in the perspective view in the top panel. *However*, this acceleration is nearly irrelevant for particles at everyday speed. If Figure 18.13 purports to represent, say, a bullet crossing the solar system, the journey would take years so the acceleration of the red path would be *tiny* in terms of m/s^2. The bullet's acceleration due to the *time* part of the metric would be much larger than this, so we would not even notice the tiny extra effect of the spatial part of the metric. Spatial curvature becomes important only at speeds near c because in that case the deflection due to spatial curvature occurs so quickly that the acceleration (in terms of m/s^2) is substantial—comparable to the effect of the time part of the metric. This gives the whole model a speed dependence that makes it testably different from Newtonian gravity (Section 18.6).

I stress again: Figure 18.12 must *not* be confused with the potential, despite their similar shapes. The potential represents a relationship between time and space, while Figure 18.12 represents a relationship between two dimensions of space. Because all particles move through time, the potential is important for all particles, which are attracted to lower regions. Curvature of space has approximately zero effect on low-speed particles, and when we do use Figure 18.12 the idea of attraction is misleading. Rather, a particle simply follows a straight path along the curved surface as would a taut string.

I also stress that Figure 18.12 merely helps you visualize a relationship between radius and circumference that exists in nature regardless of how we choose to visualize it; *curvature of space does not require extra dimensions to "curve into."* Observers in this space have no sense of being in the landscape you see in these figures—they cannot see the vertical dimension on the page, which is not a real physical dimension, so they cannot see the geometry at a glance. They can only infer the geometry from measurements as described in the following section.

Check your understanding. Compare the smallest circle in Figure 18.12 with the next (larger) circle. According to my graphics software, the increase in circumference is equal to about two of the heavy black meter sticks. *(a)* In a flat space, how much would the circumference increase? *(b)* Take the ratio of actual circumference increase to your answer in part (a); this is $\frac{2\pi \Delta r}{2\pi \Delta r_m} = \frac{\Delta r}{\Delta r_m}$. Try to locate your value of $\frac{\Delta r}{\Delta r_m}$ on Figure 18.10; is your value typical in the solar system?

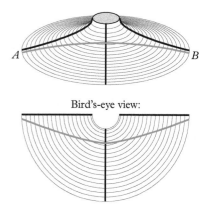

Figure 18.13 Top: *the shortest distance or straightest path in a curved space is best visualized by imagining a taut string between two points A and B; the distance is longer than if the space between A and B were flat.* Bottom: *a bird's-eye view of the same path reveals acceleration toward the center of curvature (the source mass). The magnitude of this acceleration is many m/s^2 if the path is traversed quickly (e.g., by a photon), but tiny if traversed slowly (e.g., by a rocket).*

Think about it

See Section 18.7 for further contrasts between the effects of the time and space parts of the metric.

Box 18.2 Types of curvature

As emphasized in Section 18.3, general relativity is based on a *local* relationship between the mass at a point and the curvature there. This locality is the only way to respect *c* as a speed limit; if a source mass changes, long-distance effects are felt only after rippling through a series of local relationships. Curvature is therefore defined by a local measurement: the convergence of local inertial worldlines. This box looks at curved *surfaces* as concrete analogies.

Let us contrast our local criterion for curvature with an intuitive *global* criterion: circumnavigation. On Earth, we can head in one direction and eventually approach our starting position from the other side, without ever changing course. This test is easily visualized and seems to prove curvature but is less useful than it seems, for several reasons. First, it is too global: we cannot deduce "how much circumnavigation" happened at each locality visited along the way. Second, some spaces that can be circumnavigated are nevertheless locally flat! Curl a sheet of paper with parallel lines into a cylinder: it can now be circumnavigated even as the lines remain parallel (it is still locally flat). As general relativists we are committed to a local criterion for curvature, so we must follow the parallel-line criterion and consider a cylinder to be locally flat. In other words, *circumnavigation measures something that is irrelevant to gravity.*

If calling a cylinder locally flat really bothers you, it may help to consider more descriptive names for the two different things being tested here. The type of curvature identified by the parallel-line test may be called *intrinsic* curvature—it is woven into the fabric of the space—whereas circumnavigation may reveal *extrinsic* curvature, which stands apart from the local geometry. Intrinsic curvature is the only type relevant to gravity, so "curvature" in this book always refers to intrinsic curvature.

The intrinsic category is further subdivided into positive (initially parallel lines eventually converge) and negative (initially parallel lines eventually diverge) types. We associate gravity with positive curvature because it makes initially parallel *worldlines* converge (on net). Even considering the spatial part alone, we can see that the curvature is positive by imagining parallel lines entering the left side of Figure 18.12: the funnel shape will make them converge as they approach the center. Positive curvature can also be defined as the radius of a circle being more than its circumference divided by 2π (which is how we constructed Figure 18.12), or the angles of a triangle adding to more than 180°; all these criteria are tightly linked. The sphere is a familiar example of positive curvature, but fails to represent gravity in one important respect: the sphere is equally curved everywhere. Spacetime, in contrast, is most highly curved near a source mass, and becomes nearly flat far away. This makes the funnel a much better icon for the curvature pattern caused by gravity.

18.6 Observable consequences of the Schwarzschild metric

There are several practical tests of spacetime curvature in our solar system, and general relativity has passed all these tests to high precision.

Deflection of light. In Section 18.1 we reviewed the Newtonian prediction for how gravity affects light in the solar system: light passing very close to the Sun should be deflected by a mere 0.00024 degrees. We also saw that a metric model of gravity based only on the time part of the metric can be functionally equivalent to the Newtonian model. Indeed, by 1911 Einstein was also predicting a deflection of 0.00024 degrees based on a metric model of the time part. He encouraged astronomers to test this prediction, but they could do it only during a total solar eclipse, when the glare of the Sun itself would not be a factor. Astronomers did set out for an eclipse in Russia in 1914, but were stymied by clouds and the outbreak of World War I.

Einstein then perfected the Einstein equation and found that the Sun should affect the spatial part of the metric just as much as the time part. The spatial part has a negligible effect on particles at everyday speeds because those particles travel through time much more than through space. Light, however, travels equally through time and space, so the spatial part of the metric really matters for predictions involving light. Einstein's final prediction, then, was that light should be deflected by *twice* the Newtonian value (Figure 18.14) because the time and space parts of the metric *each* cause a deflection of 0.00024 degrees. This was confirmed by measurements taken during the 1919 eclipse, making Einstein world famous as the man who proved Newton wrong and founded a new vision of space and time.

However, eclipses never became a precise test of general relativity because eclipse observations are strictly limited in time and are at the mercy of the weather. The modern, precise version of this test uses distant sources of radio waves called quasars, which can be observed at any time because the Sun is not a blinding radio source. By tracking the apparent separation between quasars as the Sun passes between them, astronomers have confirmed the general relativity prediction for the deflection of light to a precision of one part in 10,000.

Recently, observations of normal stars in visible light have made a comeback in terms of testing the curvature of space in the solar system. In the 1990s the Hipparcos satellite mapped stars all over the sky, determining their positions very precisely to provide a foundation for many other kinds of precision astronomical research. Hipparcos repeatedly mapped the entire sky without ever pointing near the blinding Sun (this is possible because the apparent position of the Sun changes over the course of a year). The Hipparcos star-position measurements were so precise, though, that they revealed the tiny deflection light experiences when entering the solar system and reaching Earth even *without* passing near the Sun. Hipparcos was able to confirm the Schwarzschild deflection of light to a precision of one part in 1000. An even higher-precision satellite called Gaia was launched in 2013, and will soon measure the curvature of space 1000 times more precisely than Hipparcos. This is so precise that Gaia may be able to test minute details of the model, such as effects on the solar system metric due to planets, rotation of the Sun, and so on.

Geodetic precession. In the Newtonian model, an arrow that points to a distant star and orbits the Sun will always point to the *same* distant star. This is illustrated in the lower part of Figure 18.15, where the arrow is attached to a nonrotating test mass to make it a bit more concrete. In the Schwarzschild model, curvature of space causes the arrow to shift direction ever so slightly as it travels through space. The funnel from Figure 18.12 is reproduced at the top of Figure 18.15 to remind you of this curvature. Focus on the outer part of the

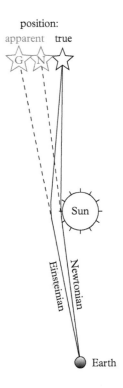

Figure 18.14 *Figure 18.1 is adapted here to compare the deflection of light from a* single *background star in Newtonian (N) and general relativity (G) models. The latter includes the Sun's effect on the spatial part of the metric, and thus doubles the predicted deflection. The more sharply bent path appears not to graze the Sun only because the angles are highly exaggerated (the apparent positions really shift by only 0.05% and 0.1% of the Sun's apparent diameter) and the figure is not to scale.*

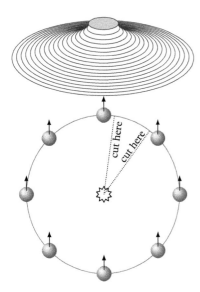

Figure 18.15 *Paper model of geodetic precession. Cut the lower diagram—showing a hypothetical flagpole orbiting the Sun—along the dotted lines and seal the cut to make a cone roughly approximating the curved-space model in the upper diagram. Does the flagpole still start and end the orbit pointing in the same direction, as it does in flat space? Inspired by Epstein's* Relativity Visualized.

Think about it

The Moon is not the best test of geodetic precession because its orbit is not very elliptical, and is complicated due to Earth not being a sphere. But the orbit of the Moon is *very* well studied, and it matches the curvature-of-space prediction to a precision of about 1 part in 200.

funnel, where all the planets orbit: we can model this part of the funnel, at least approximately, with a simple cone. Make that cone by cutting the wedge out of the page and pulling the cuts together; the paper naturally forms a cone. Now you can see that the arrow at the end of the orbit does not point in the same direction as the arrow at the start; in fact, the shift in direction accumulates slowly throughout this orbit and successive orbits. This shift is called **geodetic precession** because it is due to the shape of space.

In practice, gyroscopes are used in place of flagpoles; the rotation of a gyroscope stabilizes it for a long time, and the axis of rotation defines a direction in space. This effect has been confirmed in the space around Earth to a precision of 1 part in 350 by the Gravity Probe B satellite. The anomalous precession of Mercury is conceptually related: the long end of Mercury's highly elliptical orbit defines a direction in space that (absent perturbations by other planets) remains constant in a Newtonian potential but shifts continually in general relativity. You can repeat the modeling exercise in Figure 18.15 without the flagpole by drawing an elliptical orbit, and you will see that the orbit cannot repeat exactly. While a substantial part of Mercury's anomalous precession is due to the curvature of space, be aware that much of it is due instead to the time part of the Schwarzschild metric differing from the Newtonian approximation. Therefore, gyroscopes provide the most targeted way to measure curvature of space alone.

An even subtler effect on Earth-orbiting gyroscopes is due to the *rotation* of the Earth. The Schwarzschild metric makes no attempt to model the effects of source mass rotation, so we will return to this topic in Chapter 19.

Time delay of light. The astrophysicist Irwin Shapiro (b. 1929) realized in the 1960s that the extra path length from A to B in Figure 18.13 could be measured via the extra travel time for a light ray. Imagine A is Earth, B is an interplanetary spacecraft, and the red line is a light ray between them. The spacecraft is at times in a position where the light ray crosses only fairly flat space far from the Sun, and at other times in a position where the light ray crosses the more highly curved space near the Sun. A series of experiments with different spacecraft (and planets, which reflect radar signals back to us) has confirmed that the latter type of path takes more time, by about 100 microseconds each way when passing very close to the Sun. This is a brief delay, but modern clocks are extremely precise—and experimenters have worked very hard at reducing all other uncertainties in this experiment. The most precise measurement to date, from the *Cassini* spacecraft, agrees with the prediction of general relativity to a precision of 1 part in 100,000. The time delay has also been verified using regular radio pulses from neutron stars (called pulsars in this context) outside the solar system.

To be clear, only half the time delay is due to curvature of space. The other half is due to clocks ticking slowly near the Sun. Light moves at speed c relative to *local* observers, so the fact that clocks tick slowly near the Sun implies that *distant* clocks measure a longer light travel time (review Box 18.1 if necessary). The time and space parts of the metric contribute equally to the time delay of light, so the effect of spatial curvature is to double the time delay that would have been predicted in a "Newtonian" metric model. The following section will sharpen the distinction between the effects of the time and space parts of the metric.

Check your understanding. Take the cone you made to understand Figure 18.15 and flatten it. Draw a straight line across the orbit, grazing the Sun. When you make the paper into a cone again, what does this represent?

18.7 Time versus space parts of the metric

We have seen that trajectories depend on the space part of the metric only to the extent that particles move through space—which is often very little by the standard of *c*. This section presents mental pictures to help cement this idea in place.

First, consider a particle at rest—a pen in your hand, say. This pen does not feel the space part of the metric *at all* while it is stationary. If gravity consisted of only the space part of the metric, this pen would *not* accelerate toward the center of the Earth when released from your hand. To anthropomorphize a bit, a particle has no reason to care about the relationships between different points in space if it stays at one point. It starts moving through space only because its motion through *time* results in acceleration toward regions of slower time. Once moving through space, the space part of the metric matters in principle—but at everyday low speeds its effect is so small that gravity is essentially just acceleration toward regions of slower time (Chapter 14). At high speeds the slow-time effect remains but is supplemented by curvature of space.

Second, consider the oft-used analogy of a bowling-ball source mass on a taut rubber sheet. The bowling ball stretches the rubber sheet into the shape of a gravitational potential, which then acts like a coin funnel in making particles (represented by marbles) orbit. This extends the coin funnel analogy by explicitly linking potential depth to the source mass, and by demonstrating the complicated and dynamic spacetimes created by multiple and moving source masses. However, the rubber sheet does *not* demonstrate curvature of space. As Section 15.3 noted regarding coin funnels, a curved surface *by itself* cannot make a stationary marble accelerate—the visualization works for us only because we instinctively imagine the marble "wanting" to fall down the funnel. Like the coin funnel, the rubber sheet hides the key role of *time*: an apparently stationary particle is actually hurtling through time, so its trajectory is determined by its local "time landscape" rather than by curvature of space. Thus, as with the coin funnel, the low-lying parts of the rubber sheet must be interpreted as slower-time regions. The rubber sheet analogy is redeemed somewhat by the fact that slower-time regions *also* tend to host more spatial curvature, but true analogies for spatial curvature—such as the taut string in Section 18.5—simply do not involve time.

Next, we run some numbers to compare the metric time and space effects on Mercury, the fastest planet in the solar system at 47 km/s. Although fast by everyday standards, Mercury moves only about 0.00016 of a light-second per

Confusion alert

See Chapter 14 to review how accelerating *toward* slow-time regions actually maximizes proper time.

second (another way of saying that $\frac{v}{c} \approx 0.00016$). Mercury's speed through time (about one second per second) is therefore thousands of times greater than its speed through space. Thus, although space and time are equally affected by the Schwarzschild metric, Mercury is thousands of times less sensitive to the spatial part. This is why Mercury's orbit was modeled to better than 99.9% accuracy using the time part of the metric alone.

Finally, we assess the time and space effects on the travel time (measured in coordinate time) across the central billion kilometers of the solar system, for both slow and fast inertial particles. We begin with light. Curvature of space implies (for any particle) about 15 km of extra distance across the solar system when passing near the Sun. For light, this costs about 50 extra microseconds of travel time. The metric time coefficient dictates that clocks near the Sun tick slowly; because light moves at *c* relative to *local* clocks, this costs the light an additional 50 microseconds of travel time as measured by the distant clocks that keep coordinate time. Thus, the time and space parts of the metric each *increase* the travel time for light, but by a tiny percentage of the roughly one hour that light would take to cross a billion kilometers of empty space.

The effects play out differently for slowly moving particles. A bullet fired at 500 m/s initial speed across the solar system has to travel the extra 15 km near the Sun due to spatial curvature, just as light does. As with light, this is a small fraction of the billion or so kilometers traveled, so spatial curvature by itself would result in a tiny fractional increase in travel time for the bullet. The time part of the metric has a more dramatic effect: it makes the bullet accelerate to hundreds of km/s as it approaches the Sun. The bullet slows down again as it leaves the vicinity of the Sun, but at all times throughout this journey the bullet has a higher speed—often *much* higher—than its initial 500 m/s. This reduces the travel time far more than the extra 15 km increases it. The net effect is that gravity reduces the bullet's travel time substantially compared to the same bullet crossing the same distance of empty space.

Check your understanding. (a) Would the prediction for the time delay of light in the solar system change substantially if we used the Newtonian rather than the Schwarzschild time coefficient? Why or why not? (b) Repeat for the deflection of light.

CHAPTER SUMMARY

- The Einstein equation is a prescription for relating the metric to sources of gravity. It is difficult to solve (i.e., to find a metric given a description of the sources) but solutions have been found for certain specific cases.

- The Schwarzschild metric is an exact solution of the Einstein equation for the spacetime around a static, spherical source mass. Trajectories in this metric match those of the Newtonian model of gravity in the limit of slow orbits far from the source mass.

- The Schwarzschild metric has a time coefficient that matches the metric equivalent of the Newtonian model at large r, but predicts stronger gravity at small r.

- A conceptually new feature of the Schwarzschild metric is curvature of space: the coefficient on the space part of the metric is a function of r. The space coefficient is important for high-speed particles (especially light) but has very small effects on low-speed particles.

- Orbiting gyroscopes test spatial curvature specifically through an effect known as geodetic precession. Most other experiments (deflection of light, time delay of light, and anomalous precession of Mercury) test some combination of the time and space parts of the metric. In all cases the Schwarzschild metric has been confirmed to high precision, and that gives us confidence that the Einstein equation accurately captures the relationships between sources and metrics.

 FURTHER READING

Was Einstein Right? by Clifford M. Will clearly articulates Einstein's key insights while unfolding the story of the race to test general relativity in the twentieth century. *Gravity from the Ground Up* by Bernard Schutz discusses those insights at a more technical but still accessible level. *Einstein's Mistakes* by Hans Ohanian describes Einstein's struggles with these concepts and how Einstein learned from his own mistakes. Einstein also learned from others; see "History: Einstein was no lone genius" by Michael Jansses and Jürgen Renn in *Nature* magazine (http://www.nature.com/news/history-einstein-was-no-lone-genius-1.18793). The story of the 1919 eclipse observations that made Einstein famous is well told in a *New York Times* article by Dennis Overbye (https://www.nytimes.com/2017/07/31/science/eclipse-einstein-general-relativity.html).

Parts of Section 18.3 were inspired by John C. Baez and Emory F. Bunn's short, self-contained treatment of the meaning of the Einstein equation, posted online at http://arxiv.org/pdf/gr-qc/0103044v5.pdf. They also explain how curved spacetime makes general relativity thinking tools differ from those of special relativity. Mathematically adept students may try *A General Relativity Workbook* by Thomas Moore for more on the Einstein equation.

The comparison of travel time for light vs. a bullet in Section 18.7 was inspired by Lewis Carroll Epstein's *Relativity Visualized*. Epstein offers a unique visualization by drawing a proper time dimension through which the bullet advances but light does not. Epstein also goes beyond the taut-string analogy for curved space by teaching you how to push a vector "straight ahead" even in a curved space. For those interested in learning more about curved spaces, Rudolf v. B. Rucker's *Geometry, Relativity and the Fourth Dimension* is a good resource for the thoughtful beginner.

CHECK YOUR UNDERSTANDING: EXPLANATIONS

18.1   We have not yet examined what gravity could do to the spatial part of the metric; our model so far (Equation 18.1) is limited to one frame; Equation 18.1 does not account for source momentum; and Equation 18.1 cannot deal with dynamic situations (sources that change with time).

18.2   In the symmetric frame the velocities of particles F, O, and B look like this:

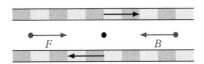

In this frame B *still* must accelerate upward (transverse distances are not affected by the frame change). The two source densities are equal, so we must attribute the net acceleration entirely to the effects of source momentum depending on the test particle velocity. Particle B moves parallel to the lower current and so is repelled from it, while B moves antiparallel to the upper current and so is attracted to it.

18.3   Assuming they are initially at rest relative to each other, net geodesic convergence will cause the particles in the cloud to converge and the cloud will shrink in volume. If geodesics converge in some directions but diverge in others, the cloud will shrink in some directions but expand in others—leaving the volume of the cloud unchanged.

18.4   (a) At $r = 100$, according to the figure, $\frac{\Delta t}{\Delta \tau}$ is barely more than one, maybe 1.01 ($\frac{\Delta \tau}{\Delta t}$ is perhaps 0.99). (b) Gravitational redshift is the standard way. A photon will arrive at a distant observer with only 99% of its original frequency and energy. (c) The difference between models is so small as to be negligible here. For a substantial difference, one would have to look closer to the compact object.

18.5   (a) The circumference is equal to $2\pi r$, so in a flat space increasing the radius by 1 m would increase the circumference by 2π m. (b) The requested ratio is $\frac{2}{2\pi} \approx 0.3$. This is off the charts on Figure 18.10, which shows that such severe effects are possible only at very small r—assuming an entire solar mass is contained inside that r. Space in our solar system is curved much more gently than this.

18.6   This should look like the deflection of light in Figures 18.14.

18.7   (a) No, there is little difference between the time coefficients in the two models except at extremely small values of r that are inaccessible in the solar system. (b) The same reasoning applies just as well to the metric effect on the deflection angle.

EXERCISES

18.1   On Figure 18.1, add a background star on the *right* side of the Sun; trace the path of light to Earth and then trace that path back to see the apparent position. Repeat for a second star. Convince yourself that any given star appears farther from the Sun than it would in the absence of the Sun's gravity.

18.2   Do parallel mass currents attract each other more, less, or the same as you would expect if source momentum had no effect? *Hint:* consider several frames including the rest frame of the parallel currents.

18.3   Do antiparallel mass currents (meaning each moves upstream of the other) attract each other more, less, or the same as you would expect if source momentum had no effect?

18.4   Equation 18.3 lists all possible metric terms in in a (t, x, y) coordinate system. How many terms are possible in a (t, x, y, z) coordinate system? *Hint:* find all unique pairs of coordinates.

18.5 In the solar system, where do we find net convergence of geodesics? Are these the same places where spacetime is curved?

18.6 Draw a spacetime diagram with a stationary mass. Show that geodesics converge on the mass.

18.7 Explain the conceptual difference between the Einstein equation and solutions to that equation.

18.8 How does Figure 18.9 show that the time coefficients of the Newtonian and Schwarzschild metrics are about the same everywhere in the solar system?

18.9 Find the ratio $\frac{\Delta r}{\Delta r_m}$ for the largest pair of circles in Figure 18.12. Find that ratio in Figure 18.9 and identify how far from a solar-mass object you would be if you measured this ratio. Why does no place in the solar system have such a ratio?

18.10 If the circles in Figure 18.12 represent all points equidistant from the Sun, what shape do they really represent? Why have I not drawn that shape in Figure 18.12?

18.11 How could you use a triangle to test for spatial curvature? *(a)* Describe how you could use the right triangle formed in the top panel of Figure 18.13 by half of the red trajectory and two heavy black lines. *(b)* Now sketch a triangle far from the source mass in Figure 18.13; describe how and why it differs from the first triangle.

18.12 Why is Mercury the best planet to look for an anomalous precession due to curvature of space?

18.13 Why does light take longer to cross the solar system if the light happens to cross very close to the Sun? How much of this effect is predicted by the Newtonian metric?

18.14 Why does light take longer to cross the solar system if the light happens to cross very close to the Sun, while a bullet takes *less* time if it does so?

18.15 Without looking at Figure 18.15, recreate the idea on paper yourself. How would you create a similar model for the part of the deflection of light due to spatial curvature?

+ PROBLEMS

18.1 Consider the Sun moving in front of two distant stars:

Draw a series of sketches illustrating the apparent positions of the stars and the Sun as the Sun approaches the stars, passes in front, and continues to the right. *Hint:* review Figure 18.14; see also Exercise 18.1. (This experiment has actually been done with distant pulsars, and the data agree with the Schwarzschild prediction.)

18.2 If contact with three GPS satellites would be required to find your location in a two-dimensional space (Figure 7.13), how many satellites are required in three spatial dimensions? In a four-dimensional spacetime, how many are required to locate you in all four coordinates (t, x, y, z)? *Note:* some implementations of GPS may use the fact that you are on the surface of Earth to substitute for information from one satellite, so in practice you may be able to get a GPS location with one fewer satellite.

18.3 Draw the acceleration field in and around a spherical shell of mass. Argue that this field has a net convergence only where there is mass (i.e., at the shell itself).

18.4 *(a)* Draw a spacetime diagram illustrating two balls dropped simultaneously from different heights on Earth. The height difference should be enough to notice a difference in acceleration; place the surface of the Earth to the right of your diagram.

(b) Redraw this diagram in a freely falling frame. Do the worldlines converge or diverge? Explain how your diagram makes sense in light of the rule that mass causes net convergence of geodesics.

18.5 Does curvature of space affect gravitational redshift? Explain why or why not.

18.6 Carefully compare Figures 15.4 and 18.12. Why does the latter have rings that are separated by consistent distances as measured along the surface of the funnel, while the former does not?

18.7 If you have studied optics, you know that light always propagates perpendicularly to its wavefront. Use this to explain how a beam of light would deflect when part of its wavefront is in a region of slower time than other parts of the wavefront.

18.8 Classify all the tests of general relativity discussed in this section into three categories: testing the time part of the metric alone, the space part of the metric alone, or testing a combination of the two.

18.9 What kind of situation would increase the precession due to spatial curvature well beyond the small amount exhibited by Mercury? Specify the kind of orbit, orbiting body, and/or source mass—anything you think is relevant.

18.10 Why is it slightly unfair to compare the time coefficients of the Newtonian and Schwarzschild metrics at a given r?

18.11 Professor Zweistein offers a competing theory of gravity that has curvature of space but not curvature of time. How could you determine which theory is correct? List as many ways as possible, but list the most obvious and definitive tests first.

18.12 Consider an observer in a rotating merry-go-round where the rim moves at nearly c. *(a)* How does length contraction affect, if at all, the circumference as measured by placing meter sticks along the rim? *(b)* How does length contraction affect, if at all, the radius as measured by placing meter sticks from center to rim? *(c)* Does this frame exhibit spatial curvature?

18.13 *(a)* Use the metric to calculate $\frac{\Delta\tau}{\Delta t}$ for a stationary clock on the surface of the Sun. If your calculator does not have enough precision, think about how to use the Newtonian approximation to the time coefficient to compute the *difference* between $\frac{\Delta\tau}{\Delta t}$ and 1—and why that approximation is justified here. *(b)* Why is "compute $\frac{\Delta\tau}{\Delta t}$ for a stationary clock at the center of the Sun" a much more complicated question?

18.14 Since Earth formed 4.5 billion years ago, how much more time has passed at its surface than in deep space? *(a)* Argue that the Newtonian approximation for $\frac{\Delta\tau}{\Delta t}$, $1 - \frac{GM}{c^2 r}$, is justified here. *(b)* Use this ratio to calculate the *difference* in time. *(c)* Argue that this should also be an order-of-magnitude estimate for the difference in time between the center and surface of Earth. You may wish to review how the potential depth varies inside and outside a ball of mass (Figure 16.8). *Note:* in the 1960s, Richard Feynman gave a lecture in which the difference is quoted as "one or two days" and, perhaps due to Feynman's eminence, this estimate was not checked and corrected until 2016. This illustrates the importance of thinking critically about all statements, even those made by trusted sources (see *The young centre of the Earth* by Uggerhøj *et al.*, *European Journal of Physics* vol. 37, no. 3. for more context).

18.15 The "Newtonian metric" (Equation 18.1) has a time coefficient of $\left(1 - \frac{GM}{c^2 r}\right)^2$ while the Schwarzschild time coefficient is $\left(1 - \frac{2GM}{c^2 r}\right)$. Why are they numerically so similar? Multiply out $\left(1 - \frac{GM}{c^2 r}\right)^2$ and show that it equals the Schwarzschild coefficient plus an additional term, $\left(\frac{GM}{c^2 r}\right)^2$. Show that this additional term is much smaller than the other terms except near compact objects.

18.16 There is no theoretical basis to predict the presence (or absence) of extrinsic curvature in our universe, but we can look for it empirically. Research how this has been done without sending a robot to circumnavigate the universe, and what the results are.

Beyond the Schwarzschild Metric

19

General relativity explains much more than the spacetime around static spherical masses. This chapter provides brief summaries and starting points for further exploration of selected topics.

19.1 General relativity in context

The physicist John Archibald Wheeler (1911–2008) summarized general relativity most succinctly: mass-energy tells spacetime how to curve, and curved spacetime tells mass-energy how to move. As mass and energy move around, they cause the metric to change with time. General relativity thus allows us to model complicated and dynamic situations such as the collapse of a star or the merger of two compact objects. In most situations, the Einstein equation is too difficult to solve for a single metric expression such as Equation 18.3. Instead, computers are used to solve the equations numerically—that is, starting with an initial mass-energy distribution an initial metric is computed, which is used to predict how particles move in the next instant, which is then used to find the metric in the next instant, and so on. There are a few cases, however, where time dependence of the metric is very important and the basic behavior can be understood without computers; these are examined in Sections 19.3 and 19.5.

General relativity is not the only possible metric theory of gravity, but it is the simplest. An example of a more complicated approach is that the "constant" G may vary in space and/or time. Such variations, if they exist, would also be governed by relativistic principles, so this hypothesis does lead to specific equations and predictions for observable consequences. Experiments guided by these predictions, however, have not found any deviation from general relativity. General relativity thus reigns as the least complicated model that is consistent with all the data. General relativity can also be tested without reference to any other specific metric theory: if the deflection of light, for example, is not *exactly* twice the Newtonian value then general relativity is wrong and we should seek some other metric theory. Again, general relativity has comfortably passed many such tests.

We can also step back from pitting general relativity against other metric theories and test the foundation of all metric theories, which is the equivalence

The Elements of Relativity. David M. Wittman, Oxford University Press (2018).
© David M. Wittman 2018. DOI 10.1093/oso/9780199658633.001.0001

principle. Violations of the equivalence principle, if found, would suggest that the metric theories may be good approximations but cannot tell the whole story. To date though, experiments confirm the equivalence principle to better than one part in one trillion.

Although the metric approach and general relativity in particular have passed all these tests to date, some very large questions remain. A key assumption behind the mathematical tools involved is that the metric has a definite value at any event, and varies smoothly over space and time. Yet quantum mechanics— the physics of the very small—establishes a limit on how definite the value of anything can be when evaluated at a precise place and time. For example, the smaller the time window in which we look at a particle, the more its energy appears to fluctuate rather than maintain a definite value. The standard tools of general relativity ignore this quantum uncertainty, so they are ill suited to model the unfolding of microscopic events. For everyday work such as modeling the solar system this is not a problem, as quantum effects average out to zero on scales much larger than an atom. This is also not a problem for quantum scientists studying subatomic particles, because gravitational effects are negligible for such studies. A problem arises only when modeling something extremely small *and* extremely massive, such as a star that has collapsed to smaller than an atom. Even in that case, though, the effect of such an object *on its larger-scale spacetime surroundings* can be modeled well with general relativity (Section 20.1).

Nevetherless, the tension between general relativity and quantum mechanics spurs scientists to look for new theories that could unify both (see *Further Reading*). Some of these theories do predict small departures from the equivalence principle, so it is possible that in coming decades a departure will be detected and point the way to a deeper theory. If so, general relativity would remain a very good approximate description of nature, just as the Newtonian model provides an approximate description that is good enough for most purposes.

General relativity is a comprehensive model for how mass-energy affects spacetime and vice versa. The rest of this chapter and Chapter 20 describe the *consequences of this model in specific situations.* Keep this context in mind as we tour the applications.

Check your understanding. Is it accurate to say that tests of the equivalence principle confirm general relativity?

19.2 Gravitomagnetism

The Schwarzschild metric applies to the spacetime around *motionless* masses. Masses can move in many different ways so there is not one specific metric for spacetimes around moving masses, but we can already predict a general feature of such metrics. Section 18.2 showed that source motion causes mixed terms—

for example a $(\Delta x)(\Delta t)$ term—in addition to the "pure" $(\Delta x)^2$ and $(\Delta t)^2$ terms. To recap that section, mixed terms are distinctive because the *sign* of Δx matters when Δx is not squared; therefore, the *direction* of test particle motion affects the proper time. In turn, the maximum proper time path, and thereby the gravitational acceleration, now depend on the velocity of the test particle in addition to the standard factors. This phenomenon is called **gravitomagnetism** because the acceleration of electric charges in magnetic fields is similarly velocity-dependent. For the idealized linear mass currents studied in Section 18.2, gravitomagnetism repels particles moving in the same direction as the source and attracts particles moving in the opposite direction. At everyday speeds these are very small effects in addition to the normal attractive force, but nevertheless interesting because they do not exist in the Newtonian model.

Nature does not provide us with linear mass currents, but spinning masses are ubiquitous in nature and can be modeled as mass currents that circle back on themselves as in Figure 19.1. There, a particle orbits parallel to the nearby mass current (known as the *prograde* direction) so the net attraction is less than with a nonrotating source mass. Orbits in the opposite (retrograde) direction yield a net attraction above and beyond the standard gravitational attraction. Sufficiently close to a massive, rapidly rotating source, retrograde orbits are impossible because they would have to be faster than c to overcome this attraction—but prograde orbits can still be maintained at substantially lower speeds.

Physicists can make more complete and quantitative predictions thanks to the mathematician Roy Kerr (b. 1934), who solved the Einstein equation for the spacetime metric around a spinning spherical mass and produced what is now known as the **Kerr metric**. Figure 19.2 shows how source mass spin affects inertial trajectories that are initially pointed directly toward or away from the source mass: an infalling particle begins to move in the direction of spin (by an amount highly exaggerated in this illustration). The box around the infalling particle represents its local inertial frame, which is "dragged" into a new orientation. For this reason, gravitomagnetic effects are often called "frame dragging" or "dragging of inertial frames." Frame reorientation is worth noting because it lends itself to precise measurement with a gyroscope—which spins in a constant direction relative to its local inertial frame. Nevertheless, "frame dragging" is unfortunate jargon because it is easily misinterpreted as "spacetime itself is forced to move in the spin direction." To see why that would be a misinterpretation, consider the outbound particle in Figure 19.2: reversing the particle velocity reverses the direction of the effect so the deflection is now *opposite* the spin direction. This makes perfect sense if your key thinking tool for the effect of source motion is "velocity dependence" but makes little sense if you think "spacetime itself moves in the spin direction." Velocity dependence is really the core idea here and provides a more flexible thinking tool.

For testing these predicted effects, astronomers can look near rapidly-spinning massive objects, but we defer that discussion to Section 20.5. The remainder of this section focuses on tests of gravitomagnetism around Earth, where we can

Think about it

See Problem 19.1 for a thought experiment involving the mixed-term effect on clocks.

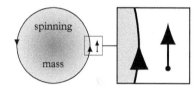

Figure 19.1 *A spinning mass can be modeled as a mass current that loops back on itself. Here, a nearby orbiting particle sees the spinning source mass as a parallel mass current. The far side of the source mass does form a current in the opposite direction, but has less effect due to its greater distance.*

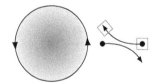

Figure 19.2 *A spinning source mass causes an infalling particle to begin moving in the same direction as the rotation of the source mass. A particle launched directly outward, in contrast, is deflected in the opposite direction; it is also slowed by the usual gravitational attraction that applies regardless of source mass spin.*

Newtonian model

General relativity

Figure 19.3 *Gravitomagnetic effect of Earth's rotation (white arrow) on polar satellite orbits. In the Newtonian model there is no effect. In general relativity, orbits advance in the spin direction; the size of the effect is greatly exaggerated here.*

build precise experiments even if gravitomagnetic effects are small around the slowly spinning Earth.

Figure 19.3 shows the gravitomagnetic effect of Earth's spin on satellite orbits that repeatedly cross over the poles: they advance in the direction of spin. This type of orbit does not lend itself to the parallel mass current thinking tool, but was selected here because it provides the cleanest comparison with the Newtonian model. You may recall (Section 16.1) that Newtonian orbits in a spherical potential always repeat exactly, so you may think that any orbit could be used for the comparison. Earth, however, bulges at the equator, and this causes most orbits to advance even in the Newtonian model—by much more than the additional small advance predicted by general relativity. The exception: polar orbits are unaffected by the equatorial bulge. Therefore, a satellite trajectory test would ideally use a polar satellite with precise tracking capability. As it happens, the satellites best equipped for precise tracking are not in polar orbits so the large effects of Earth's asphericity must be carefully accounted for. The LARES (Laser Relativity Satellite) team nevertheless claims to have detected the gravitomagnetic effect to a precision of 1 part in 25. Other experts, however, argue that uncertainty in Earth's asphericity is too large to claim this degree of precision, and do not consider this a definitive detection of gravitomagnetism.

Earth's asphericity matters much less if, instead of studying trajectories, we study frame orientation as indicated by a gyroscope. Recall from Section 18.6 that gyroscope orientation is also used to measure curvature of space (geodetic precession). If a gyroscope precesses, can we determine whether the cause was geodetic or gravitomagnetic? Yes, because only the latter depends on the gyroscope's orientation relative to Earth's spin. Both effects can be measured by launching a spacecraft with multiple gyroscopes, some oriented in directions sensitive to the spin effect and others, insensitive. This was the idea behind Gravity Probe B, the satellite mission we first encountered in Section 18.6 where it confirmed the geodetic effect around Earth to a precision of 1 part in 350. The mission also had the more heroic goal of detecting the much smaller gravitomagnetic effect, which it did to a precision of one part in five. Although this is a less precise measurement than originally hoped for, Gravity Probe B nevertheless confirmed that source mass rotation really does have a gravitomagnetic effect on the surrounding spacetime.

Gravitomagnetism more generally encompasses the effects of *any* type of source mass motion, not just rotation. Figure 19.4 shows how the Moon's motion can be viewed as an orbit around the Sun perturbed by the "mass current" provided by the moving Earth. Astronauts planted on the Moon an ingenious device called a retroflector, which reflects light directly back to its source. With this device in place, physicists routinely shoot laser beams at the Moon and receive the return signal (called *lunar laser ranging*). By precisely measuring the time for the round trip, the distance to the Moon at any given time is measured to within ±1 cm. These data are so precise that if gravitomagnetism were to be removed from the general relativity model for the Moon's orbit, the model would fail to match the data—by about 600 times the uncertainty in the data. This is an independent confirmation of gravitomagnetism—due to Earth's bodily motion rather than its spin—to a precision of 1 part in 600.

Take a step back from these details and remind yourself that gravitomagnetism is a distinctive outcome of applying frame-based thinking tools to gravity, as in Sections 18.1–18.2. The confirmation of gravitomagnetic predictions thus represents a triumph of frame-based reasoning, and the principle of relativity, over Newton's absolute time and space.

Check your understanding. Venus is about the same size and mass as the Earth but rotates much more slowly (once every 243 days). *(a)* Will the gravitomagnetic effect on gyroscopes around Venus be smaller than, the same size as, or larger than around Earth? *(b)* How would orbits around these planets compare in terms of geodetic precession?

19.3 Gravitational waves

The mass currents we used to think about gravitomagnetism typify the effect of a source mass moving at constant velocity. In this section we focus on *accelerating* source masses, which reveal a conceptually new aspect of gravity.

Imagine an initially static source mass that is pushed from one position to another nearby position, where it stays. Before the push, the spacetime is described everywhere by the Schwarzschild metric centered on the original position. Long after the push, the spacetime must be described by a Schwarzschild metric centered on the new position. The details of this transition can be computed with the Einstein equation (Section 18.3), and it turns out that changes in the metric ripple out from the source mass at speed *c*. Soon after the change in source position, the metric near the source has already been "updated" while distant regions are still unaffected. This results in a mismatch between the metric near the source and the metric further away. Over time, the "updated" region grows and pushes the mismatch further out into the "outdated" region, much as slapping the water in one corner of a pool causes a mismatch in water levels that ripples outward as a wave. In fact, the gravitational ripple has mathematical properties just like those of a wave: a **gravitational wave**.

Although the one-time slap provides a vivid picture, *any* source mass acceleration generates gravitational waves, and the most common cause of acceleration is that a source mass is itself orbiting another source mass (recall from Section 16.1 that circular motion always involves acceleration toward the center of the circle). Mutually orbiting pairs of stars—called binary stars—thus continually send gravitational waves rippling outward. Gravitational waves, like most other types of wave, carry energy outward from the source of the disturbance. Imagine standing in a pool, locking hands with a partner, and twirling around in an "orbit." You will tire quickly, because much of your energy goes into generating waves that

Figure 19.4 *The Moon's motion is mostly determined by the Sun; this one-month segment of Earth's orbit around the Sun is, on this scale, indistinguishable from the Moon's path over the same month. In the frame of the Sun, the moving Earth forms a "mass current" with gravitomagnetic effects on the nearby Moon.*

Confusion alert

In many contexts such as Earth science, a "gravity wave" is any wave in which gravity plays a role, such as an ocean wave. The waves discussed here—subtle disturbances in the metric itself—are conceptually quite different but have a similar name: gravitational waves.

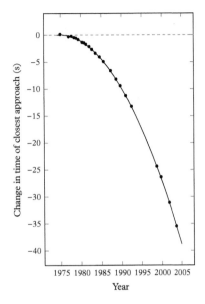

Figure 19.5 *The orbital period of the Hulse–Taylor binary neutron star system is decreasing rapidly due to energy loss to gravitational waves. The data (points) match the general relativity prediction (curve) extremely well. Adapted from a figure in Relativistic Binary Pulsar B1913+16: Thirty Years of Observations and Analysis by J. M. Weisberg and J. H. Taylor.*

Think about it

Binary pulsars test general relativity in multiple ways. Their orbits precess due to curvature of space, for example, tens of thousands of times faster than that of Mercury. They also demonstrate that the "speed of gravity" (the speed with which changes in the metric propagate) is really c, because the predicted energy drain due to gravitational waves depends on this speed.

ripple across the pool. Similarly, gravitational waves drain energy from an orbit; this causes the orbiting masses to spiral in over time, much like coins in a coin funnel. This orbital decay is a fundamentally new prediction of metric theories of gravity, compared to the Newtonian theory. However, the waves are extremely weak—making the orbital decay extremely slow to unfold—unless the masses and accelerations are very high. The predicted effect on solar system orbits is far too small to ever detect. This makes our planet quite stable, but forces us to look outside the solar system to test this prediction.

To produce non-negligible amounts of gravitational waves, a binary star must have an orbit only a few million km across. Two ordinary stars approaching this closely would simply form a single large ball of gas rather than maintain a tight orbit. Only the compact remnants of stars—neutron stars or black holes—could enter such small orbits. Neutron stars are ideal because some of them (also called pulsars) emit regular pulses of radio waves, thus allowing astronomers to track their motions precisely. General relativity predicts that a binary neutron star system losing energy by emitting gravitational waves will spiral in closer and closer together, yielding a shorter and shorter orbital period. Such a system was observed starting in the 1970s by Joseph Taylor and Russell Hulse; the quickening of the orbital period over time beautifully matches the general relativity prediction as shown in Figure 19.5. Astronomers have since found additional binary pulsars—some accelerating even more rapidly in even tighter orbits—and the energy loss rate continues to agree precisely with general relativity, leaving little room for alternative metric theories that may predict slightly different loss rates.

Hulse and Taylor won the Nobel Prize in 1993 for establishing this technique and proving, albeit somewhat indirectly, the existence of gravitational waves. Physicists and astronomers nevertheless sought to detect gravitational waves more directly, in order to learn more about the astrophysical events that generate them. As a start, the frequency of the waves tells us the frequency of the orbit that generates the waves, and this could be the *only* way to observe a mutually orbiting pair of black holes. Two black holes in a tight orbit could go around each other ten or twenty times per second, a frequency that, if it were a sound wave, would be heard as a very low bass hum. As the orbit of such a system shrinks and speeds up, the wave frequency increases. The amplitude or "loudness" of the waves would also increase due to the stronger accelerations in a tighter orbit. This process would culminate in a final dramatic increase of frequency and amplitude—called a "chirp" in analogy with sound waves—as the black holes merge. A passing gravitational wave alternately stretches and squeezes space, so we can detect it by monitoring the lengths of perpendicular rulers. The amount of stretching and squeezing is tiny—a 4-km "ruler" changes in length by much less than the width of an atomic nucleus—but gravitational wave observatories are now able to detect this.

In 2015 the newly upgraded Laser Interferometer Gravitational-Wave Observatory (LIGO) clearly detected a gravitational wave for the first time (Figure 19.6). The wave was clearly detected in two independent detectors

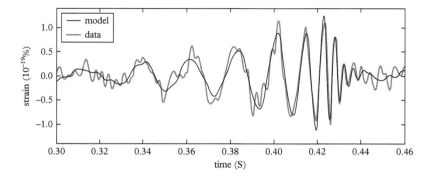

Figure 19.6 *Ripples in the metric passing the LIGO detector in Hanford, Washington. Each cycle of the wave corresponds to half an orbit of two black holes spiraling in around each other; the cycles become more frequent and more violent as the orbit becomes smaller, until the black holes merge and settle down into a single black hole.*

thousands of kilometers apart, so there is no doubt that this was a real astrophysical event. The two detections occurred 7 ms apart, thus allowing LIGO to infer at least roughly the direction of the event in the same way that humans (unconsciously) use the difference in reception time between two ears to roughly locate the direction of a sound. Even better, the frequency, amplitude, and increase in frequency and amplitude are so clearly measurable that the LIGO group was able to infer the masses of the progenitors (about thirty solar masses each), the distance of the event (about 1.2 billion light-years), and the amount of energy released in gravitational waves (the equivalent of about three solar masses). Since then LIGO has detected additional events and has even been able to infer the spin of some of the merging black holes. Detection capabilities are continuously improving, so we can expect to learn much more from this entirely new way of listening to the universe—not only about black hole populations but also about violent and dynamic aspects of spacetime. For example, a clearer picture of the very end of the wave will tell us more about how the spacetime around the newly merged black hole settles down into a Kerr spacetime.

Gravitational waves passing by Earth could be detected in other ways as well, for example by a specific pattern of small changes in the arrival times of pulses from pulsars in different parts of the sky, as seen by radio telescopes at different locations on Earth. Networks of radio telescopes set up for this purpose are called pulsar timing arrays. These networks are sensitive to black hole pairs in larger orbits, well before their final merger event; so far, they have only been able to put upper limits on the gravitational waves generated by supermassive black hole pairs. Even greater sensitivity to many sources of gravitational waves could be achieved by placing detectors millions of km apart in space.

Returning to the big picture of relativity, gravitational waves are the clearest proof that spacetime itself is a dynamic physical entity, not just a featureless stage on which other physical events play out. Although we routinely accept this dynamism today, it was a striking departure from the Newtonian conception of absolute space and time when Einstein formulated it a century ago. Remarkably, these waves are so subtle that scientists needed a full century to develop the experimental technology to confirm their existence directly. To quote the writer

Think about it

Rainer Weiss, Kip Thorne, and Barry Barish were awarded the 2017 Nobel Prize in Physics for their contributions to and leadership of the LIGO effort.

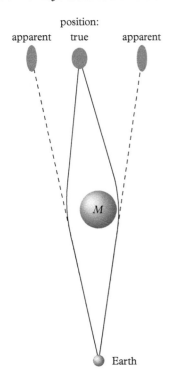

Figure 19.7 *A mass always deflects light toward it, which makes the light converge downstream as if the mass were a lens. In this case the deflection allows light from a distant galaxy to reach Earth along two different paths, and we see two images of that galaxy. This figure is very much not to scale—vertical distances have been shortened much more than the horizontal distances, so the bending angles are highly exaggerated.*

Eden Philpotts, "The universe is full of magical things, patiently waiting for our wits to grow sharper."

Check your understanding. Why are gravitational waves not important in the evolution of solar system orbits?

19.4 Gravitational lensing*

In Figure 18.1 we saw how the presence of the Sun changes the apparent position of background stars as seen from Earth. Figure 19.7 shows that, if a mass M provides a great enough deflection, light from a single source (an elliptical galaxy in this case) can actually take two *different* paths to Earth; we then see two copies of the background object in two different places. This is a striking example of a more general phenomenon: light rays tend to converge "downstream" of a mass. Because this bears some similarity to the way light converges after passing through an ordinary glass lens, this phenomenon is called **gravitational lensing**.

Gravity bends light in a particular pattern best exemplified by the base of a wine glass, which is shaped somewhat like a gravitational potential. In Figure 19.8 a wine glass is placed on a regular grid of circular "galaxies." The galaxy most closely aligned with the center of the lens is distorted nearly into a ring. This is dictated by symmetry: if a lens is perfectly aligned light from the source can take any way around the lens and still arrive at the observer. Perfect symmetry is rare, however, and in Figure 19.8 the off-center view breaks the ring. Similarly, Earth is never exactly aligned with a gravitational lens (a massive galaxy or cluster of galaxies) and a background source of light, so we too see broken rings in the sky. Broken rings are actually composed of multiple (highly distorted) images of the same galaxy. Figure 19.9 shows a beautiful example, with a massive yellow galaxy lensing a blue background galaxy. The symmetry of such images can be further broken when the lensing mass itself is asymmetric; the lumpy mass distributions in clusters of galaxies yield a rich variety of distorted images.

Light sources substantially further off axis are distorted merely into ellipses that face the center of the lens. Figure 19.8 has many such distortions, and we can understand them by referring back to the left galaxy image in Figure 19.7. Compared to the left side of the source galaxy, light from the right side of the source galaxy passed closer to the lensing mass M and suffered greater deflection; this squeezes the sides of the galaxy image closer together.

Astronomers have found hundreds of examples of "strong" gravitational lensing, meaning we see multiple images of a single source, or something close to a ring. The ring size (or separation between images of the same source) is proportional to the deflection angle, which tells us the mass of the lens. More detailed analyses of the distortion pattern allow us to map out how the mass is distributed, so lensing is a key tool for studying dark matter (Section 17.5) and many other aspects of astrophysics. We use this tool with confidence because lensing-based mass estimates are double-checked against other methods of inferring mass where possible, and the agreement is good. The general relativistic relationship between mass and deflection of light is thus used every day by astronomers.

Astronomers have developed a number of additional ways to use gravitational lensing:

Microlensing. For stellar-mass lenses the diameter of the ring may be so small that we do not see it as a ring—it may cover only a few pixels in our best cameras. However, the fact remains that more total light reaches us when the alignment between Earth, lens, and background source is good. When a faint, compact object crosses between us and a background star, it goes from poorly aligned to well aligned and back to poorly aligned in a matter of months; this causes an increase followed by a symmetric decrease in apparent brightness of the background star—even if we never see the faint compact object itself. This is the primary way in which astronomers survey our galaxy for faint compact objects. In addition, any planets around those objects will cause an additional brief boost in apparent brightness. This is actually the best technique for finding extrasolar planets far from their host stars, because reflex motion (Section 17.7) diminishes with star-planet separation.

As a corollary, in practice we do not see rings or arcs around compact objects such as neutron stars and black holes; we are simply too far away to see the tiny arcs that should be there. Seeing such features in detail, as we often do in artists' conceptions, would require traveling much closer to those objects so that they loom much larger in the sky.

Time delays. In Figure 19.7, the two paths taken by light are not of equal length. If light can take the long way around and still hit Earth, it must be deflected

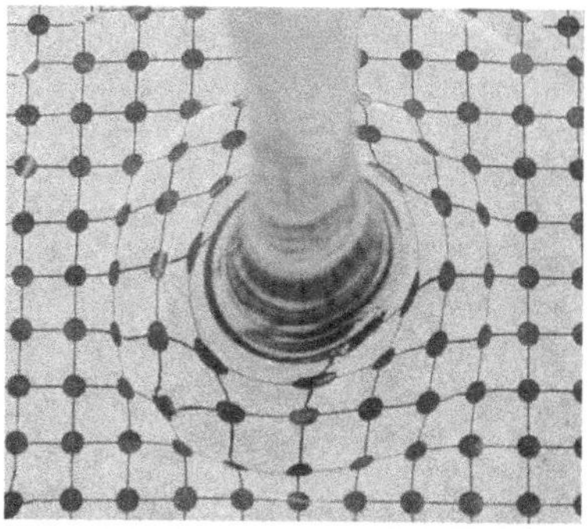

Figure 19.8 *The base of a wine glass bends light much like a gravitational lens.*

Figure 19.9 *Gravitational lensing of a distant (blue) galaxy by an intervening (yellow) galaxy. If the symmetry were perfect we would see a ring, but the symmetry is broken because the true position of the blue galaxy is not exactly behind the center of the yellow galaxy. Image credit: ESA, Hubble Space Telescope and NASA.*

Figure 19.10 *Time delays in action: red circles indicate three images of the same spiral galaxy in the background of a cluster of yellow elliptical galaxies. The white circle highlights how one spiral arm of the southern image is further lensed nearly into a ring by one specific galaxy in the cluster. Around that ring we see four images (yellow points) of a supernova explosion in that spiral arm. Each image has a unique time delay so this snapshot shows the explosion at four different stages of brightness. The image in the central red circle has a longer time delay and showed the explosion a full year after this snapshot. Image credit: NASA, ESA, S. Rodney (John Hopkins University, USA) and the FrontierSN team; T. Treu (University of California Los Angeles, USA), P. Kelly (University of California Berkeley, USA) and the GLASS team; J. Lotz (STScI) and the Frontier Fields team; M. Postman (STScI) and the CLASH team; and Z. Levay (STScI).*

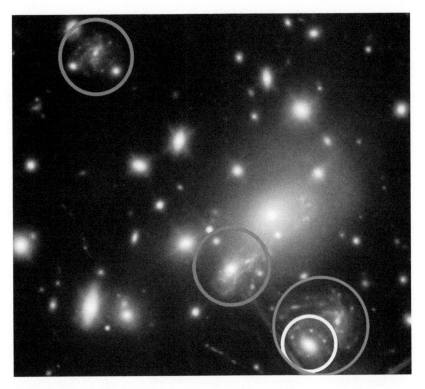

by a larger angle, which means that it must pass closer to the lensing mass M. That light requires a longer travel time because of the greater Shapiro time delay (Section 18.6) deeper down in the gravitational potential. If the background source of light changes with time, we see that change register first in the outermost image and then later in the more central image. The length of the delay is determined by the distribution of mass in the lens—weeks if the lens is a single galaxy, up to years if the lens is a massive cluster of galaxies. Well-measured time delays thus help astronomers characterize the distribution of mass in the universe.

A spectacular case is the galaxy cluster MACS J1149.5+2223, which bends the light of a particularly pretty spiral galaxy into three separate images (Figure 19.10). Additionally, one of the spiral arms in the southern image is nearly perfectly placed to form a ring around one of the galaxies in the cluster. A star in that spiral arm exploded and is observed as a bright point of light, seen in four different places along that ring in the southern image. The time delay differences between the four images in this smaller-scale lens are on the order of weeks, so the unequal brightness of the four points reflects different views slightly before, during, and after the peak brightness of the explosion. Even more remarkably, we can zoom out to the larger-scale lensing caused by the entire cluster of galaxy and ask about the much longer time delays between the three widely separated images of the pretty spiral galaxy. As mentioned previously, the outermost image experiences the least delay, so we would expect to see the explosion first in the northern image, then the southern image, then the central image. But the explosion was discovered in the southern image in fall 2014; astronomers calculate that the explosion would have been visible in the northern image sixteen years prior to that, but no one was looking at the time. They also calculated that the explosion would appear in the central image in fall 2015, and indeed it appeared at the predicted time and place. This is a beautiful example of how well we can understand nature.

Weak lensing. Far from the lens axis in Figure 19.8 the distortion becomes quite weak; yet most galaxies *are* far from the lens axis so collectively they still encode much information about the lens. Weak lensing analysis techniques enable us to study mass in lower-density regions, which constitute the vast majority of the universe. Thus, while strong lensing is best for studying specific massive lenses, weak lensing is better suited to studying the mass distribution throughout the universe more generally.

Regardless of the specific technique, gravitational lensing uses background sources of light to characterize the lensing mass—it does *not* require us to observe light from the lensing mass itself. Because most of the mass in the universe is dark (Section 17.5) lensing is a go-to tool for studying the distribution of mass in the universe and the behavior of dark matter. This tool works because general relativity taught us that a given mass bends light *twice* as much as in the Newtonian model (Section 18.6).

Check your understanding. (a) Does gravitational lensing allow astronomers to measure the collective mass of stars in a galaxy, or the collective mass of all forms of matter in that galaxy? *(b)* Which galaxy's mass is measured—the source galaxy or the lens galaxy?

19.5 Cosmology*

General relativity provided, for the first time, a theoretical framework capable of modeling the universe as a whole. Observations of distant space reveal a sea of galaxies extending more or less uniformly in every direction, with no hint of a boundary. Newtonian thinking tools fail in this case, as follows. Given the symmetry of a borderless universe with a uniform density of mass, the acceleration field at any point cannot have a preferred direction, so the acceleration must be zero everywhere. Poisson showed, however, that local variations in acceleration are tied to the local mass density (Section 18.1); with no such variations, the mass density must be zero. Therefore, if the cosmos is symmetric the only self-consistent Newtonian model is an empty one. General relativity has no such weakness because spacetime curvature, rather than acceleration, is the fundamental quantity. Curvature can be the same everywhere without causing particles to

Think about it

Calling the cosmos uniform is a simplification that overlooks local variations in density, just as "the Earth is round" is an approximation that overlooks local features. But just as a zoomed-out picture of the Earth *is* round, uniform models of the cosmos do extremely well in matching a diverse array of data on large scales.

accelerate relative to the coordinate system; we need only find the metric that does this.

The first attempts to solve the Einstein equation to model a uniform sea of galaxies exposed a new issue: the result depends on how much energy is in each cubic meter of empty space. You might expect that that this **vacuum energy** is zero, but no law of physics proves this. In the solar system, observations show that vacuum energy is small compared *to other forms* of energy via the following argument. The Einstein equation shows that if vacuum energy exists, it would cause a repulsive effect that would compete with the usual gravitational effect of mass. We can examine solar system orbits for any signs of this repulsion; each planet will be affected in proportion to the amount of vacuum between it and the Sun, so let us look at the distant planet Neptune. If Neptune accelerates toward the Sun less than expected, this would indicate that some of the attractive part of gravity was reduced by the repulsive effect of vacuum energy. We see no such thing, and after accounting for observational uncertainties we can say that the density of vacuum energy, if any, must be less than 10^{-15} kg/m^3—negligible for everyday purposes.

However, there is so much more space between our galaxy and other galaxies that attraction or repulsion between *galaxies* provides a much more sensitive test for vacuum energy. Because data on galaxies were not available at the time, Einstein fell back on a preconception that the universe as a whole must be static and unchanging. Knowing that the universe does contain mass that is attractive, the only way to construct a static model was to assume that the vacuum energy (also called the cosmological constant) must have just the right density to exactly balance intergalactic gravity and provide a static universe. This density was not in conflict with observations available at the time; it was well below the upper limit that could be tested with solar system orbits.

Nevertheless, Einstein was wrong on two counts. First, the hypothetical balance between attraction and repulsion was shown to be unstable: motions and interactions of stars and galaxies would end up tilting the balance one way or the other. So, general relativity yields the important theoretical result that the universe *must* be dynamically evolving. Second, observations eventually showed that the universe is not static at all. The idea that vacuum energy could be very small but not quite zero was soon dropped in favor of the simplifying assumption that it was exactly zero.

You might think that (in the absence of vacuum energy) gravitational attraction implies universal collapse, but this really depends on the initial conditions. If you walk into a room and see a ball moving upward through the air, you do not say that gravity makes upward motion impossible, merely that it will slow the initial upward motion. Similarly, with the universe as a whole gravity will slow any expansion that exists as a result of an initial condition; depending on the average mass density of the universe, gravity may or may not be strong enough to reverse the expansion. In the 1920s, Alexander Friedmann (1888–1925) and Georges Lemaître (1894–1966) independently produced solutions of

Think about it

Why is vacuum energy repulsive? The Einstein equation shows that, in addition to energy and momentum, *pressure* is a source of gravity. You can picture this as an effect of source motion, as pressure in a box is typically due to particles inside hitting the box walls. Strictly speaking, though, pressure measures how much the energy of the box increases when the box is squeezed. A box of vacuum energy would have negative pressure because compressing it—removing some of the energy-carrying space—would *decrease* its energy. Plugging this negative pressure into the Einstein equation then yields negative gravitational attraction; i.e., repulsion.

the Einstein equation describing an expanding universe. Lemaître showed how this model would appear to an observer attached to one galaxy. All other galaxies would appear to be receding in a particular pattern: the recession velocity would be *proportional to the distance*. This proportionality is a hallmark of uniformly expanding models because, when all distances are expanded by the same factor in the same time, larger initial distances yield more kilometers of expansion in that time (Figure 19.11). This model was confirmed in 1929 when Edwin Hubble (1889–1953) published observations showing that recession speed (measured via the Doppler effect) is indeed proportional to distance from us—an effect now called Hubble's Law but first predicted by Lemaître.

This relationship between recession speed and distance ensures that if we "run the movie backward" we will eventually see all galaxies converge on each other simultaneously. We use this event to define $t = 0$; running the movie forward again we will see particles expand away from each other in what is now known as the **Big Bang**. However, the Big Bang is unlike any explosion because *space* (as quantified by the metric) is expanding and galaxies are just along for the ride. Furthermore, the cosmic metric is approximately the same everywhere so the Big Bang had no center; it happened everywhere (Figure 19.11 should help convince you that there is no center of expansion). The Big Bang happened at the precise value of x, y, z that you are sitting at right now! It also happened at every other value of x, y, z; what makes this possible is that all these coordinates were, at $t = 0$, "in the same place" in the sense of having zero meter-stick distance (also called proper distance; Section 18.5) between them. Because the explosion image is misleading, there have been attempts to rename the Big Bang model—*Minute Physics* has suggested "the Everywhere Stretch"—but the name has stuck.

"Stretch" is in fact a good mental image: if we attach galaxy stickers to an expanding balloon, the stickers expand away from each other but have no motion relative to a balloon-based coordinate system. Furthermore, the stickers are bound by local forces, just as solar systems and galaxies remain bound by their own gravity. Only *intergalactic* space expands, even if we say informally that "the universe" expands. Intergalactic photons lack local binding forces and stretch along with space. This yields a new interpretation for the observation that light from distant galaxies is received at low frequencies: rather than a Doppler effect due to motion of the source galaxy through space, it is a stretching of space itself.

A crucial prediction of the Big Bang model is a *hot* early universe, because expansion has a cooling effect: photons are diluted over a larger volume, and each photon has lower frequency (less energy). Twentieth-century astronomers discovered two key pieces of fossil evidence confirming Big Bang predictions that the universe was hotter in the past: the cosmic microwave background radiation and the abundances of the lightest elements (which match a predicted process called Big Bang nucleosynthesis; see *Further Reading*). The Big Bang model became so successful in explaining a diverse body of evidence that there are no longer any seriously competing models (Box 19.2).

Figure 19.11 *The distances between the lower set of galaxies have been expanded to produce the upper set. To see that each galaxy sees uniform recession, draw the worldline for any galaxy and think in that frame. Observers in that galaxy consider themselves stationary and see other galaxies receding, with more distant galaxies receding more quickly (Hubble's Law). Thus, there is no particular center of expansion. (Ignore the cues from the edge because in reality there is no known edge to the galaxy distribution.) This exercise is more visually striking if you cut out the upper set and slide it horizontally to make "your" galaxy appear stationary.*

Think about it

The "Big Bang" name was initially pejorative, coined by adherents of a model called steady state, who were skeptical that the cosmos could have had a definite beginning. "Big Bang" lost its pejorative sense when steady state was contradicted by observations that the early universe was very hot.

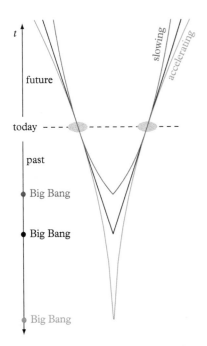

Figure 19.12 *Ellipses represent galaxies moving apart since the Big Bang, either at a constant rate (black worldlines), at a rate that slows over time (blue worldlines), or a rate that accelerates over time (red worldlines). The higher the average density of mass in the universe, the more slowing we expect and thus the younger the universe. The higher the density of vacuum energy, the more acceleration we expect and thus the older the universe.*

Astronomers have worked diligently to measure the quantities that affect the details of the model. Chief among these is the current expansion rate; all else equal, a faster expansion rate implies a younger universe, as galaxies would need less time to reach their current separations. Next, how does the expansion rate change with time? Figure 19.12 shows two galaxies on worldlines where they have a particular separation and relative velocity today. Three models extrapolate from this into the past and future. The black model has the galaxies coasting at constant velocity since the Big Bang (where the worldlines intersect). The blue model slows over time, which means past expansion was faster and required less time. Not shown in the figure is the cause of slowing: mass and gravity. We predict slowing in proportion to the energy (mass) density of the cosmos, due to gravity. The red model accelerates, which means slow past expansion and an older universe, and could only be due to a repulsive effect like vacuum energy.

To find which model represents our universe we could watch a galaxy for millions of years to see how its speed changes. But we are impatient, so we use the fact that wavelengths of light expand along with space. If we receive a photon (call it P) that has tripled its wavelength since the time it was emitted (as measured by the "bar code" discussed in Section 9.3), we know that intergalactic space tripled since P was emitted. Look again at Figure 19.12 and find when each model had one third the current intergalactic separation—you will find that the slowing model puts this milestone relatively recently, while the accelerating model puts it about twice as far in the past. The models thus make very different predictions about the time P spent in transit, and therefore about the *distance* to P's source. We identify the correct model by measuring the distance to that source. This is a challenge that took astronomers much of the twentieth century to master, but it is now proven beyond reasonable doubt that an accelerating model best fits our universe. Thus, vacuum energy does appear to be nonzero, but just enough to have a noticeable effect over billions of light-years.

This discovery has stimulated theoretical physicists to explore whether other physical mechanisms may mimic vacuum energy in causing acceleration (Box 19.3). Nevertheless, any alternative explanation would have to behave much like vacuum energy. To illustrate this, the red model in Figure 19.13 shows what happens in a universe like ours, with both vacuum energy *and* mass. We see slowing early on, as vacuum energy was less important then (there was less space between galaxies) and normal attractive gravity was more important (mass density was higher, with the same stuff squeezed into less space). At later times, the mass density is dilute, so vacuum energy and acceleration take over. Detailed observations of the expansion history indeed confirm an early era of *de*celeration followed by a later era of acceleration.

Note that the red model in Figure 19.13 happens to have about the same age as the black coasting model. This illustrates why acceleration had actually been predicted by some cosmologists. The ages of the oldest stars had been known independently to be about 13 billion years old, matching the coasting model. But the cosmos *cannot* be coasting because we know it has mass and gravity, which cause slowing. And if current expansion is slower than past expansion, our model is too young to accommodate the known ages of the oldest stars. The age problem is solved if at some point there is acceleration *in addition to* the slowing that we know must be there. Cosmologists who made this argument were vindicated when direct measurements of the expansion history confirmed the acceleration—which dominates only at late times, as one would expect with vacuum energy. While vacuum energy may not be the final word on the cause of the acceleration, it has proven to be another of Einstein's enduring legacies.

This section has compressed many ideas into a brief summary, so let us return to the big picture. Our cosmos appears to have a uniform density with no edge, which makes it unsuitable for Newtonian thinking tools. General relativity thus enabled, for the first time, quantitative modeling of the cosmos as a whole. A metric model with just a few ingredients is consistent with a vast array of data, but research continues on understanding the exact physical origin of some of the ingredients. Much of this chapter has emphasized the dynamic nature of spacetime in general relativity, and the Big Bang model is perhaps the simplest example of that—it evolves in time but is the same everywhere in space.

Check your understanding. Other galaxies recede from us, so it looks like we are at the center of expansion. Extrapolate Figure 19.11 back in time to explain how this view is inaccurate, and develop an accurate response to the question "Where did the Big Bang happen?"

Figure 19.13 *Simplified version of Figure 19.12 comparing the coasting model to a deceleration-plus-acceleration (mass-plus-vacuum energy) model.*

Think about it

Currently, the most precise Big Bang model has an age of 13.80 ± 0.02 billion years.

Box 19.1 Limits of the Big Bang model

The Big Bang metric model really deals with the expansion after $t = 0$ rather than what happened *at* $t = 0$—or before, whatever that may mean. The fossil evidence goes all the way back to the first second, and well-known laws of physics allow us to extrapolate back to the first microsecond or nanosecond, but we cannot yet rigorously extrapolate all the way back to $t = 0$. This is because at $t \approx 10^{-43}$ s quantum effects become important, and (Section 19.1) it is difficult to simultaneously deal with quantum and gravitational effects. This difficulty does not mean that we can make no models of the earliest moments and the ultimate origin of the universe—it means that a variety of models compete with each other to extend the basic model in different ways.

Another limit worth reiterating is that cosmological models deal only with the largest scales; they make no attempt to model specific galaxies, much less the internal workings of galaxies or their constituent solar systems. A good analogy is that we can model the overall shape of the Earth and its causes (spherical due to gravity, plus an equatorial bulge due to rotation) without accounting for individual mountains. The global model may provide useful context for studies of mountains (or galaxies) and vice versa, but "local vs. global" remains a remarkably useful distinction.

Box 19.2 Alternatives to the Big Bang model

Physicists in the twentieth century were very creative in building alternative models to compete with the Big Bang metric model in explaining the apparent recession of galaxies. Only the Big Bang survived further experimental tests, but the creation and testing of these alternative models illustrate how science works. Here we briefly examine one such model, called tired light, and one experiment that tests this model alongside the metric model.

The tired light model postulated that light loses energy (thus losing frequency and appearing redder while still traveling at c) just from traveling enormous distances through space. In the Big Bang model, the stretching of the light wave comes instead from the expansion of space itself while the light is in transit. Therefore, the distance between two flashes of light emitted—say, one second apart—will stretch to more than one light-second as the flashes travel to a distant galaxy. This is similar to a Doppler effect, but is caused by metric expansion rather than actual motion. A key prediction of the Big Bang model, then, is that *all* time-dependent processes in the distant galaxy will be seen in slow motion, while tired light predicts that only the wavelength of light is affected, as illustrated below.

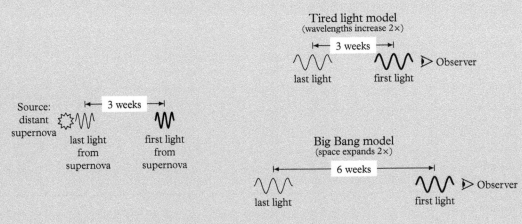

What do observations tell us? Supernova explosions that take three weeks in nearby galaxies are indeed observed to take much longer in distant galaxies, thus ruling out the tired light model.

Box 19.3 What causes cosmic acceleration?

The main text focuses on vacuum energy for its simplicity, but experts have been creative in asking what else could drive acceleration. One option—modified gravity—is that general relativity is incorrect on the largest scales. Alternative models involving extra dimensions, for example, can accelerate without any need for vacuum energy. However, so far such models do not match the data as well as general relativity with vacuum energy.

Another model, called dark energy, extends the idea of vacuum energy by allowing its density to dilute somewhat as the cosmos expands. Vacuum energy can thus be considered the simplest example of the class of dark energy models. Do not confuse dark energy with dark matter! Think of dark energy as shorthand for "the unidentified cause of

Box 19.3 *continued*

the acceleration of the cosmic expansion," whereas dark matter is far more extensively characterized (Section 17.5). Furthermore, while modified gravity definitely cannot explain all the observations that point to the existence of dark matter, modified gravity may yet explain cosmic acceleration.

CHAPTER SUMMARY

- General relativity is a comprehensive framework for predicting how mass and energy curve spacetime. The following four outcomes result from applying that framework in different contexts.
- Gravitomagnetism—velocity-dependent gravitational acceleration caused by the motion of a source mass—is a key prediction of general relativity and has been confirmed.
- Source masses that accelerate produce gravitational waves. This prediction was long ago confirmed indirectly with binary pulsars. In a technical tour de force by the LIGO group and a triumph for the metric model, these waves were finally seen directly in 2015.
- Gravitational lensing—the deflection of light in a wide array of astrophysical contexts—is a widely used tool for revealing the distribution of mass in the universe.
- General relativity provides the first self-consistent physical framework for cosmology, and implies that the cosmos itself must be dynamic. This leads directly to the Big Bang model of the expanding universe. The physical origin of vacuum energy is not yet understood, but models including it are consistent with a vast array of cosmological data.

FURTHER READING

The Confrontation between General Relativity and Experiment by Clifford M. Will, part of the *Living Reviews in Relativity* series, is a thorough and current review of experimental tests of general relativity and is available at http://relativity.livingreviews.org/Articles/lrr-2014-4/. The discussion is aimed at experts, but the figures wonderfully illustrate the precision of the evidence. The book *Was Einstein Right?* by the same author is an excellent treatment of the same topic for general audiences, and also explains in more detail the motivation behind one alternative metric theory of gravity.

A brief (six page) and readable description of the main contenders for a theory of quantum gravity can be found in the paper *Quantum Gravity for Dummies* (no relation to the book series), available at https://arxiv.org/abs/1402.2757.

Einstein's Unfinished Symphony by Marcia Bartusiak is a readable history of physicists' efforts—and false starts—to detect gravitational waves on Earth. Published in 2000, this book conveys a good sense of a decades-long effort that had yet to pay off. The LIGO detection in 2015 will make you appreciate this even more.

Among many other educational resources, the LIGO website allows you to "listen" to gravitational waves at https://www.ligo.caltech.edu/video/ligo20160211v2.

For a good selection of stunning images of gravitational lensing, search Astronomy Picture of the Day (http://apod.nasa.gov) for "gravitational lens."

You can visualize that the Big Bang has no center with an interactive tool that far surpasses Figure 19.11, available at http://bit.ly/1qBwzpL.

Big Bang by Simon Singh is an excellent history of cosmology, exploring the ideas as well as the characters. This is a good place to start for a readable account of evidence briefly cited here, such as the cosmic microwave background and Big Bang nucleosynthesis.

CHECK YOUR UNDERSTANDING: EXPLANATIONS

19.1   The answer is more nuanced than a simple yes or no. Equivalence principle tests point to the need for a metric model of gravity. Because general relativity is the leading metric theory, equivalence principle tests definitely support general relativity, but they do not necessarily support general relativity over other metric models. (Physicists do strongly prefer general relativity over other metric models though, because it is the simplest such model; we prefer models that are no more complicated than necessary.)

19.2   (a) Because Venus spins much more slowly, gyroscopes orbiting there will have much smaller gravitomagnetic precession. (b) The geodetic effect will be about the same for both planets because this effect is based solely on mass.

19.3   Planets have very small accelerations because they slowly swing around in very wide orbits. (Reminder from Section 16.1: the magnitude of the acceleration in circular motion is $\frac{v^2}{r}$.) The gravitational waves that result from such small accelerations are incredibly weak, and drain energy from the orbit at a negligible rate.

19.4   (a) All forms of matter (and energy). (b) The lens galaxy.

19.5   The Big Bang happened everywhere. Imagine a microscopic sheet of graph paper expanding to everyday size—*all* coordinates on that graph paper were present from the beginning. In terms of Figure 19.11: if we extrapolate sufficiently far back in time all worldlines intersect at one point, so all galaxies were present at the $t = 0$ event.

❓ EXERCISES

19.1 Summarize general relativity in one sentence.

19.2 Why is combining general relativity with quantum mechanics so difficult?

19.3 Why do physicists continue testing general relativity even after it has been so well confirmed?

19.4 *Mythbusters* visited a telescope firing a laser at a retroreflector on the Moon and measuring the round-trip travel time. A crew member bumped the telescope slightly and the return of the laser was no longer seen. How does this refute the myth that NASA never actually went to the Moon?

19.5 Would you feel a gravitational wave passing through your body?

19.6 How is gravitational lensing like an everyday lens, and how is it different?

19.7 Explain how dark energy is different from dark matter.

19.8 Draw a model on Figure 19.12 that slows enough to recollapse. What can we infer about the matter and vacuum energy content of that model?

➕ PROBLEMS

19.1 Section 19.2 argues that a mixed term like $(\Delta x)(\Delta t)$ in the metric means that the proper time of a test particle depends on its direction of motion. Assume that motion parallel to the mass current contributes positively to proper time, and motion in the opposite direction contributes negatively. Use this idea to create a new type of twin paradox story. To help Alice and Bob meet at the end, consider the idea behind Figure 19.1.

19.2 Could a metric with mixed terms allow a person to travel through the time coordinate in a direction of his choosing, and perhaps meet an earlier version of himself? Do an Internet search on "closed time-like curves" and report what experts have found regarding this question. Keep in mind that not all mathematically possible scenarios are physically possible.

19.3 Does the speed of a wave depend on the force or speed of the initial disturbance that causes the wave? (You may find it helpful to think about making waves in a swimming pool.) What *does* the speed of a wave depend on? Relate this to the fact that gravitational waves travel at c.

19.4 Research the story behind the discovery of Supernova Refsdal, the supernova shown in Fig-

ure 19.10. Who is Refsdal and what was his contribution?

19.5 Strong lensing—giant arcs and/or multiple images—happens only along the densest lines of sight in the universe. *(a)* Explain why the density *along a line of sight* should be measured in units of kg/m^2 or g/cm^2. *(b)* Do an Internet search to find published estimates of these densities. You may find the answer surprising.

19.6 Research and explain the fossil evidence that the early universe was hot. How does this confirm the Big Bang model?

19.7 Do an Internet search for the latest results on the cosmic expansion history using supernovae. At what confidence do they confirm the acceleration? Are the supernovae distant enough to clearly reveal the era of deceleration before the current era of acceleration?

19.8 It seems like a coincidence that our cosmos has a mixture of vacuum energy and mass that gives roughly the same age as a coasting model. Another way to state this is that the acceleration picked up only rather recently; acceleration would have been negligible for observers billions of years in the past, and will be blindingly obvious to observers billions of years in the future. Research what cosmologists think about this coincidence, and what models they are exploring as a result.

19.9   Do religion and science conflict? You may have heard that the Bible supports a cosmic age of about 6000 years (the "Ussher chronology"), which certainly seems to conflict with the Big Bang's 13.8 billion years. Test your preconceptions by researching the story of how Pope Pius XII endorsed the Big Bang in the 1950s, and how some atheists of that time created the steady state model in opposition to the Big Bang. What were the motivations of each?

How, if at all, does this challenge your preconceptions of relationships between science and religion?

19.10   Research an alternative to the Big Bang model—steady state, tired light, or something else—and describe specific predictions of the alternative model that have been contradicted by data. Also describe how and why the Big Bang model made the correct prediction in each case.

19.11   Is energy conserved in general relativity? Research what experts say about this. You may find some disagreement on the fine points, but focus on describing the basic problem.

Black Holes

Black holes seem like something out of science fiction, but they are real. The popular picture of black holes does, however, include some myths. Now that we understand the relevant properties of gravity, we can separate fact from fiction.

20.1 What is a black hole?

We examine first the Newtonian model of a "dark star": an object with escape velocity (from its surface) greater than the speed of light. Following the discussion of Section 16.6, we know that such an object must be compact in the sense of having a large $\frac{M}{r_{\text{surf}}}$ ratio. For concreteness, let us imagine an object with a fixed mass and ask how small the radius must be. We do this by setting the escape velocity (Box 16.2) equal to c:

$$c = \sqrt{\frac{2GM}{r}} \qquad (20.1)$$

$$c^2 = \frac{2GM}{r}$$

$$r = \frac{2GM}{c^2} \qquad (20.2)$$

Use caution here: *all* the mass M must be contained within the radius r, as in the top panel of Figure 20.1, for this equation to apply. For example, if we use the mass of the Sun for M we find $r = 3$ km; this means that *if* a solar-mass star were compacted into a radius of 3 km, light leaving the surface would be starting so far down the potential that it could never get all the way out. As early as 1783, John Michell (1724–93) recognized that if such astonishingly compact objects—packing the mass of the Sun into a radius 250,000 times smaller—exist, we would not see them because light would not escape. In the Newtonian conception, $r = \frac{2GM}{c^2}$ is a boundary defining what can be observed by *distant* observers, but does not necessarily have implications for local observers or for the structure of the star itself. In general relativity, $\frac{2GM}{c^2}$, called the **Schwarzschild radius** (r_S), still defines a region from which light cannot escape—but the conceptual differences from the Newtonian "dark star" are so great that a new term was coined: **black hole**.

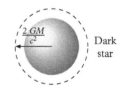

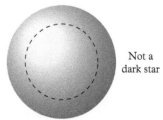

Figure 20.1 Top: *Newtonian "dark star." The dashed sphere indicates the radius from which light cannot escape to infinity. Because the star's surface lies within this radius, distant observers cannot see it.* Bottom: *if the same mass is more spread out, the star can be seen.*

The Elements of Relativity. David M. Wittman, Oxford University Press (2018).
© David M. Wittman 2018. DOI 10.1093/oso/9780199658633.001.0001

We begin with one feature of the Newtonian model that will roughly carry over to black holes. The region from which light cannot escape is defined based on the *depth* of the potential rather than its slope (acceleration). You can see this in Equation 20.1 by noting that the right-hand side of that equation is proportional to $\sqrt{\frac{GM}{r}}$, which is the depth of the potential at r. The definition of r_S has nothing to do with acceleration and everything to do with where the potential reaches a specific depth. Thus, it is conceptually possible to build a black hole without enormous acceleration at the point of no return.

Unlike a black hole, the Newtonian "dark star" does allow for escapes. Recall that escape velocity is the initial speed required to escape for *inertial* particles, which can climb the potential only by converting their kinetic energy; if they run out of kinetic energy before reaching the top of the potential they fall back in. Such calculations do not apply to an object that can fire an engine or climb a rope upward, so these objects *can* escape. Inertial objects (or light) launched with $v < v_{esc}$ could coast *very* far out before running out of kinetic energy and falling back in. Furthermore, escape could be aided by orbiting space stations that help rockets refuel, or retransmit faltering radio and television signals from the surface. In all these ways, both objects and information could leave the "dark star."

The modern understanding in the context of general relativity differs greatly. The definition of the Schwarzschild radius comes from the Schwarzschild metric (Equation 18.3, repeated here):

$$c^2(\Delta\tau)^2 = c^2\left(1 - \frac{2GM}{c^2 r}\right)(\Delta t)^2 - \left(1 - \frac{2GM}{c^2 r}\right)^{-1}(\Delta r)^2 \tag{20.3}$$

Let us recap what this means (see Section 18.4 for the full discussion). The coordinate time t is kept by stationary clocks at such a large r that Equation 20.3 reduces to the special relativity metric. Now consider the time τ on a stationary clock at some smaller r. Stationary objects by definition have $\Delta r = 0$ because they never change their position, so Equation 20.3 reduces to $\Delta\tau = \sqrt{1 - \frac{2GM}{c^2 r}}\,\Delta t$. This means that for a given amount of coordinate time, Δt, the elapsed time on the clock at r, $\Delta\tau$, is smaller by the factor $\sqrt{1 - \frac{2GM}{c^2 r}}$, which you can visualize in Figure 20.2. Light emitted by the clock at r will be received by a distant observer with its frequency and energy reduced by the same factor (gravitational redshift; Section 13.4). The distant observer sees events at a smaller r play out in slow time. For clocks at the Schwarzschild radius r_S, we plug $r_S = \frac{2GM}{c^2}$ into Equation 20.3 and find that they tick *zero* times per tick of coordinate time (see Figure 20.2 where the curve goes to zero). Light emitted at r_S thus delivers zero energy to an outside observer, rendering this region unobservable.

It is sometimes stated that "time stops at r_S" because the reasoning here gives the impression that a space probe could stay at r_S and observe eternity unfold around it with no ticks elapsing on its own clock. However, no probe can actually stay at r_S. This is because only light can follow worldlines with $\Delta\tau = 0$; these worldlines are unattainable by particles with mass. Thus, "freezing time"

Think about it

The black hole concept does not automatically imply high mass, only that $r < \frac{2GM}{c^2}$. That said, black holes in nature do tend to have masses similar to or greater than those of massive stars; see Section 20.3.

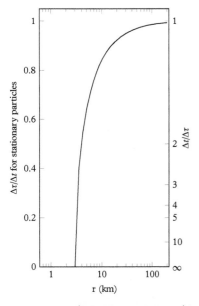

Figure 20.2 $\frac{\Delta\tau}{\Delta t}$ *(left axis) and* $\frac{\Delta t}{\Delta\tau}$ *(right axis) near a one-solar-mass black hole; this echoes Figure 18.9 but follows the curve down to* $\frac{\Delta\tau}{\Delta t} = 0$. *The zero means that light emitted from this region delivers zero energy to outside observers; it is unobservable.*

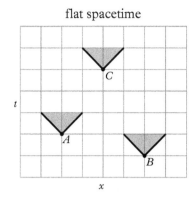

flat spacetime

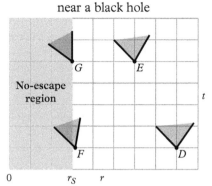

near a black hole

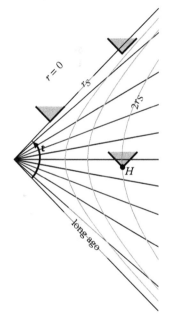

Figure 20.3 *Light cones indicating the local direction of "future" are attached to various events in gravity-free (left) and Schwarschild (right) spacetimes. Gravity bends spacetime so the cones tilt toward the black hole. The Schwarzschild radius r_S defines the tipping point where the future begins to point more toward* r = 0 *than toward a larger* t. *Due to curvature the coordinate grid should not really be drawn as square; Figure 20.4 shows an alternate view that warps the grid rather than the light cones.*

by placing a stationary clock at r_S is just as impossible as "freezing time" by moving a clock at speed c. Our probe is necessarily infalling at r_S. The probe's view of a mother ship that stays stationary outside r_S is then affected not only by gravitational blueshift but also by motion-related redshift. The net result is that the probe does *not* see ship time unfolding infinitely rapidly, as might have been predicted from the clock tick rates considered in the previous paragraph. (Section 20.2 considers in turn how the ship sees the probe.)

The fact that *light* can stay at r_S leads to another conclusion. On a spacetime diagram, worldlines of massive particles tilt less than worldlines of light rays; this defines the future light cone (Section 6.4) of each event. In the absence of gravity, spacetime is equally flat everywhere so all events have identically oriented light cones as in the left panel of Figure 20.3. But here, gravity is bending spacetime so that one edge of the light cone *remains* at r_S. This yields the picture in the right panel of Figure 20.3; the bending of the light cone must decrease with distance from the black hole and eventually return to normal sufficiently far away. At r_S the bending is just enough to align one wall of the light cone with the $r = r_S$ grid line. Thus, a particle at $r \leq r_S$ can never escape—and a particle with mass can only proceed to a smaller r, regardless of its previous trajectory or the power of its engines. Furthermore, no event at or inside $r = r_S$ can send a signal to any event at a larger r. The Schwarzschild radius thus defines a one-way boundary called an **event horizon**.

Figure 20.3 is misleading because the grid is drawn as square everywhere. The Schwarzschild t and r coordinates form a regular grid far from the black hole but curve through each other more and more as we near r_S. Figure 20.4 illustrates how we might draw this, with space and time grid lines perpendicular at one time but curved in a way that puts r_S in the future. The grid curvature has been chosen to respect special relativity locally everywhere: inertial worldlines are straight and light worldlines tilt by $\pm 45°$. A particle released from rest at H will drift straight upward through time and inevitably cross $r = r_S$ (with $r = 0$ soon after that); to

Figure 20.4 *In the presence of gravity, drawing light rays at $\pm 45°$ and inertial worldlines as straight requires warping the coordinate grid. A particle released from rest at H will proceed straight upward, initially through time but increasingly through space until it crosses r_S—from which even light rays are unable to escape because r_S itself is a 45° line.*

maintain fixed r a particle must accelerate to the right. The grid becomes so curved as it approaches r_S that events there (and at smaller r) cannot really be assigned a meaningful t coordinate—a phenomenon completely obscured by the regular grid in Figure 20.3. Thus, there is no meaningful answer to the question "at what time t did the probe hit $r = 0$?" Section 20.2 explains how this problem stems from the definition of Schwarzschild coordinates, rather than a rift in spacetime at r_S. An infalling clock just keeps ticking at one second per second in its local frame, with no glitch as it passes through the event horizon.

What does our probe encounter inside the event horizon? If the mass inside has a surface as in the Newtonian dark star model, we must suppose there is a force supporting the surface against the crush of gravity—but no known force is capable of this. Any putative structure (an iron core, say) holding up the surface must itself be proceeding toward smaller r according to the dictates of spacetime. Our probe therefore reaches $r = 0$ along with everything else that ever fell in, including the original source mass. If all the mass is really at $r = 0$, gravity becomes infinitely strong there, a situation called a **singularity**. General relativity by itself leaves no alternative to this conclusion, but quantum effects may also come into play so no one knows quite what form this central mass takes.

In summary, the spacetime around an arbitrarily compact mass is much more interesting than a Newtonian model would suggest. Newtonian thinking implies only that below a certain potential depth, light cannot climb out all the way to a very distant observer. The metric model predicts curvature is so strong that inside a certain radius, all paths forward in time point toward $r = 0$. As a result, no particle or signal inside that radius can ever get out, not even a little bit.

Check your understanding. Consider Figure 20.3. *(a)* Draw the light cone attached to an event between G and E. *(b)* Draw a light cone attached to an event (call it H) at $r = r_S$ but at a later time than event G. Can an observer travel from event G to event H?

20.2 A closer look at the horizon*

A general feature of spacetime metrics is that the sign of the $(\Delta t)^2$ term differs from that of all the spatial terms. We discussed at length in Chapter 11 how this distinguishes time from space. Consider again the Schwarzschild metric

$$c^2(\Delta\tau)^2 = c^2 \left(1 - \frac{2GM}{c^2 r}\right)(\Delta t)^2 - \left(1 - \frac{2GM}{c^2 r}\right)^{-1}(\Delta r)^2 \qquad (20.4)$$

and imagine first that we are far from a black hole, so the coefficients in parentheses are approximately one. As usual for the convention followed in this book, the $(\Delta t)^2$ term is positive and the $(\Delta r)^2$ term is negative (along with the other spatial terms not written here). As we approach the horizon the time coefficient becomes smaller, and it plummets to zero exactly at the horizon (Figure 20.2). Inside the horizon the time coefficient actually becomes negative. You can see this with algebra by noting that inside the horizon the term $\frac{2GM}{c^2 r}$ is greater than one. Therefore, the time coefficent, $\left(1 - \frac{2GM}{c^2 r}\right)$, becomes *less* than zero, and the space coefficient, $-\left(1 - \frac{2GM}{c^2 r}\right)^{-1}$, becomes *greater* than zero. In other words, the r and t coordinates have switched roles: r displacements now contribute positively to proper time and t displacements contribute negatively! Outside the horizon, any advance in proper time τ *required* a displacement in the t coordinate because t was the only coordinate with a positive coefficient. For the same reason, *inside* the horizon any advance in proper time τ requires a displacement in the r coordinate. The required displacement turns out to be toward a *smaller r*. In other words, *the future leads to $r = 0$ inside the horizon*. This is the natural extension of the light cone trend in Figure 20.3: light cones fully in the no-escape region should point *mostly to the left*.

This is mind-blowing at first—how can r become a time coordinate and t become a space coordinate? We must remember that merely naming a coordinate "r" or "t" does not determine its physical meaning or how it behaves in an equation—that can be determined only from the equations themselves. We assigned the names r and t because *far from the central mass* they behave like the space and time coordinates in special relativity. But the equations show that they do *not* behave like this inside the horizon. Do time and space completely swap meaning inside the horizon? No—we can construct alternative coordinate systems that exhibit no such behavior when crossing the horizon, indicating that the underlying geometry is smooth. It is the Schwarzschild *coordinate system*—the particular way of defining r and t—that has a problem at the horizon. Nevertheless, this coordinate system is still most often used when, say, modeling the solar system because it is most convenient for regions *far* from the horizon. Latitude-longitude coordinates on Earth provide a good analogy: if you fly over the North Pole your longitude instantaneously changes by 180°, but this is a quirk of the coordinate system at that location rather than physically real teleportation. Despite this quirk, latitude-longitude coordinates continue to be used because they have very convenient properties in most places on Earth. Regardless of coordinate choices, *the underlying physical behavior is the same*; it is simply a question of how hard we must work to interpret the coordinates. See *Further Reading* for more about alternative coordinate systems; here, we focus on learning what we can using Schwarzschild coordinates.

You may have noticed that the r coefficient in the Schwarzschild metric becomes infinite at the horizon because $\left(1 - \frac{2GM}{c^2 r}\right)$ becomes zero there. Commonly held misconceptions related to this are that gravity becomes infinitely strong there, that crossing the horizon takes infinitely long, and that distant observers forever see an infalling particle apparently frozen at the horizon because time stands still there. None of these is true. The metric tells us the ratio of proper displacements (using local clocks and meter sticks) to displacements in the global coordinates t and r. We must add many small displacements to find the total distance, and integral calculus gives us the tools to do that. Figure 20.5 shows

> **Think about it**
> _____
>
> The sign of the Δr displacement required to move forward in proper time cannot be read directly from the metric, because Δr is squared in the metric. Mathematically, the conclusion that motion is toward a smaller r comes from plugging the metric into the geodesic equation. Conceptually, it is easy to generalize from freely falling particles, which clearly move to a smaller r.

> **Think about it**
> _____
>
> Examples of physical behavior around a black hole that cannot depend on coordinate system choice: nothing can escape from the horizon; particles inside the horizon must proceed to the very center; and there is infinitely strong gravity at the very center.

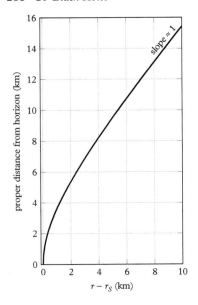

Figure 20.5 *Relationship between proper (meter-stick) distance from the horizon and r-coordinate distance from the horizon of a solar-mass black hole. The slope of this curve is obtained from the metric coefficient on r and is infinitely steep at the horizon, but yields a finite total distance. The slope of one at large distances indicates nearly flat spacetime: proper displacements are nearly the same as coordinate displacements.*

a curve with slope equal to the $\frac{\Delta r_m}{\Delta r}$ given by the Schwarzschild metric for a solar-mass black hole. This slope *is* infinitely steep at $r = r_S$, but the *height* of the curve—the cumulative proper distance "built up" by this slope—remains finite. (Think of this as the region of infinite slope being so small that the total effect is finite.) In mathematical jargon, the metric coefficient on r is *integrable*.

A similar process shows that the total proper time experienced by an infalling observer, from just outside the horizon to $r = 0$, is in the range of mere microseconds for solar-mass black holes. Coordinate time—time as measured by distant observers—is a bit more tricky. One cannot compute a coordinate time to $r = 0$, for the same reason we did not compute a proper distance to $r = 0$: the coordinates do not *mean* the same thing on both sides of the horizon. Just as we cannot have stationary meter sticks to measure r_m inside the horizon, neither can we have stationary clocks to measure time there. It is often thought that a distant observer sees an infalling particle "frozen in time" at the horizon, but this is an error caused by imagining a (physically impossible) *stationary* particle at the horizon. If we imagine instead the taillights emitted by an *infalling* particle, the energy received by a distant observer decreases dramatically with time due to both gravitational redshift (the slowness of time at the point of emission) *and* the increasing infall speed. For all practical purposes the taillights are gone within milliseconds for a solar-mass black hole. This is a thousand times longer than the elapsed time measured by the infalling particle, so it *is* like super-slow-motion video—but one that is over in milliseconds rather than frozen forever.

What if you follow a course that takes you *near* the horizon without crossing it? When you return to your base you will find that much less time has passed for you than for others who remained at the base. This is an efficient way to travel into the future, because such a course need not require much in the way of engines—on a highly elliptical orbit you can coast alternately close to and far from the slow-time region. But plan wisely—there is no going back in time if you happen to travel farther than planned into the future. And be *very* careful not to accidentally cross the horizon.

Check your understanding. Use Figure 20.5 to answer these questions. *(a)* To move from an r coordinate infinitesimally above the horizon to an r coordinate 2 km higher, how far must you move in meter-stick distance? *(b)* To move from an r coordinate 6 km above the horizon to an r coordinate 2 km higher, how far must you move in meter-stick distance?

20.3 Black holes in nature

Nature can form black holes in at least two ways. **Stellar-mass black holes** form when massive stars collapse under their own gravity, and **supermassive black holes** are found at the center of nearly every galaxy, where they formed so long ago that astronomers are still working out the details of how they formed.

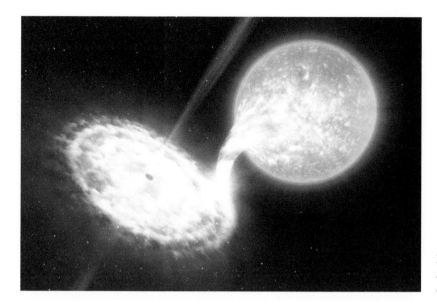

Figure 20.6 *Artist's conception of a stellar-mass black hole accreting gas from a companion. Credit: ESO/L. Calçada.*

In either case, astronomers prove that something is a black hole by first measuring its mass M and then proving that its radius is no larger than $\frac{2GM}{c^2}$. For this, they need to observe material orbiting the black hole, so stellar-mass black holes show themselves only if they reside in a binary star system. The orbit of the companion star around the black hole determines the mass of the black hole, but that is the easy part. It is more difficult to prove that an event horizon exists, because we cannot get close enough to take a clear picture. We can, however, develop strong evidence that a horizon exists if we find gas orbiting close to the horizon. This can happen if the companion star is in a phase where it becomes very puffy; gas in its outer layers then falls off and swirls in toward the black hole. This swirling gas is called an **accretion disk**. Light from the accretion disk is really what allows astronomers to test whether the companion is a black hole, so the dozens of stellar-mass black holes identified in our galactic neighborhood are a small subset of the total: only those with companion stars puffy enough to donate gas. Figure 20.6 shows a model of such a system.

In each case, the entire system—black hole, accretion disk, and companion star—is so far away from us that we see all the light blended into a single point. So, how can we possibly deduce a model with so many details? Despite the pointlike appearance, we can still examine the distribution of energy in the light (called its spectrum), and how that changes with time. Let us first consider how much information we can get from a normal binary star this way. Each star emits its own specific spectrum that depends mostly on its temperature—the hotter the temperature, the higher the energy of the typical photon emitted. The total amount of energy emitted depends on this as well as the surface area of the star. Thus, even if a hot and a cool star are so close that they look like a single point, the spectrum

---

Confusion alert

Accretion disks can be extremely luminous because swirling down a deep potential releases a great deal of energy (the coin funnel analogy is apt here). Thus, accretion disks *around black holes* are some of the most luminous objects in the universe, even if the black holes themselves are not luminous.

of that point still reveals the existence of two bodies with different temperatures and sizes. We can confirm this model in more detail if the orbit is oriented so that one occasionally eclipses the other from our view: any spectrum taken during the eclipse will be of the eclipsing star alone, and the speed with which the other spectrum winks out is another clue to the size of the other star. And, of course, cyclical changes in the Doppler effect tell us about the orbital speeds and thus the masses of the stars (Section 17.6). The same analysis works when one of the two "stars" is actually a compact object with an accretion disk; in fact it is easier in some respects because an accretion disk is distinctively hotter and smaller than a star. An accretion disk releases a great deal of gravitational potential energy, so it becomes far hotter than a star and emits extremely energetic (high-frequency) light called X-rays. Its small size is then apparent when the X-rays wink out very quickly during an eclipse.

A compact object that passes all these tests is close to being declared a black hole, but we must first rule out the possibility that it is a neutron star. A neutron star has a radius only a few times larger than r_S, so it too can form a similar accretion disk if a companion donates gas. A neutron star, however, has a hot surface that radiates light that affects the total observed spectrum. Furthermore, a blob of gas falling onto the neutron star surface emits a burst of X-rays, in sharp contrast to gas "silently" crossing an event horizon. Models of neutron stars show that they cannot be more massive than about three solar masses without collapsing. Therefore, compact objects more massive than this should not exhibit X-ray bursts—and in fact they do not, marking a substantial success for these models. Compact objects above three solar masses should thus be considered black holes by default, but the highest standard of proof is still a detailed analysis of the spectrum.

A subtler factor affecting the spectrum is gravitational redshift. The outer edge of the accretion disk is too far from the center for its light to be strongly redshifted, but the *inner* edge of the disk is more strongly affected by slow time—and more so if the inner edge approaches a black hole rather than a neutron star. Therefore, any given accretion disk shows a *range* of gravitational redshifts, with disks around black holes ranging toward greater redshift than disks around neutron stars. In another substantial success for the black hole model, this is indeed what is observed.

Astronomers have gathered similar evidence for supermassive black holes residing at the centers of galaxies. The one at the center of our own Galaxy has been particularly well studied because astronomers can follow the orbits of individual stars very close to the Galactic center (Figure 20.7). These stars orbit something that has about 4 million times the mass of the Sun yet does not emit enough light to be seen in Figure 20.7. One star approaches within 7 billion kilometers of the center of mass—about the size of Pluto's orbit around the Sun—thus proving that the 4 million solar masses are packed into a very small volume, astronomically speaking. The low luminosity of the central area indicates that little material is accreting, but there is some accretion activity at times. Studies of this activity show that the region inside the accretion disk can be no larger

Think about it

A neutron star packs its mass into a radius not much larger than a black hole, so its gravity is quite strong and its internal pressure is immense. Partly because pressure itself is a form of energy that contributes to gravity, general relativity predicts that a neutron star attempting to support more than about three solar masses will collapse and become a black hole.

Think about it

The inner part of the disk in Figure 20.6 should perhaps be shown as redder due to gravitational redshift, but the inner disk is also hotter and therefore would be bluer in the absence of gravitational redshift. The net effect is difficult to portray in a painting, but can be quantitatively modeled and matched to observations.

than 44 million km, which is smaller than Mercury's orbit around the Sun. For comparison, only about 250,000 Suns would fit inside this volume even if they were stacked like oranges at the grocery store. This is an extremely strong case for a supermassive black hole: 4 million solar masses are packed into a small, dark volume. Now that the basic properties of this black hole have been well established, astronomers are gearing up to use this black hole to test general relativity in more detail.

Other galaxies have central black holes up to 4 *billion* solar masses—1000 times more massive than the one in our galaxy. Although these galactic centers are too distant for us to pick out individual star orbits, many of them are accreting copious amounts of gas; this makes a very bright accretion disk that allows astronomers to tease out many details. Motions of gas in the accretion disk reveal both the amount of mass in the center, and that the mass is concentrated in a very small region. Furthermore, the light from these galactic centers matches accretion disk models rather than stars, which have a very different spectrum. The conclusion that black holes lie at the heart of these objects is unavoidable, but the origin of these supermassive black holes is still an area for research. Centers of galaxies are dense places, of course, but the black holes there seemed to have formed so long ago that reconstructing the details of their formation is a challenge. (Stellar-mass black holes are easier to study in this regard, because stars continue to be born and die all around us.) This is an exciting area of research, but the questions are less about relativity than about the physical conditions and processes at the hearts of galaxies long ago.

What would a black hole look like if we *could* get much closer and see details? Look back at the artist's conception in Figure 20.6 and you can now appreciate that every detail is backed up by observations and physical laws, and that much of it applies as well to supermassive black holes accreting gas from their environment. As gas falls from far away it gains kinetic energy, which turns into heat when the gas collides with the accretion disk. The disk is thus blue-hot and glows even more brightly in ultraviolet and X-ray light. Seen from a great distance, all the light in this illustration blends into a single dot, but its spectrum still tells us that the system contains a star and a small, very hot object. (The accretion disk appears large in Figure 20.6 only because it is in the foreground.) The accretion disk is so hot and luminous that it actually heats up the near side of the companion star— this too is seen in the spectrum. The spectrum reveals no evidence of a star or other light-emitting object at the center of the disk, and gravitational redshifts confirm that some of the gas in the disk is in a region of very slow time. Orbital speeds of the companion star (or, for supermassive black holes, orbits of other nearby stars or of gas in the disk) reveal the mass of the black hole. The combined spectrum of the binary system changes as one object blocks light from the other; this confirms the spectral models of each object and reveals the size of the disk. Finally, the beam apparently emerging from the black hole is actually a jet of charged particles accelerated by the twisted and changing magnetic field in the disk. Radio telescopes see this jet, and in binary systems the jet brightness waxes and wanes as the orbit points the jet alternately toward and away from us.

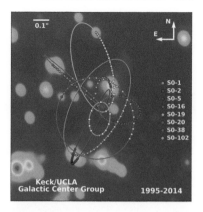

Figure 20.7 *Motion diagram of stars orbiting the Galactic center, with positions indicated yearly. Stars move very quickly near the center, indicating the presence of a very massive, compact object that does not emit much light. Image created by Prof. Andrea Ghez and her research team at UCLA from data sets obtained with the W. M. Keck Telescopes.*

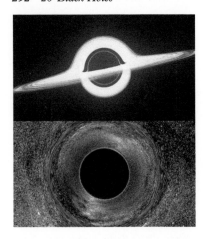

Figure 20.8 *Top: deflection of light near a black hole allows multiple perspectives on the accretion disk. Bottom: view of a hypothetical isolated stellar-mass black hole with the Milky Way galaxy serving as background. Both images were generated with raytracing code you can run on your own computer (see* Further Reading*). Image credit: Riccardo Antonelli.*

One thing missing in Figure 20.6 is the bending of light by the gravity of the black hole. Figure 20.8 shows two views of this phenomenon. The top panel shows the accretion disk alone, with no background sources of light.

We see the near part of the accretion disk nearly edge-on (like Saturn's rings) but we also see some light from the *far* side of the disk bent toward us: light from the "top" of the far side bends around the top, and light from the "bottom" of the far side bends around the bottom. The top is a little more prominent because our vantage point is slightly above the plane of the disk. In the bottom panel, we see how a black hole with *no* accretion disk distorts a background image of the Milky Way. Any actual black hole is so far from us that the dark spot (tens of km across for stellar mass black holes) is a mere speck lost in the vastness of space, so we cannot take actual pictures like this. A notable effort to overcome this challenge is the Event Horizon Telescope (EHT), which is actually a worldwide network of radio telescopes that can take very sharp pictures when used together in a technique called interferometry. The EHT may make it possible to see the "shadow" of the supermassive black hole at the center of our Galaxy. This would have been unthinkable with the technology of even a decade ago; the possibility of such a detection in this decade is a testament to the EHT team and the enduring power of black holes to motivate physicists and astronomers.

Check your understanding. If a compact object has the mass of the Sun and a photon with energy E is emitted from a spot on the accretion disk at $r = 10$ km, what energy does the photon have when it enters a telescope on Earth? *Hint:* use Figure 20.2.

20.4 Facts and myths about black holes

Myth: black holes suck. Science fiction movies and TV documentaries often give the impression that black holes suck in everything around them. Compact objects do expose (small) regions of very high acceleration. Even a relatively boring white dwarf packs about 300,000 Earth masses into the size of the Earth—and therefore has 300,000 times Earth's surface gravity. But compact objects have the same potential at large r as any other mass with the same M; they are special *only* in exposing regions of very steep potential at small r. So, if the Sun were replaced by a black hole of the same mass for example, Earth's orbit would be entirely unaffected. Spacecraft can thus orbit black holes quite easily, *unless* they approach too closely.

There are, however, a few kernels of truth in this myth. First, in the Newtonian model of gravity a highly elliptical orbit can swoop in arbitrarily close and swoop out again in a completely symmetric path. (*Orbit* here means an inertial path: no engines required.) This is *not* true in general relativity: swooping in to r_S leads irreversibly to $r = 0$. Second, even circular orbits differ at sufficiently small r. There, gravity is stronger than Newton predicted, for two reasons: the steeper-than-Newtonian decline in the time coefficient (Figure 18.9), and the fact that high-speed (i.e., small) orbits "feel" curvature of space (Section 18.5). Stronger gravity means that higher speed is required to maintain a circular orbit; inertial particles may not circle at $1.5r_S$ or less, because their speed would have to be c or greater. (Light may be found orbiting here in the so-called *photon sphere*.) A more detailed analysis shows that even between $1.5r_S$ and $3r_S$ inertial orbits are unstable: the tiniest bump from a random hydrogen atom will lead to bigger and bigger changes over time, ultimately leading either to falling in or to being ejected. This is very different from orbits in Newtonian gravity, which are stable against small bumps, and does make black holes suck more than expected based on Newtonian intuition. Nevertheless, a ship with engines could perform course corrections as needed and remain in orbit here.

These conclusions about orbits are based on the Schwarzschild metric, which does not take into account the spin of the black hole. Black holes in nature do spin, and orbits around them can be more complicated (Section 20.5). But in either case, radically non-Newtonian effects are evident only within several Schwarzschild radii, say within 10–20 km for a solar-mass black hole. No space traveller from millions of kilometers away could get this close without purposefully designing a trajectory that takes him there. Black holes do not suck on innocent travelers.

Myth: tides rip you apart at the event horizon of any black hole. Stellar-mass black holes indeed have strong tidal accelerations near their event horizons. If you go in feet first, your feet will be accelerated more than your head, so you will be stretched lengthwise—but you will also be squeezed at the sides because your shoulders must converge at $r = 0$. The stretching along the vertical direction and squeezing along the horizontal direction is so strong that it has earned the well-deserved name *spaghettification* (Figure 20.9).

Supermassive black holes may be 10^9 times more massive, but that puts the event horizon 10^9 times further from the center of mass. As a result, the acceleration at r_S is 10^9 times smaller, and the tidal field 10^{18} times smaller, than in the stellar-mass case. An astronaut will thus float through the event horizon comfortably. There is no sign or other barrier at the event horizon, but this hypothetical astronaut could nevertheless determine where the horizon is by observing the deflection of light and the gravitational *blue*shift of light from distant stars.

Myth: black holes must be massive, and their precursors must be dense. All you need to make a black hole is to squeeze an object down so that its radius is less than $\frac{2GM}{c^2}$. Any amount of mass will do, as long as the radius is appropriately small. Earth, for example, would have to be squeezed down into several millimeters and the Sun into a few kilometers.

Black holes in nature, though, *are* massive because the nature's squeezing mechanism is gravity itself. A gas cloud initially distributed throughout a very large volume of space may have enough self-gravity to collapse down further, at which

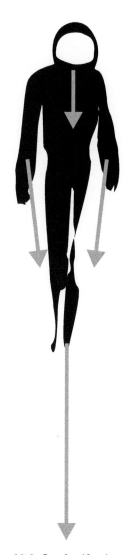

Figure 20.9 *Spaghettification: arrows show the accelerations of different parts of the astronaut. As a result, feet and head are stretched apart and opposite arms are squeezed together. The astronaut as a whole is freely falling, so she feels only these relative (tidal) accelerations. Tidal accelerations* always *have this pattern, but it is much more pronounced at a small* r.

point it will have even more self-gravity, so it will collapse down even further, and so on. This process of gravitational collapse is how stars form, starting from a size of many trillions of r_S. Most stars, for most of their lives, generate internal forces that resist further gravitational collapse. For the most massive stars these forces finally fail, and a stellar-mass black hole results.

Supermassive black holes must also have formed through gravitational collapse, but they did *not* need a particularly high density to form. Imagine two gas clouds about to collapse into black holes: Alice has a mass equal to that of the Sun and Bob has a million times that. Alice's black hole will have $r_S = 3$ km, so Bob's will have $r_S = 3$ million km. Bob is 1 million times larger in each spatial dimension, so he is 10^{18} times larger in volume. Although Bob has more mass, it is spread over such a large volume that his density is much lower. In that sense it is easier to form a supermassive black hole—but astronomers still do not know how much of a typical supermasive black hole's mass was present at its initial formation and how much was accreted later.

There may also be *micro* black holes, with roughly the mass of an atom, formed through quantum processes. There is no proof that such black holes exist, but they could conceivably be made in sufficiently energetic collisions of subatomic particles. This is not considered a likely outcome at the Large Hadron Collider, but the prospect is exciting because it could tell us something about the connection between gravity and quantum forces. There is no need to worry that Earth will be swallowed by a black hole formed in such a collision. Every day, energetic particles from space—far more energetic than humans can possibly make—bombard the top of Earth's atmosphere, and Earth is still here. *If* micro black holes are made in such collisions, they "evaporate" quicky (see below). It is also possible that micro or mini black holes were created in the Big Bang, but so far searches have not revealed any such primordial black holes.

Fact: black holes evaporate. The physicist Stephen Hawking (1942–2018) showed that, astonishingly, black holes leak energy (mostly in the form of photons) through a quantum process now called Hawking radiation. Given enough time without additional accretion, a black hole will leak all its energy and "evaporate." This is one of the fascinating processes at the juncture of gravity and quantum mechanics, but there are a few caveats in terms of our ability to ever observe it.

First, more massive black holes leak more slowly, and even stellar-mass black holes have expected lifetimes of order 10^{67} years. This is unimaginably long even compared to the current age of the universe, which is about 10^{10} years. If we counted all the grains of sand on Earth while waiting 10^{10} years between grains, we still would not make a dent in the lifetime of one of these black holes. Practically speaking, this means that ordinary black holes leak photons so slowly that there is no hope of detecting such photons directly. On the other hand, evaporation gives us a way to look for micro black holes—either in the wild or created in particle accelerators—as follows. Because lower-mass black holes leak more rapidly, they rapidly become even *lower* mass and leak even *more* rapidly. This leads to a burst of photons at the end as the final bit of mass is converted into photons quickly. (No

Think about it

Because stars are well-understood, the existence of stellar-mass black holes—the final stage in the life of a massive star—was predicted before they were identified in nature. Supermassive black holes were not predicted. Rather, supermassive models were developed to explain observations of galactic centers.

one knows if some kind of remnant would be left, or what that would look like.) Physicists have searched for such bursts but have found none, and that allows us to place upper limits on the number of micro black holes created in colliders and in the Big Bang.

Second, black holes in our universe continue to *gain* mass—even if they are not being fed by a companion star—because passing photons fall into them at a much higher rate than they emit Hawking radiation. It will be about 10^{18} years before the universe expands enough to reduce the photon supply to the level at which stellar-mass black holes begin to experience a net loss of energy.

Although Hawking radiation is not likely to be observed directly, it has proven fruitful in stimulating thinking about how to combine quantum mechanics and general relativity, and how to describe black holes in terms of thermodynamic concepts such as temperature. Readers wishing to learn more about black hole thermodynamics are enouraged to consult the *Further Reading* list at the end of this chapter.

Check your understanding. Identify something you have seen in a science fiction story that you now know is false or implausible. Explain your reasoning.

20.5 Spinning black holes

Just as a figure skater spins faster by bringing her arms in, a star spins up rapidly as it collapses to form a black hole. Black holes found by astronomers are therefore expected to spin quite rapidly, and the surrounding spacetime is thus described by the Kerr metric introduced in Section 19.2. Far from such a black hole, or around a (hypothetical) slowly rotating black hole, the Kerr metric is approximately the same as the Schwarzschild metric. We will therefore focus on effects at small r around rapidly rotating black holes.

Section 19.2 explains how particles in orbits near the equator "see" a mass current flowing nearby. To recap, a prograde orbit (in the same direction as the spin) is analogous to a parallel mass current as shown in Figure 20.10, which results in less net gravitational attraction due to the repulsive effect deduced in Section 18.2. A retrograde orbit, in contrast, is analogous to an antiparallel mass current whose motion increases the net gravitational attraction. To balance this attraction, a retrograde orbit needs higher speed (and a prograde orbit, lower speed) than is the case around a nonspinning black hole. Retrograde orbits thus run into the limit where the required circular speed would be c, at some r larger than $1.5r_S$ (the limit for nonspinning black holes) while prograde orbits run into this limit at some r *smaller* than $1.5r_S$. This creates a region where particles can *only* move in a prograde direction. In this region, prograde is the only way forward in proper time. This effect also allows prograde orbits to reach closer to $r = 0$ than in the Schwarzschild case, which means that accretion disks can reach closer to rapidly spinning black holes than to slow spinners or nonspinners.

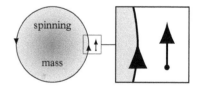

Figure 20.10 *Figure 19.1 is repeated here for your convenience. A spinning mass can be modeled as a mass current that loops back on itself. Here, a nearby orbiting particle sees the spinning source mass as a parallel mass current. The far side of the source mass does form a current in the opposite direction, but has less effect due to its greater distance.*

Can we verify any of this? Yes! If the inner edge of the accretion disk is closer to $r = 0$, it experiences a region of slower time, and its light experiences a greater gravitational redshift on its way to us. Thus, the range of gravitational redshifts measured in an accretion disk (Section 20.3) is greater for more rapidly spinning black holes. Observations of accretion disks indeed match the rapidly spinning model, thereby confirming that black holes are rapidly spinning *and* that the Kerr metric provides a good model for the spacetime around spinning black holes.

Thus, we have high confidence that rapidly spinning black holes do have a "future is prograde" region (outside of and in addition to the "future is $r = 0$" region). The prograde-only region *does* allow particles to escape. Recall that the faster the prograde motion, the stronger the gravitomagnetic repulsion. With speeds as fast as c, prograde particles can escape from as close as $r = \frac{GM}{c^2}$, or half the Schwarzschild radius.

The mathematician and physicist Roger Penrose (b. 1931) has shown how this feature of spacetime would allow an advanced civilization to extract rotational energy from a spinning black hole. Imagine a garbage truck passing through the region where the velocity-dependent repulsive effect is strong. When the garbage is pushed out the back, by Newton's third law the rest of the truck receives a push *forward*. The now-slower garbage experiences less repulsion and falls in—but the now-faster truck can use the increased repulsive effect to escape with *more* energy than it entered with. This process reduces the rotation of the black hole, but a great deal of energy could be extracted before the black hole would slow down substantially.

Check your understanding. How can we use observations to determine how rapidly a black hole is spinning?

Think about it

A flash of light emitted in the region $\frac{GM}{c^2} < r < \frac{2GM}{c^2}$ can escape *only if* it is emitted in the direction of the black hole spin. This would drastically affect a picture of a prograde accretion disk, as we would see only the side that is rotating toward us. We are too far away to see this effect directly in a picture, but it does affect the spectrum of the disk.

Box 20.1 Wormholes

We can imagine interestingly curved spacetimes that satisfy equations but do not actually occur in nature. For example, in Chapter 16.4 we deduced that the interior of a massive hollow spherical shell has some counterintuitive properties because the potential there is both low and flat. But massive shells do not occur in nature—and even if we tried to construct one we may not find a material strong enough to support an extremely massive shell against its own gravity. Skeptics can call such a shell impossible, while optimists can call it not-yet-possible.

The wormhole—wherein space is curved in a way that provides a short passage between widely separated regions of space—is much closer to impossible. It is relatively easy to write an equation describing such a spacetime (think of two copies of Figure 18.12 with their throats connected), but to call it "possible" we must also understand how to generate this geometry using some arrangement of matter and energy, and how to keep it from collapsing. Investigations in this regard have not been promising: the opening of a wormhole where none existed previously is probably forbidden, and any wormhole that does exist would probably collapse before a particle could traverse it. The word "probably" is necessary here mainly because quantum effects are not yet well understood. Quantum effects have at times revolutionized our understanding of gravitational systems—witness Hawking radiation and

Box 20.1 *continued*

black hole evaporation (Section 20.4). Yet, in the same way that Hawking radiation has no practical impact in our universe, it seems likely that macroscopic wormholes will remain science fiction even if quantum effects somehow allow wormholes to exist in principle.

CHAPTER SUMMARY

- General relativity offers a black hole model far more interesting than the Newtonian "dark star" whose only feature is light not escaping. Close enough to a black hole, spacetime curves so much that moving forward in proper time requires decreasing r. Not only is escape impossible for any particle inside the Schwarzschild radius $r_S = \frac{2GM}{c^2}$, the future there always leads to $r = 0$. The Schwarzschild radius therefore defines an important one-way spacetime boundary called the event horizon.

- Black holes don't suck: at large r the spacetime around a black hole behaves just as the Newtonian model predicts.

- Black holes have been observed by astronomers: stellar-mass black holes (with masses from several to tens of solar masses) form through the collapse of massive stars, and supermassive black holes (with masses from millions to billions of solar masses) lurk in the centers of most galaxies. Black holes with other masses are allowed in principle, but may or may not be formable through natural processes. Slow time just outside black holes has been observed in the form of gravitational redshifts.

- Black holes in nature spin rapidly, creating gravitomagnetic repulsion for prograde orbits. Prograde orbits are thus possible closer in than in the Schwarzschild case.

☰ FURTHER READING

Gravity from the Ground Up by Bernard Schutz discusses black hole thermodynamics (Hawking radiation for example) and spinning black holes at a level that nonexperts can follow. For college physics students, Thomas Moore's *A General Relativity Workbook* also contains clear discussions of both topics, as well as a discussion of alternative coordinate systems around static black holes.

Many images and animations of stellar motions around the black hole at the center of our Galaxy can be found at http://www.galacticcenter.astro.ucla.edu/multimedia.html.

The site http://spiro.fisica.unipd.it/~antonell/schwarzschild lets you visualize the bending of light around a Schwarzschild black hole (as in Figure 20.8) in real time in your browser. You can control your position and motion relative to the black hole as well as the direction in which you look. The author of this site, Riccardo Antonelli, describes his code in the April 2015 issue of *Hacker Monthly*. Tracing light rays in a Kerr spacetime is more complicated, but was done for the 2014 movie *Interstellar*. *The Science of Interstellar* by Kip Thorne may be a useful resource for students wishing to explore behind the scenes of this movie.

The *New York Times* profiled the challenges facing the Event Horizon Telescope team in a June 8, 2015 article titled *Black Hole Hunters*. This is a highly recommended exposition of the process of science in action.

To learn more about how the modern understanding of the black hole evolved, see *Black Hole: How an Idea Abandoned by Newtonians, Hated by Einstein, and Gambled On by Hawking Became Loved* by Marcia Bartusiak. This book examines personalities from John Michell to Stephen

Hawking and illuminates the interaction between theory and observation.

Visualizing Interstellar's Wormhole by Olivier James *et al.* (*American Journal of Physics* vol. 83, p. 486, 2015) provides more details on wormholes at a reasonably accessible level. There is math at the university physics level, but less specialized readers will still enjoy the introduction and the visualizations. The *Interstellar* team's website http://www.dneg.com/dneg_vfx/wormhole/ has additional images and videos.

CHECK YOUR UNDERSTANDING: EXPLANATIONS

20.1   (a) This cone should be tilted toward $r = 0$ by an amount greater than shown at E but less than shown at G. (b) This cone looks the same as the one at G. The only thing that could travel from G to H would be light, as illustrated by the vertical right edge of the cone at G.

20.2   (a) About 5.5 km; (b) about 2.4 km.

20.3   The curve in Figure 20.2 goes through $\frac{\Delta\tau}{\Delta t}$ of about 0.8 at $r = 10$. This tells us that a photon leaving

$r = 10$ km with energy E will arrive at Earth (which is far to the right where $\frac{\Delta\tau}{\Delta t} = 1.0$) with an energy of about $0.8E$.

20.4   Answers may vary.

20.5   The more the black hole spins, the closer the innermost orbit can get to $r = 0$ and therefore the greater the gravitational redshift we can observe.

EXERCISES

In all exercises and problems, assume a Schwarzschild black hole unless otherwise specified.

20.1   *(a)* How can you escape a black hole in the Newtonian model? *(b)* Why is the general relativity model different?

20.2   Explain how you could use a black hole as a time machine. (This refers to the usual practical sense that when you step out of the machine you find that much more time has elapsed outside the machine, rather than the strange behavior of the t coordinate discussed in Section 20.2.)

20.3   Building on Exercise 20.2, can you use this machine to perform time travel in either direction? Compare to the special relativity forward-only time machine described in Exercise 10.7. Does gravity allow you to do anything different?

20.4   *(a)* Why does Section 20.4 say that a horizon-crossing observer will observe gravitational *blue*shift of light from distant stars? *(b)* Does an observer need to cross the horizon to observe this blueshift?

20.5 Explain how the behavior of the r coordinate near the horizon is related to Figure 18.12 being cut off at the top.

20.6 Look again at the color of the accretion disk in Figure 20.6. Describe how the Doppler effect due to motion of gas in the disk would affect its color.

20.7 What is the evidence that black holes exist in nature?

20.8 *(a)* How do astronomers determine the mass of a black hole? *(b)* What masses do they typically have, and what does this say about where they came from?

20.9 *(a)* What must be true about a black hole if astronomers are able to find it? *(b)* What does this imply about the number of black holes in our Galactic neighborhood, compared to the number that we know about?

20.10 If a solar-mass black hole with no accretion disk entered our solar system, would we notice? Explain how we would notice, or why we would not.

20.11 Answer this question posted on an internet discussion board: "If sound waves have no mass, can they pass through a black hole?"

20.12 When discussing orbits around spinning black holes, the text assumed orbits in the equatorial plane. Would polar orbits be different? Justify your answer.

20.13 How is a black hole like Vegas?

➕ PROBLEMS

20.1 Use Newtonian gravity to find the radius at which a circular orbit around a mass M must have speed c. You should find that this is inside the Schwarzschild radius. Explain conceptually why general relativity would predict that orbits become impossibly fast at a larger r than Newtonian gravity predicts.

20.2 (You may wish to pave the way for this problem by doing Exercise 20.4.) *(a)* How much is light from a distant star blueshifted when observed by a hypothetical observer stationary at the event horizon? Even if your prediction sounds absurd, proceed to the next part of this question. *(b)* Imagine we send a probe in to measure the blueshift as it crosses the event horizon. Explain why the absurd prediction is not realized.

20.3 Explain the premise behind the comic in Figure 20.11. Discuss ways in which this comic is accurate or inaccurate.

Maybe it wasn't such a good idea to locate a restaurant at the edge of a black hole.

Figure 20.11 *Analyze this comic in Problem 20.3.*

20.4 Explain the relationship between Figures 20.5 and 18.10.

20.5 Find your own example of a black hole appearing in popular culture, and discuss ways in which this example is accurate or inaccurate.

20.6 To estimate the size of an unresolved object, astronomers often use the argument that if its light varies on timescales of, say, one minute then the object can be no more than one light-minute across. Investigate the basis of this argument and explain it in your own words.

20.7 Watch the *PBS Spacetime* video at https://www.youtube.com/watch?v=vNaEBbFbvcY. Write a paragraph explaining why you do or do not think events inside black holes really happen.

20.8 In the 2014 movie *Interstellar*, astronauts visit a planet near a black hole where one hour corresponds to seven years on Earth (a ratio of about 1:60,000). *(a)* Explain why the planet has not been ripped apart by tidal forces. (*Hint:* the name of the black hole is Gargantua.) *(b)* The astronauts want to minimize the Earth-years elapsed during this mission. They plan to park the mother ship in a black hole orbit just larger than that of the planet and make only a brief shuttle visit to the planet itself, reasoning that they will pay the 1:60,000 time penalty only for time spent on the planet. Explain why this should not be an effective plan. *(c)* In the movie the mother ship maintains a roughly 1:1 clock tick ratio with Earth. Roughly how many Schwarzschild radii from the center of the black hole must the mother ship be if this is true: less than 1, exactly 1, a bit more than one, a few, or many? *(d)* Choosing from the same options, roughly how many Schwarzschild radii from the center of the black hole is the planet, given the 1:60,000 clock tick ratio between the surface of the planet and the mother ship? *(e)* Given that large clock tick ratio, comment on the energy required to set the shuttle down on the surface and bring it back to the mother ship.

20.9 If there is a planet where, hypothetically, one hour elapses on the surface for every two hours that elapse aboard an orbiting space station, *(a)* how much energy would be required to launch a 10,000

kg ship from the surface to the station? *(b)* How much energy would be required to *lower* the same ship *from* the station—without crashing?

20.10 Imagine that you are doing some black hole explorations with a very narrow, 1-km long spaceship. *(a)* You keep the long end pointed toward a solar-mass black hole, which has an event horizon radius $\frac{2GM}{c^2} = 3$ km. You slowly move toward the black hole until one end of your ship is just touching the event horizon. How much stronger is the gravitational force on that end of the ship, compared to the end of the ship furthest from the black hole? Phrase your answer in terms of "X times stronger" rather than an actual number of pounds. *(b)* How would your ship "feel" this difference? In other words, you would want to build your ship to resist what? *(c)* Could you reduce the stress on your ship by rotating your ship 90° about its center, so that it's lying "across" the black hole rather than pointing to it? Why or why not? Draw a picture if you have any doubts. *(d)* Now you explore a black hole that is a billion times more massive, one of the supermassive black holes that lie at the centers of galaxies. You do the same thing as in part (a), point one end of the ship so that it just touches the event horizon. How much stronger is the gravitational force on that end of the ship, compared to the end of the ship furthest from the black hole? *(e)* Which of the two black holes is more likely to spaghettify you as you cross the event horizon?

20.11 Would a spinning black hole make a better time machine than a stationary black hole of equal mass? Explain why or why not.

20.12 Astronomers say that gravity is a more powerful engine than nuclear fusion: 1 kg of hydrogen fuel releases more energy by accreting onto a black hole than by undergoing fusion. Investigate this claim and explain it in your own words. Be sure to address the following points: *(a)* If an object is arbitrarily compact, what limits the amount of energy gravity can release? *Hint:* the event horizon is not the only limiting factor. *(b)* Would nuclear processes outperform gravity if your kilogram of fuel was half matter and half antimatter?

20.13 If a wormhole is essentially two copies of Figure 18.12 with their throats connected, does it necessarily cause gravitational acceleration? Think carefully about the meaning of Figure 18.12 and the curvature of space vs. time.

20.14 Do some research to find out why: *(a)* singularities are considered problematic; *(b)* how black holes cloak the problematic aspects of singularities; and *(c)* whether other situations may produce uncloaked or "naked" singularities.

Index